Stochastic Climate Theory

Springer
Berlin
Heidelberg
New York
Barcelona
Hong Kong
London
Milan
Paris
Singapore
Tokyo

Serguei G. Dobrovolski

Stochastic Climate Theory

Models and Applications

With 167 Figures and 13 Tables

Springer

Prof. Dr. Serguei G. Dobrovolski
Water Problems Institute
Russian Academy of Sciences
Gubkina 3
Moscow 117971
Russia
email: sgdobro@orc.ru

ISBN 3-540-66310-X Springer-Verlag Berlin Heidelberg New York

Library of Congress Cataloging-in-Publication Data applied for
Die Deutsche Bibliothek – CIP-Einheitsaufnahme
Dobrovolski, Serguei G.: Stochastic climate theory: models and applications / Serguei G. Dobrovolski. – Berlin; Heidelberg; New York; Barcelona; Hong Kong; London; Milan; Paris; Singapore; Tokyo: Springer, 2000
ISBN 3-540-66310-X

Cover design: Design & Production, Heidelberg
Typesetting: camera-ready by the author

SPIN 10644717 32/3136xz-5 4 3 2 1 0 – Printed on acid-free paper

This book is dedicated
to the memory of my father

Preface

All attempts to study climatic variations show that climate can only be interpreted as a stochastic system. However, no monograph specifically devoted to the stochastic theory of climate is available, either in Russian or in any other language, and it was my intention to remedy, at least partially, this gap in the literature. I also wanted to popularize models and approaches, which are closely related to the climate theory: from the Bernoulli symmetric random walk through Hasselmann's stochastic climate models theory. The reader is assumed to have at his command little more than the rudiments of probability theory: in this work I try to describe the theory of random functions (with emphasis on climate variations) from its very beginnings.

The author started his first experiments on stochastic climate models almost 25 years ago, at the Water Problems Institute of the Russian Academy of Sciences. Frequent discussions with A.I.Duvanin, V.E.Privalsky, L.P.Kuznetsova, R.K.Kliege, and V.P.Yushkov have had a considerable influence on my work. Later contacts with the Department of Atmospheric Physics of the Moscow State University, the Hydrological Sciences Branch of the NASA Goddard Space Flight Center, and the Center of Atmospheric Sciences of the University of Mexico were very fruitful.

The material of this book partly stems from a set of lectures given at the Moscow State University. Also several research projects were focused on stochastic climate models of the atmosphere, hydrosphere, and cryosphere and financially supported (partly) my work: Grants 97-05-64427 and 97--05-65105 of the Russian Foundation for Fundamental Research, INTAS Project No. 93-0979, and some others.

The author is grateful to J.Adem for many fruitful discussions. Special thanks are due to B.J.Choudhury for remotely sensed land surface hydrological parameters, which are the basis of Section 3.2.

The author would appreciate comments and suggestions sent to the Email address: sgdobro@orc.ru

1999 S.G.Dobrovolski

Contents

Introduction

Almost all discussions about potential anthropogenic change as well as attempts to forecast climatic developments lead back to climate variability. The collection of observations shows atmospheric and oceanic variations on all time scales, ranging from turbulent fluctuations to glacial cycles.

Great success has been achieved in formulating deterministic principles (balances of energy, mass, impulse and tracers, astronomical cycles, etc.) and in implementing them in dynamic models [models for the general circulation of the atmosphere and the ocean (GCMs) and others]. Nevertheless, all attempts to forecast future developments and to reproduce past data - a process necessary to gain understanding of relevant mechanisms - show that *climate can only be interpreted as a stochastic system*. Several reasons add to this:

1. Climate is a result of a high-dimensional, complex system. Even simpler subsystems exhibit nonlinear, chaotic dynamics. Models therefore involve simplifications, discretization, and parametrizations. Their effects in the model output can only be expressed as random with respect to reality.
2. Unknown external factors influence the system in a temporally varying manner.
3. Observations contain errors, and represent only a limited picture in space and time. Consequently, definitive causal inferences cannot be made, and only probabilistic statements can be achieved.

Thus, a key task in climate research is the development of "conceptual models" to describe the most important temporal or spatio-temporal features of observational records. These "reduced" models not only help to gain a better understanding, but they can also be used as a tool for forecasting; the quality of the forecast, in turn, can be taken to validate or improve the understanding.

Furthermore, experiments with sophisticated models such as GCMs produce an output of many variables for tens to hundreds of thousands of gridpoints; this output does not immediately yield insight because of the enormous amount of data and the chaotic nature of the dynamics. Hence, the behavior of this "virtual climate system" is also interpreted in terms of conceptual models. Apart from model interpretation, also model diagnostics and validation fall back on these reduced models. Here, one is not only interested in the confirmation of certain averaged patterns; a major aspect is the reproduction of a realistic temporal variability, which may lead the way to adequate representation of certain physical or biochemical mechanisms.

Because of the above-mentioned random effects, these models possess an essentially stochastic nature. Due to the complex interaction mechanisms of climate subsystems, random input signals can be transformed to observed output with significantly modified temporal and spatio-temporal characteristics. For instance, short-term fluctuations can be converted into longer-term variations by means of inertia components, e.g., heat storage in large water masses. Even more complex phenomena can be evoked by positive feedback mechanisms (e.g., convective transport).

Stochastic models of climate variability represent an essential tool for the above tasks. First, this book gives a basic introduction to the underlying theory. Second, the identification and analysis of stochastic models is presented for the main climatic subsystems covering a wide range of time scales. Recent developments have also allowed the design of more complex stochastic dynamical models; an exemplary study for an ice age application is presented as well.

Chapter 1 of the book is introductory. An overview of the climate system is given: climatic data, processes, and spatio-temporal scales are described. A new classification of the climate system components, which is more consistent with the stochastic nature of climatic variations than previous classifications (from our point of view), is proposed. Finally, deterministic models of the climate system – their advantages and limitations – are discussed.

In the second chapter, a theoretical basis for the stochastic approach to climate studies is described. An introduction to the theory of random functions, with emphasis on climate variability, is given, and the meaning of basic notions of the stochastic climate theory are discussed: stationarity and nonstationarity of a climatic variable, the ergodicity of climatic variations, spectral decomposition of atmospheric, oceanic, and land surface parameters. Also problems of stochastic model identification and parameter estimation are treated with examples from climatology. Using the above mathematical background, a physical basis of the stochastic climate theory is discussed: the atmospheric "forcing" of the climate system, Hasselmann's hypothesis on two-scale weather-climate separation and his theory of stochastic climate models, the immediate applications of this theory.

Chapter 3 is an application of the stochastic climate theory to recent climatic variations - during the so-called instrumental period of climate studies, i.e., approximately 100 years. Variabilities of three main types of climatic subsystems are analyzed: "thick subsystems" (ocean, atmosphere, ice sheets), "thin subsystems" (sea ice, snow cover, vegetation cover), "two-dimensional" and "local objects" (rivers, mountain glaciers). Three important natural phenomena are discussed: the global climatic temporal invariance of state parameters of thick subsystems, local-global polarization of transfer-accumulation climatic processes, and local-global convergence of state parameters of thin subsystems. It is supposed that these three natural phenomena form the basis for the general stochastic pattern of climatic variability.

Finally, Chapter 4 is dedicated to a stochastic study of the paleovariability of global and regional climates. Both stochastic and stochastic dynamical models are used to analyze existing paleorecords and to reconstruct important features of glaciation-deglaciation processes.

1 Climatic System: Data, Processes, Scales, and Deterministic Models

Roughly speaking, the climate system consists of the atmosphere, ocean, ice sheets, and land surface. In order to give a more detailed definition and classify the climate system's components, we must take into consideration the ideology of climate research. During the past 30 years, general circulation models (GCMs, see Sect. 1.4.1), together with more or less detailed models of the ocean, ice sheets, and land, were considered as the most powerful and almost unique instrument in climate modeling. These models include and describe very many components, objects, and processes, and their temporal and spatial resolution is rather detailed. Thus, the idea of many investigations related to climatic changes is the following: everything in the climate system is interrelated; feedbacks are everywhere; the larger the number of processes and components of the climate system we take into consideration, and the more detailed their description, the better are our results. In accordance with this concept, the notion of the climatic system constantly "grew". Step by step new components were included into it: stratosphere, vegetation cover, even the earth core and the mantle (in paleoclimate studies), etc. So the description of the climate system in climate-change studies often resembles lists of contents of manuals on atmospheric, ocean, hydrological, and other sciences put together.

However, in reality, the set of components of the climate system strongly depends on the type of problem we are investigating, on the specific goals of the research, and especially on the scale of the processes. At a glance it seems that the larger the temporal and spatial scales under consideration are, the more complete must be the list of processes and components of the system that we must include into the models. However, we often meet just the opposite situation: for an effective description and understanding of large and extra-large temporal scales of the processes, strongly simplified ("aggregated" or "conceptual") models are needed with a restricted set of climatic components and processes.

We can illustrate this problem by analogy with Brownian motion. The speed and coordinates of Brownian particles depend on the motion of molecules. However, nobody would consider describing the motion of each specific molecule in order to describe the character of the motion of Brownian particles. Moreover, such an approach would be absolutely impossible because Brownian motion has a probabilistic (statistical) nature.

Consequently, when we study the evolution of the climate system from century to century or from one glacial cycle to another, it is clear that we must not (and cannot) describe the behavior of each *real* cyclone and anticyclone. Moreover, it is not evident at all that in these cases we must describe the behavior of each *model* cyclone and anticyclone and the evolution of all components of the atmosphere, ocean, land, and ice sheets. Perhaps, just like Brownian motion, the behavior of the climate system at large temporal scales is governed by the laws of statistics rather than by the laws of theoretical mechanics. So the task of a researcher is to establish some optimal set of components of the climate system rather than to try to consider as many objects and processes as he can. Below, we will attempt a new classification of the components of the climate system, which could be an appropriate first approximation while studying natural climatic changes.

1.1 Main Components of the Climate System

Figure 1.1 illustrates our approach to the composition of the climate system from the point of view of changes at climatic temporal scales, i.e., more than approximately 1 month (see discussion in Sect. 1.3). The components of the climate system can be divided into three groups.

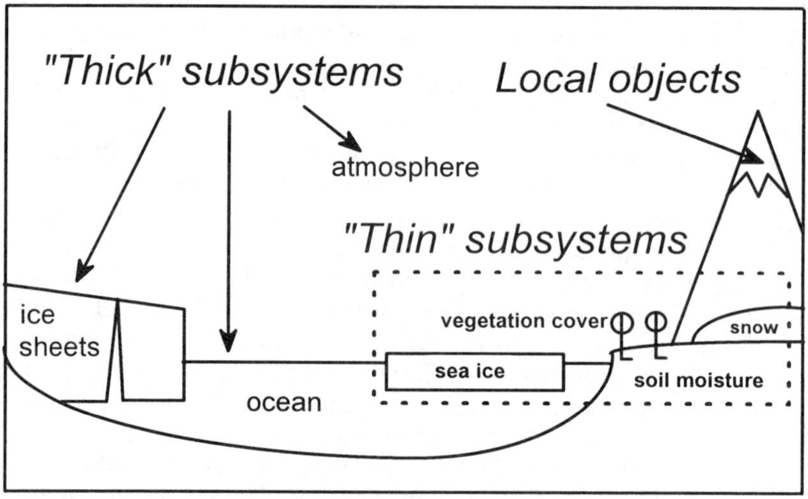

Fig. 1.1. Main components of the climate system

1.1.1 "Thick" Subsystems

Three main subsystems form this group: the atmosphere, ocean, and ice sheets. Their common features are the following:

1. These subsystems are essentially three-dimensional: their thickness is considerable and has an order of kilometers (about 2 km and more for ice sheets, about 4 km for the ocean, more than 10 km for the troposphere). The thickness of these subsystems does not change considerably at time scales comparable with the period of observations, i.e., about 100 years. For instance, the relative change in ocean thickness (global mean sea level) during the past century was only 0.004%. Moreover, the maximum relative changes in the ocean thickness during the last six glaciation-deglaciation periods were only around 2.5%. Seemingly, appropriate changes in the thickness of the atmosphere were also not very large. However, variations in the ice sheets were great at the time scales of ice ages; so the ice sheets are the less stable subsystem of the thick climate components.
2. The area of thick subsystems is constant or almost constant at time scales up to hundreds of years. The atmosphere does not possess lateral boundaries; its area is equal to the area of the earth's surface and does not change in time. The area of the ocean did not change considerably even during the ice ages, i.e., since 600,000 years B.P.: only the oceanic shelf, which has an average width of 70 km, was denuded during glaciation-deglaciation events. Thus, appropriate relative changes in the surface area of the oceans were not more than several percent because the total area of the oceanic shelf represents only 8.9% of the ocean surface. Here again we see that the ice sheets are the less stable component of the thick subsystem - although present-day Antarctic and Greenland sheets are comparetively stable (evidently, because they are surrounded by the ocean), 17,000 - 20,000 years B.P. ice sheets were at least twice as large as contemporary ice sheets.
3. Thus, at a time scale from months to hundreds of years, the area of thick subsystems cannot be considered as their most important, "state" parameter, which could affect the whole climate system. On the contrary, ocean and atmosphere temperature, especially sea surface and near-surface air temperature, strongly affect the climatic system. Sea level and thickness of ice sheets do not govern the system, although these parameters can be indicators of changes within it.
4. It will be shown below that the above simple, evident, and well-known properties of the thick subsystems are essential for the understanding and modeling of their changes at climatic time scales, as well as for understanding differences between the behavior of the state parameters of "thick" and "thin" subsystems.

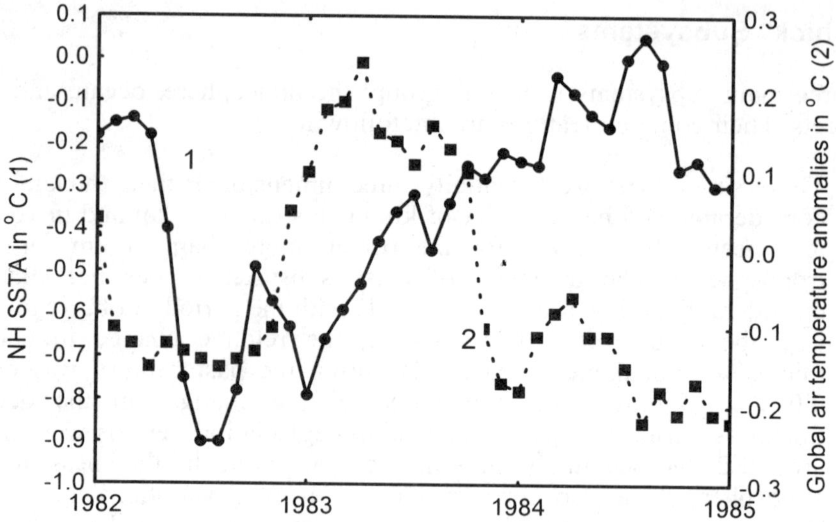

Fig. 1.2. Typical changes in spatially averaged "state" parameters of "thick" subsystems: *1* Monthly mean sea surface temperature anomalies averaged over the Northern Hemisphere (0 - 60°N); *2* globally averaged (85 °S - 85 °N) mean tropospheric temperature anomalies. (Data from D.E.Parker 1995, pers. comm., *1*; from USDC 1994, *2*)

The last point deserves some comment. It will be demonstrated in the next sections that the most important feature of the global state parameters of thick subsystems is their nonstationary, relatively slow variability, which resembles manifestations of such fundamental mathematical models as the Wiener process or the Bernoulli symmetric random walk with small grid steps. In Fig. 1.2 spatially averaged temperatures of the sea surface (1) and the troposphere (2) are presented. As well as other records of state parameters of thick subsystems at time scales from months to dozens of thousands of years, these curves do not seem to be stationary. In other words, there is no fixed mean value, which "attracts" the curve. (Also note that the two curves in Fig. 1.2 do not resemble each other).

This property of thick subsystems is related to the enormous volume, heat, and water capacities of the ice sheets and the ocean: the volume of the oceanic water is equal to approximately 1,400 million km^3, the volume of the ice sheets is about 30 million km^3. Also the absence of stationarity in the state parameters of thick subsystems seemingly is due to the multiplicity of mechanisms and processes within the ocean and the atmosphere. These processes can easily redistribute enormous impulses of heat fluxes between the ocean and the ice sheets and accumulate these additional portions of anomalous heat within them (see Sect. 1.2).

1. 1. 2 "Thin" Subsystems

The most important "thin" subsystems of the climate system are snow cover, sea ice, vegetation cover, and soil moisture. The differences between thin and thick subsystems are the following:

1. Unlike ocean-atmosphere-ice sheets, thin subsystems are essentially two-dimensional structures. Their characteristic thickness is 10^0-10^1 m, i.e., thousands of times less than that of thick subsystems, whereas their area is comparable to that of the ocean and ice sheets. Thus, thin covers represent a sort of unstable film, and their thickness can change by dozens of percent in several weeks or months. Although thin subsystems affect the ocean and the atmosphere, their own dependence on thick structures is much more dramatic: the very existence of sea ice, snow, vegetation cover, and soil moisture in a specific region depends on the atmospheric and oceanic (for sea ice) conditions.
2. The area of thin subsystems changes quickly: it has a pronounced seasonal dynamics. Moreover, monthly and even decadal anomalies in the global area of these covers can reach dozens of percent.
3. So at climatic time scales the area of thick subsystems (or "spatial density", in the case of the vegetation cover) is their most important state parameter.
4. The behavior of this parameter of thin subsystems differs from that of state parameters of thick subsystems. Figure 1.3. shows that changes of the thin earth covers, on the time scales under consideration, are more stationary than those of thick covers. It is evident that some mean level of the spatial extent of these earth covers exists, which attracts snow cover and vegetation cover (also sea ice area and vegetation density).

Seemingly the last property of thin subsystems is related to the existence of strong negative feedbacks at time scales of months, years, etc. In turn, the existence of feedbacks is due to the fact that these covers, unlike thick covers, can very quickly reach the limits of their extent, which are governed by planetary and astronomical factors. It will be demonstrated below that the first-order Markov process with a relatively small coefficient (which is the simplest and most fundamental manifestation of the negative feedback in nature) is a good first approximation for the description of variability in thin subsystems. Also it will be shown that, mathematically speaking, the mean characteristic time scale (correlation time) of global thin subsystems is 1 - 2 months, whereas the characteristic time scale of global thick covers is dozens of years and more. So the difference between the behavior of the two main types of climatic subsystems is dramatic. It is interesting that for larger time scales this difference gradually disappears: the subsystems are not independent.

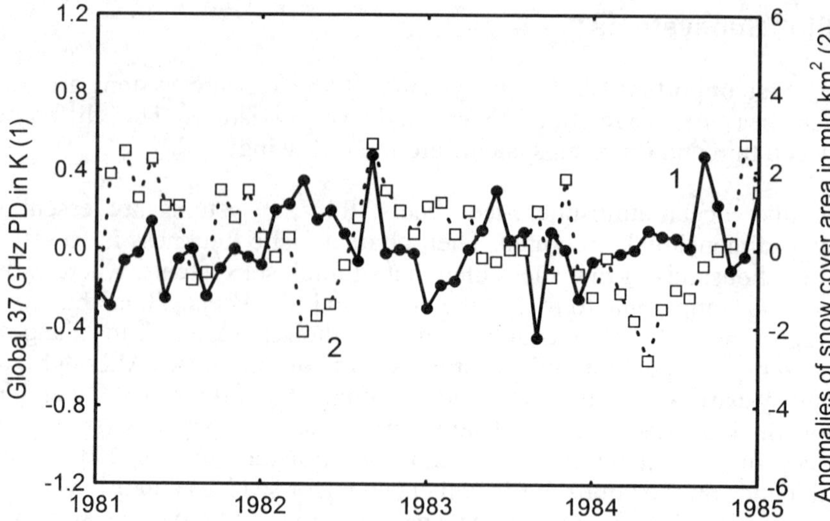

Fig. 1.3. Typical variability of *global* parameters of thin climatic subsystems: *1* monthly anomalies of 37 GHz polarization difference (40°N-40°S); this parameter describes the spatial density of vegetation cover; *2* global monthly anomalies of snow area. (Recalculated using data from B.J. Choudhury 1996, pers. comm., *1*; from IPCC 1993, *2*)

1.1.3 Local and Discrete Objects

The main natural objects of this kind are rivers, lakes, and mountain glaciers. From the point of view of the study of the whole climate system, they can be considered in a first approximation as one- or zero--dimensional structures. The most characteristic feature of local and discrete objects (as the term local indicates) is their restricted area; in turn, this feature determines the properties of the local subsystems' behavior and their role in the climate system. Unlike thin subsystems, local objects cannot significantly affect the global parameters of thick subsystems. They can however, to a certain extent, be indicators of global climatic changes. Because of the variety of local objects, their most relevant parameters are different: ice volume for the mountain glaciers, runoff discharge for rivers, and water level and surface area for lakes.

One of the larger local objects is the Caspian Sea (strictly speaking, a terminal lake). Its area is about 400,000 km^2, which is slightly more than 0.1% of the area of the global ocean. The depth of the Caspian sea (more than 1 km in its southern part) is comparable with that of the ocean, but because of its small size the relative changes in the area are considerable: more than 10% from the 1930s to the 1970s. Therefore the negative feedbacks related to the changes in the surface of the Caspian sea, through the mechanism of anomalous evaporation, are important. Thus,

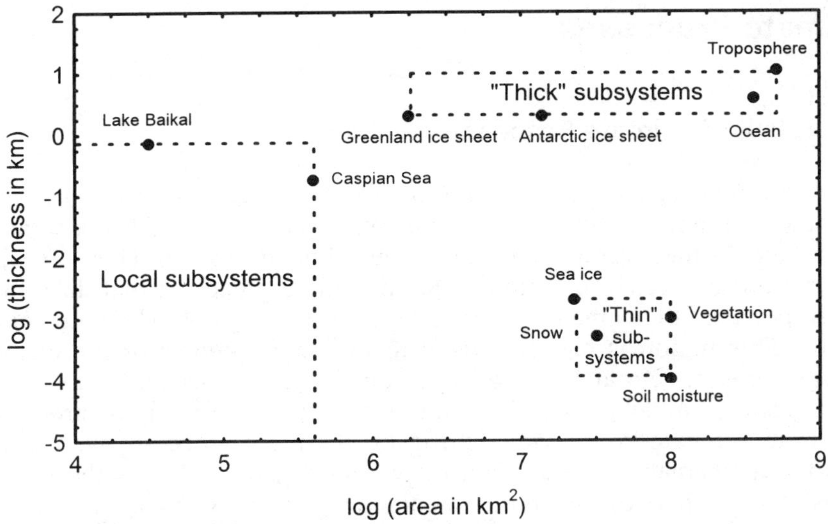

Fig. 1.4. Characteristic dimensions of the climatic subsystems in bilogarithmic scales. *Dashed lines* denote maximal and minimal values of respective parameters for each group of climatic subsystems. The thickness of the soil moisture subsystem is given in water equivalent

characteristic (integral or correlation) time of the variations in level of the Caspian Sea (dozens of years) is much less than that of the ocean.

Variations in the volume of the mountain glaciers also differ from those of thick subsystems. Unlike time series describing the evolution of thick and thin subsystems, many records of the volume of mountain glaciers contain statistically significant deterministic (at time scales of dozens of years) trends, which is seemingly due to the relatively small size of mountain glaciers and the resulting lack of stability of these objects.

One-dimensional components of the climate system, rivers, especially in the temperate and subarctic zones, have more in common with large subsystems, namely with thin subsystems. This is natural because rivers channel the water flow originating from snow melt and soil moisture. However, here again we see that restricted dimensions of the typical river channel, its one-dimensional pattern result in considerable transformation of the structure of thin subsystem variability: probability distributions of monthly and annual anomalies in river runoff are strongly asymmetrical, and runoff autocovariances are characterized by pronounced seasonal dynamics.

Finally, Fig. 1.4 illustrates differences between the main groups of climate subsystems. Note that the three groups in Fig. 1.4 form distinct domains without overlaps.

1.2
Climate Processes

1.2.1
Overview of Climate Processes

There are two approaches to the notion climate processes. From one point of view: "Climate processes are all the individual processes that separately contribute to the overall behavior of the climate system. They are also interactions and feedbacks among the individual processes that determine the response of the climate system to external forcing ... " (Houghton et al. 1996). This notion is clearly related to the "big" models of the climate system - general circulation models coupled with the ocean, land surface, and ice sheets models. From this point of view practically all the processes within the climate system are significant; the more processes we include into the mathematical description of the system, the better is the model. Thus "a myriad" (expression of the authors of the above monograph) such processes exists. These are physical, chemical, and biological processes on various spatial and temporal scales - from the scales of molecules and cloud particles up to the global and larger scale, and from partitions of one second to thousands of years.

Another variant of the climate processes is the following (we can also call these processes mechanisms in order to distinguish them from the former). Climate mechanisms are individual processes, feedbacks, and interactions, which cause changes in the climate system at climatic temporal scales. This approach to climate processes enables us to restrict, at the first approximation, the number of phenomena under consideration, and to retain only those of which could be more or less immediate causes of climatic variations and variability.

In turn, climate mechanisms can be divided into three major groups:

1. External mechanisms. These are mechanisms which include natural objects and processes of cosmic, planetary, and geological origin.
2. Internal mechanisms, i.e., natural mechanisms within the ocean--atmosphere-ice sheets-land system.
3. Man-made processes and factors.

The above simple classification depends on the temporal scales under consideration. For instance, processes in the earth core can be considered as external if we deal with contemporary climate, but they may be included into the climate system in studies of paleoclimates. The difference between cosmic and internal mechanisms is more distinct, although cosmic processes always operate through internal climate processes.

Anticipating the content of the next chapters, it is important to distinguish possible forms of responses of the climate system to different climate mechanisms. In the first approximation, four fundamental variants of such responses could take place: deterministic trends, quasiperiodical oscillations, chaotic behavior (attractors), and stochastic processes. Also various combinations of the four types can occur. Below, we will discuss possible relationships between these types of forcing-response behavior (man-made processes will be considered in Chap. 3).

1.2.2 External Climate Mechanisms

Figure 1.5 lists the external climate mechanisms which are most frequently cited. The relative importance of these processes is a subject of serious discussions today. Thus, we deal here more with hypotheses than with established facts, although the problem of external natural climate signals is extremely important.

In principle, explosive volcanic eruptions could be one of the most important potential factors of climatic variability. However, the exact mechanisms of related processes are far from being accurately studied, and in many recent works it is stressed that model answers to volcanic signals are comparable with the errors in the relevant observational data (for example Mann et al. 1995; earlier similar results were obtained by Bradley 1988, Self and Rampino 1988, and others). Perhaps the most important question is how to forecast and model volcanic eruptions themselves. Seemingly, these events have a random nature and can be described by stationary processes, which are uncorrelated in time. For instance, in a very interesting early report (EPA 1984), volcanic forcing of

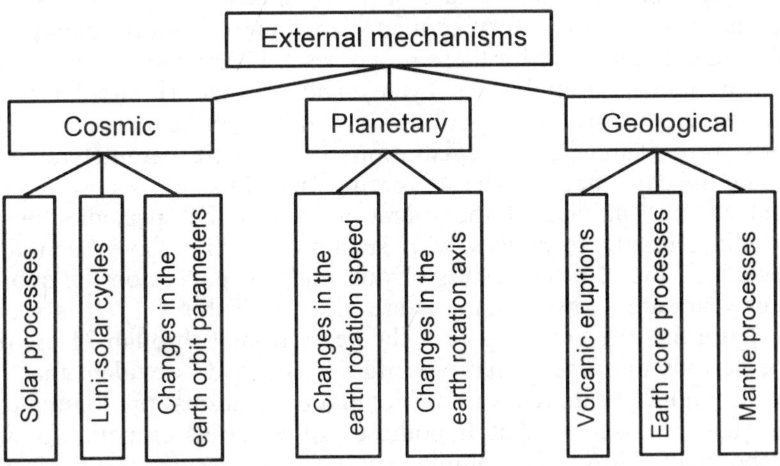

Fig. 1.5. Probable external mechanisms of climatic variations

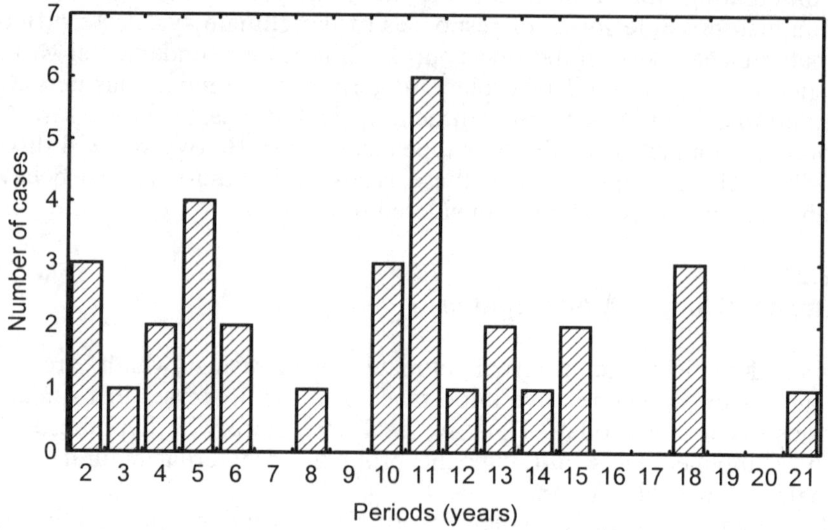

Fig. 1.6. Periods of oscillations found in climatic data and models

the atmosphere was prescribed as a white noise with time steps equal to 1 year.

Other types of geological processes possibly affecting the climate system, earth core and mantle processes, are very slow. At time scales of months to hundreds of years they can be described, in the first approximation, as processes with deterministic trends. The problems of assessing a geological signal while studying mean sea level series will be discussed in the following parts of the book.

An important common feature of possible cosmic and planetary factors of climate is the oscillatory character of their variations. Sharp spectral peaks are characteristic for these processes: 11, 22 years (solar cycles), 5.5 years (harmonic of an 11-year cycle), and 18 years (luni-solar cycle) are the most frequently cited periods of extraterrestrial processes. If these cycles were found in the registrations of climatic parameters, it would indicate the existence of external forcing mechanisms.

Indeed, an analysis of the literature shows that periodic or quasi--periodic climatic responses with such frequencies have been formally detected. In Fig. 1.6 the results of our study of papers on the subject for recent years are shown (citations are given in Table 1.1). It is seen from this figure that the greater part of the years from 1 through 21 are thought to be periods of some climatic oscillation. Also it is evident that periods of 5, 11, and 21 years (as well as biennial oscillations) predominate. Thus, at a glance, it seems that hypotheses of external climatic forcing are corroborated. However, sample analysis of results from Table 1.1 demonstrates that they are often dubious. Our calculations, using data

Table 1.1. Examples of periods of oscillations detected in climatic data and model simulations

Author	Parameter	Mechanism	Period(s)
Dymnikov and Gritsun (1996)	Climate parameters	Baroclinic oscillations	50 days
Takahashi and Zhao (1997)	Stratosphere param.	Quasi-biennial oscillations	1.5 years
Kane and de Paula (1996)	Mauna Loa CO_2	ENSO	2-6,8,14 years
Yuan et al. (1996)	Climatic parameters	Antarctic circumpolar wave	4-5 years
Currie (1996)	Cyclone parameters	Luni-solar, solar cycles	10, 18 years
Takata et al. (1997)	Permafrost param.	Ocean-land interactions	10-15 years
Carton (1997)	Sea temperature	Atlantic dipole	10 years
Currie and Vines (1996)	Australian rainfall	Luni-solar, solar cycles	10, 18 years
Babkin and Selyakov (1995)	Volga runoff	Solar cycle	11 years
Butler and Johnson (1996)	Air temperature	Solar cycle	11 years
Jovanovich (1993)	Air humidity, Sofia	Sun activity	11-12, 19-22 years
Kane (1997)	Droughts, Brasilia	ENSO	13, 26 years
Mann et al. (1995)	Climate parameters	Auto-oscillations	15-35,50-150 years
Brayan and Griffies (1996)	Climate parameters	Thermohaline convection	50-60 years
Schlesinger and Ramankutty (1994)	Global temperature	Internal variability	65-70 years
Dovgalyuk and Klimenko (1996)	Tree rings	ENSO	74, 320 years
Berger and Loutre (1997)	Ice sheet volume	Insolation changes	20,40, 100 ka
Feng (1996)	Ice sheet volume	Albedo feedbacks	20, 40, 110 ka
Chistyakov (1996)	Ice sheet volume	Insolation changes	100, 20 ka

from papers cited in Table 1.1, usually did not manifest spectral peaks (except quasi-biennial oscillations and a 100-ka periodicity of glaciation--deglaciation processes). Details of these calculations will be discussed in detail in Chapter 2.

1.2.3 Internal Mechanisms of Climatic Variations

A list of possible internal mechanisms of climatic variations (oscillatory ones) has been already included into Table 1.1. In addition, the following mechanisms, together with examples of respective recent publications, can be listed: interactions between ENSO, Hadley cell, and Walker circulation (Astafieva 1996); compensatory monsoons (Carl et al. 1995); ENSO-sea ice connections (Dugam 1995); teleconnections (Efimov et al. 1995,

Hurrell 1996); biogeochemical feedbacks (Elsen and Rotmans 1993, Puchkovsky 1996); equatorial thermocline processes (Gu and Philander 1997); Madden-Julian oscillations (Gualdi et al. 1997); global attractors for the Lorenz model (Krupchatnikov 1995) and other nonlinear structures (Sonechkin and Ivashchenko 1996); interactions between seasonal dynamics and nonlinear dynamics of the atmosphere (Kurgansky et al. 1996); volcanic aerosols-high clouds connections (Song et al. 1996); biogenic feedbacks in glacial cycles (Van der Sluijs et al. 1996); ENSO as chaotic oscillator (Wang and Fang 1996).

Thus the above list, together with Table 1.1, gives about 20 different possible mechanisms of internal climatic variations, and this list is far from being exhaustive: new hypotheses on climatic mechanisms are constantly appearing. Note that all these mechanisms are deterministic, i.e., it is supposed that exact initial and boundary conditions unequivocally determine the future behavior of the mechanism. In turn, the errors of such forecasts are supposed to be related to the lack of accurate observational data or to the existence of unknown mechanisms. In accordance with such an approach, the whole climatic system can be represented as a mechanical sum of the above (and discovered in the future) mechanisms.

However, some questions arise while considering the problem of climate mechanisms. First, what causes the existence and functioning of climatic the mechanisms themselves? Also, what are the criteria for distinguishing one mechanism from another? Finally, is there a guarantee that the number of climate mechanisms is finite?

Let us take the El Niño-Southern Oscillation (ENSO) phenomenon as an example and try to answer the above questions. ENSO is usually considered as the most powerful (or one of the most powerful) climatic mechanisms. However, many uncertainties are related with it. For instance, it is stressed in the WGI contribution to the IPCC Second Assessment Report:

> No current coupled atmosphere-ocean model simulates all aspects of ENSO events, but some (underlined by us - S.D.) of the observed interannual variations in the atmosphere associated with these events are captured (Houghton et al. 1996, p. 34). Evidence that the El Niño-Southern oscillation has varied in period, recurrence interval, and area and strength of impact is found in historical instrumental data ... The cause of these variations is not known (Ibid, p. 167).

Similar conclusions have been drawn in many other recent works:

> Relationships between ENSO and air temperature and rainfall change unpredictably (Nicholls et al. 1996). Theories of ENSO forecasts are unreliable (Pearce 1994). There are no close relationships between ENSO and precipitation persistence (Simmonds and Hope 1997). The probability of 1990-1995 ENSO is equal to 1/2,000 (Trenberth and Hoar 1996). Responses of precipitation registrations at 100 Costa Rica stations (small country - S.D.) to ENSO signal differ from station to station as for their magnitude and time lag (Waylen et al. 1996). Hydrodynamic

models could not give a forecast of El Niño return in 1993-1994 (Wright et al. 1995), etc.

The above conclusions are not surprising, because in the 1980s researchers already formulated them, applying ENSO patterns to the observational data of that time: Boer (1985), Fennessy et al. (1985), Namias (1985), Tokioka et al. (1986), Sirutis et al. (1986), Zebiak (1986), Owen and Palmer (1987), Fennessy and Shukla (1988), and many others demonstrated the small statistical relevance of the nontropical responses to the tropical sea surface temperature anomalies (El Niño). These papers also stressed the complex and heterogeneous character of such relations. Moreover, Walsh (1986), Namias (1987), Kane (1989), and others stated that the response of the tropical atmosphere itself to the El Niño phenomenon is extremely heterogeneous, and that it is difficult to use these relationships for long-term forecasts.

The above results show that manifestations of ENSO still contain a large random component and may be described by some sorts of stochastic processes. For example, Blanke et al. (1995) in their study of the effects of stochastic wind stress forcing on ENSO irregularity, demonstrated the importance of taking account of randomized atmospheric variability for the understanding of the absence of periodicities in ENSO events.

Similar features are characteristic for other possible climate mechanisms listed in this section and in Table 1.1, whose importance is perhaps less evident than that of ENSO. Thus, answers to the questions formulated at the beginning of this section can be as follows:

1. The quantity of climatic mechanisms seems to be very large (practically infinite).
2. Moreover, it is very difficult to distinguish one climate mechanism from another because there are no sharp spatial and structural boundaries between them, also because of intensive nonlinear interactions between specific mechanisms and their multiscale character.
3. The predictability of the effects of specific mechanisms is strongly limited and seemingly does not exceed a time period of a few months or a few years (we do not speak here about time scales related to glaciation-deglaciation events).
4. Processes governing climate mechanisms are likely to be random, to a large extent, as well as processes of the third, fourth, and other ranks.
5. Therefore, the effects of climate mechanisms and the mechanisms themselves are likely to be manifestations of random process models rather than deterministic processes.

1.2.4
Transfer-Accumulation Processes

The impression could be created by the previous section that everything is random and vague within the climate system. Indeed, at a glance, it seems that the system is governed by a myriad of various processes, and its resulting behavior is rather like that of a big turbulence. Classification of internal climate processes (mechanisms) seems impossible or senseless.

However, as in the case of climate components, a very simple but effective classification of climate processes is possible and necessary. Although the proposed pattern is so obvious that it seems to be a banality, it could provide a new understanding of the main features of climatic variabilities of the most important climatic parameters and processes.

Our idea is: all the incalculable variety of climate processes and mechanisms could be reduced to only two main groups of processes - (1) transfer processes, i.e., processes which are a redistribution of heat and water (and other substances) between main climatic subsystems–atmosphere, ocean, ice sheets, and others; (2) accumulation processes, which are changes in heat, water (and other substance) content in each specific subsystem.

Transfer processes between the atmosphere and each other subsystem (we can call them intrasubsystem transfers, IST) are the most important because of their intensity and variability, their important role in other transfer processes, e.g., interactions between sea ice and water, between snow and land, etc.). These processes include the following main vertical transports: evaporation, precipitation, contact turbulent heat exchange, radiation fluxes, and CO_2 fluxes. ISTs are closely related to the properties and patterns of the atmosphere: their magnitude is restricted, they are extremely unstable, but their monthly or annual anomalies are stationary, in the first approximation. The characteristic time of the transfer processes is several days. As will be shown below, the character of global-scale ISTs does not differ considerably from the local-scale.

On the contrary, properties of accumulation within subsystem (AS) are related to the features of much more inertial climatic components - thin subsystems, oceans and ice sheets, depending on their heat and water capacities. With respect to the oceans and ice sheets, these capacities are enormous, and grow almost linearly with the area of the part of climatic subsystem under consideration. As a result, the character of the ASs shows a strong dependence on their spatial scales. On the local scale there are various types of AS variability, whereas on the local scale all ASs converge to some universal type of variability. It will be shown below that ASs for the ocean and ice sheets are described by the fundamental Wiener process.

1.3 Scales of Climatic Variability

1.3.1 Spatial Scales

There are two main spatial scales in the climate system: a characteristic scale related to synoptic vortexes in the atmosphere, about 10^3 km (linear scale) or 10^6 km (characteristic area), and, evidently, a global scale of 10^4 km (10^8 km^2). Synoptic processes impose their spatial scale not only on changes with characteristic time of several days but also on month-to-month, year-to-year, and larger climatic changes because temporal averaging preserves the spatial structure of the processes. In other words, if there is a long-term anomaly of some climatic parameter, this anomaly has a horizontal size comparable with the characteristic size of cyclones-anticyclones

Spatial correlation functions of atmospheric, oceanic, and land surface parameters are a manifestation of this feature of climatic processes. As an example, spatial autocovariances of monthly anomalies of precipitation and sea surface temperature are presented in Fig. 1.7. This figure shows that correlation radius (measure of spatial structure corresponding to autocovariances equal to $1/e$, i.e., 0.368) of monthly anomalies is from 750 to 1,500 km and more. For many climatic parameters spatial

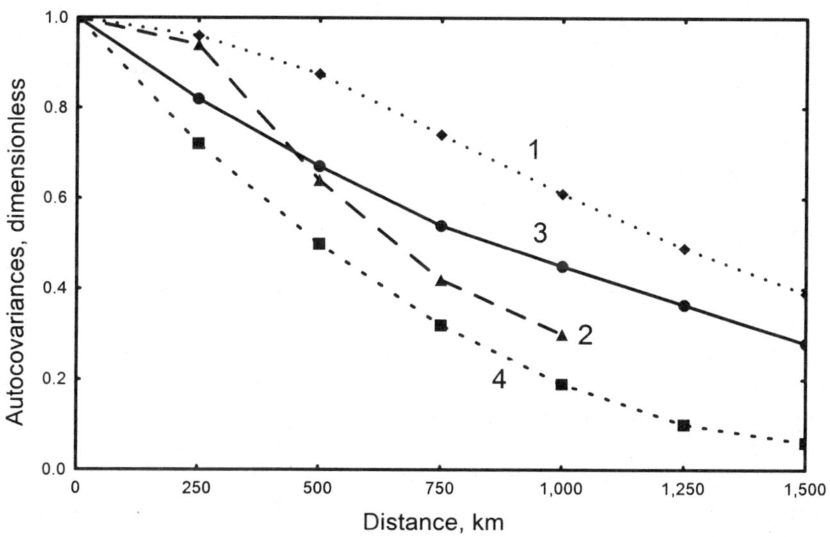

Fig. 1.7. Spatial correlation functions of monthly anomalies: *1* precipitation, longitudinal direction, (65°N, January); *2* the same but for July; *3* North Atlantic SSTA, latitudinal direction; *4* North Atlantic SSTA, longitudinal direction. (After Dobrovolski 1992)

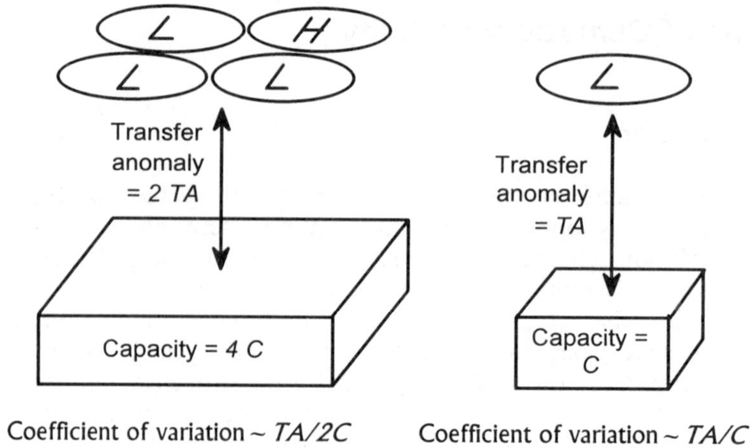

Fig. 1.8. Scheme of climatic spatial scales

correlations become zero at a distance of 2,000 - 2,500 km.

Relations between synoptic and global spatial scales are the key factor of climatic variability and variations. This important problem will be discussed in the following sections; here, we would like to demonstrate preliminarily a very simple scheme (Fig. 1.8) explaining how this mechanism works.

Let us consider a part of the land surface with a horizontal size of the order of synoptic atmospheric vortexes. Suppose that during a specific month (or year) cyclones prevail over this territory, so the amount of precipitation will be above normal. We denote the corresponding anomaly of vertical water flux (precipitation minus evapotranspiration) as *TA*, and the water capacity of the territory as *C*.

The most important mechanism of climatic variations is their feedback. In turn, feedbacks are closely related to the so-called coefficient of variation (*CV*), which is the measure of variability: mean standard variation of the process divided by its mean value. But standard variation is related to characteristic anomaly of transfers (*TA* in our case), and the mean value is closely related to the capacity of territory (water capacity in our case, *C*): *CV* is proportional to mean transfer anomaly/capacity.

Then let us enlarge the territory under consideration, say, by four times. The water capacity of the territory will increase linearly with the area, i.e., by four times. There will be place for four structures of anomalous conditions in the atmosphere, but because of the absence of horizontal correlation at these spatial scales, anomalous high and low pressure areas will be distributed randomly over the territory. On average, there will be three anomalous structures of one sign (say, cyclonic), and one structure of opposite sign (anticyclonic). So the overall anomaly of vertical flux will increase by only two times, and the corresponding coefficient of variation will decrease, on average, by two times.

The same scheme works if we consider not only land but also oceanic surface. The mechanisms of heat transfer-accumulation processes over the land and the ocean resemble that described above.

Consequently, if we take the global scale, capacities of main natural reservoirs of water and heat will increase hundreds of times whereas mean anomalies of vertical fluxes between climatic subsystems will increase only by dozens of times. So the coefficient of variation and negative feedbacks will decrease by at least an order of magnitude in comparison with local scales. Thus we deal here with a universal spatial-temporal pattern; in the next sections it will be shown that this is the basis of a very important mechanism of local-global polarization of climatic transfer--accumulation processes.

1.3.2 Temporal Scales

Roughly speaking, time scales for the atmosphere are on the order of days, for the land and upper ocean months and seasons, and for the deep ocean and ice sheets hundreds to thousands of years. To be more exact, characteristic temporal scales of the climate system are related to the following main types of processes:

1. Periodical changes caused by periodical external signals: (a) diurnal cycle; (b) annual cycle; (d) 6-month harmonic of the annual cycle. They are characterized by sharp spectral peaks and deterministic changes of phases: the summer of the appropriate hemisphere is warmer than the winter practically everywhere save in the equatorial latitudinal belt, day is usually warmer than night, etc. Therefore, the classical theory of stationary random processes is not applicable for the description of these phenomena, and a special notion of "spectrum" and "spectral peak" is needed for this case.
2. Perhaps some other types of periodical changes, which are less evident: for instance, 100-ka glacial cycle (possibly also 20-ka and 40-ka cycles).
3. A quasi-periodical process related to internal atmospheric variability: synoptic variability. It is characterized by a wide spectral peak at the periods of several days, and the phase of the oscillation is, to a large extent, random. So the theory of stationary random processes is applicable, at a first approximation.
4. Nonperiodical processes, which "occupy" frequencies between and beyond the above peaks. Seemingly, genuine climate processes are related to this type of variability.

In order to visualize relationships between variations of different scales in the climate system, it is useful to compare Figs. 1.9 - 1.14. These figures represent a sample set of temporal variability patterns of the same

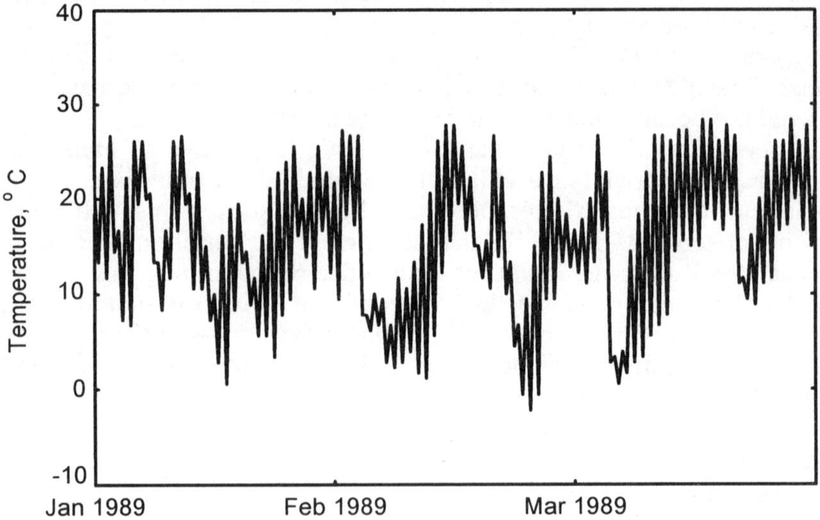

Fig. 1.9. Near-surface air temperatures in New Orleans (USA) with diurnal resolution. (Data from World Weather Disk 1994)

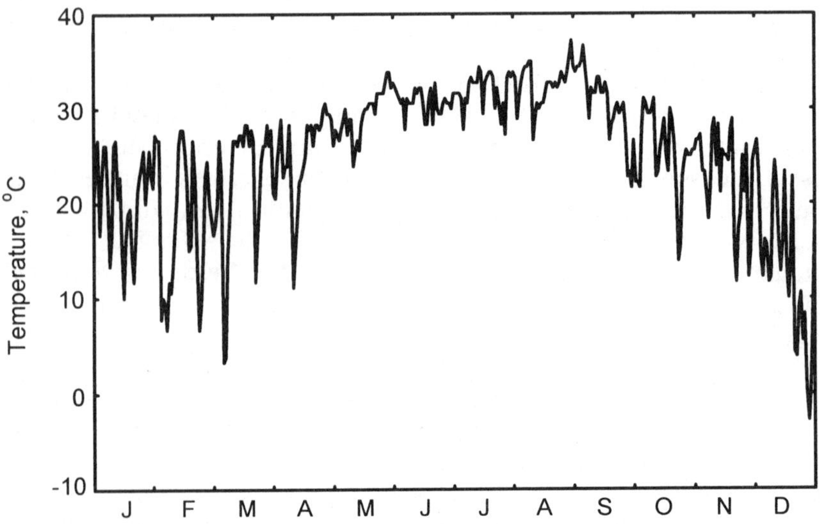

Fig. 1.10. Seasonal dynamics of near-surface air temperatures in New Orleans in 1989. Daily resolution (maximum daily temperatures). (Data from World Weather Disk 1994)

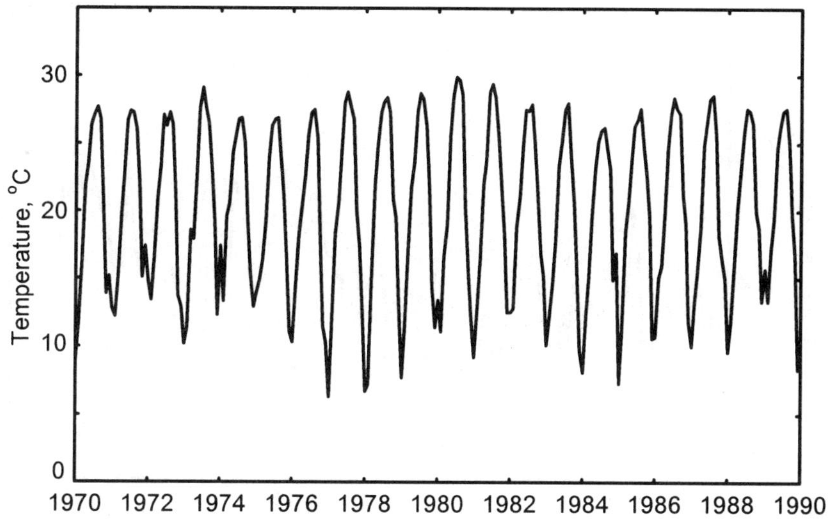

Fig. 1.11. Mean monthly near-surface air temperatures in New Orleans for 1970-1990. (Data from World Weather Disk 1994)

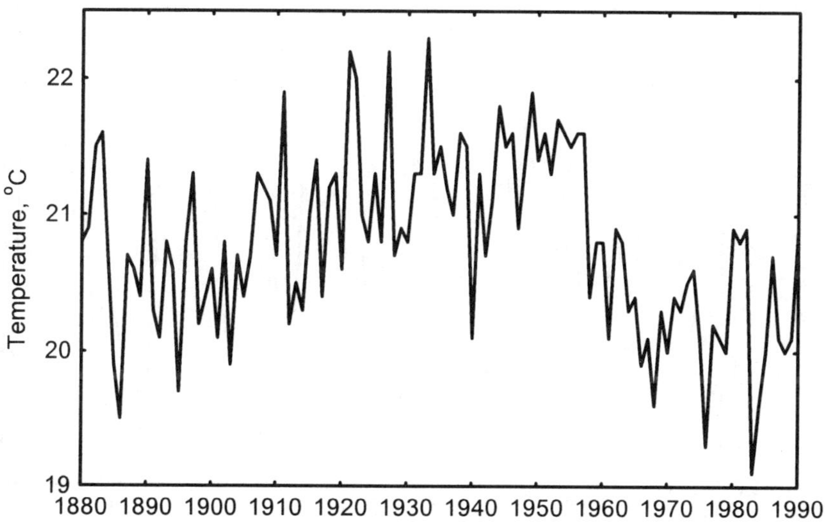

Fig. 1.12. Mean annual near-surface air temperatures in New Orleans for 1880-1990. (Data from World Weather Disk 1994)

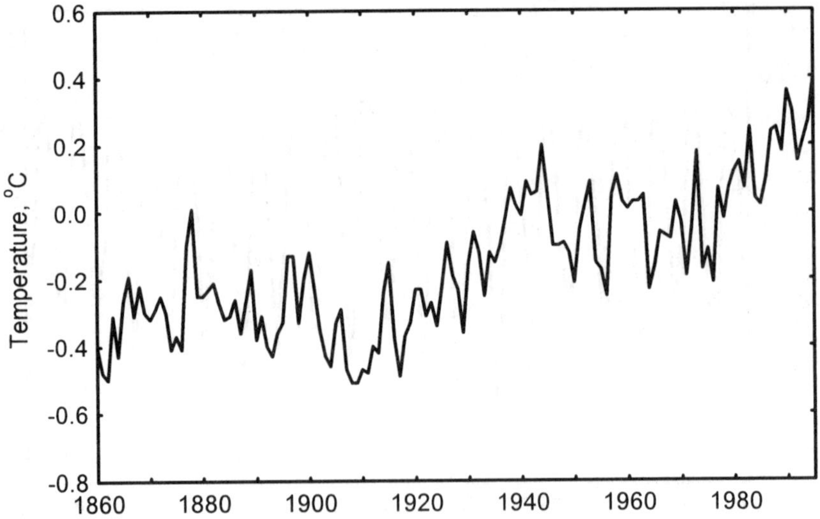

Fig. 1.13. Mean global near-surface annual air temperatures. (Data from Global Climate Observing ... 1996)

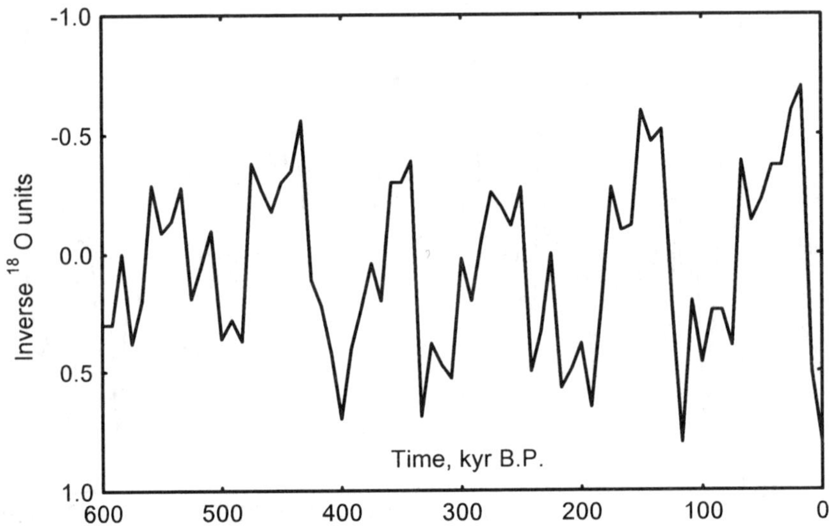

Fig. 1.14. SPECMAP ^{18}O data (related to estimations of air temperatures). (Data from Berger 1984)

parameter (near-surface air temperature in New Orleans, USA, 29°59′N, 90°15′′W) or presumably related parameters (global temperatures) plotted with different temporal resolution and scales. In Fig. 1.9 the resolution is enough to reproduce diurnal variations during several months. Figure 1.10 represents seasonal variations of daily (maximum) temperatures. In turn, Figs. 1.11 and 1.12 show changes in monthly and annual temperatures at the same site during respectively 20 and 110 years. Finally, instrumental- and paleorecords of global air temperature and related parameter (^{18}O) are demonstrated in Figs. 1.13 and 1.14.

The first impression after visual examining of Figs. 1.9-1.14 is the striking difference between temperature variability patterns at different time scales. Without figure captions, one would think that there is nothing in common between these six patterns. Indeed, variations in diurnal cycles are stationary irregular changes of relatively small-range high-frequency variability. Week-to-week and month-to-month variations of daily temperatures seem to be non-stationary with respect to both mean value and variance. The monthly temperature record resembles the stationary small-range changes of much more intensive pulsations. Finally, long-term changes in annual temperatures demonstrate no monotonous "trend" but manifest considerable low-frequency component of variations.

Moreover, global temperature patterns - both instrumental and paleoreconstructed - differ from each other as well as from the above four types of variability. The instrumental global temperature record for the past 130 years gives an impression of monotonous visible trend, and the 600-ka paleoregistration manifests more or less stationary oscillations with a very strong random component.

In the next chapters we will try to quantitatively analyze these and other registrations of the climate system parameters. Here we would simply like to stress that treatment of observational climatic data must be very accurate and must be based on an adequate understanding of the nature of processes in order to find some common features in apparently different paradigms. It is also evident, from these simple examples, that there are many difficulties in constructing deterministic models (GCMs), which claim to describe variabilities at all the above temporal scales at the same time.

One can find varying definitions of "climate", and "climate changes", with respect to time scales, in the scientific literature. For instance, it was previously thought that a frequency of 1 cycle/30 years was related to the minimum spectra of atmospheric variables, and therefore 1 cycle/30 years was considered as a high-frequency limit to climatic changes. (Later estimations of climatic spectra did not corroborate this hypothesis). Also climatic "variations" were distinguished from shorter-term climatic "variability". Finally, a vague definition of climate as "average weather" is widespread (see, e.g., Houghton et al. 1996, p. 55). We find the definition given by Klaus Hasselmann (1976, 1981), and his well-known classification of temporal scales the most adequate; he calls "climatic" temporal scales corresponding to frequencies lower than approximately 1 cycle/month.

To compare different climatic scales and illustrate definitions of climate (climatic changes), a schematic generalized spectrum of atmospheric variability is given in Fig. 1.15. It corresponds approximately to local changes in the near-surface air temperatures in moderate and subtropical zones, and is presented in bilogarithmic scales. An exact definition of spectral density will be introduced in the next chapter. Here we only note that it demonstrates the distribution of the energy of variations over cyclic frequencies (periods of time are also denoted in the figure).

There are clearly two types of spectral structures in Fig. 1.15.

1. Frequency intervals with neighboring peaks. There are three such intervals in Fig. 1.15: (a) one related to the speed of the earth rotation, namely diurnal and synoptic cycles. The latter, as it was already noted above, is much less sharp. (b) A group of peaks related to the earth's revolution round the sun, 12 and 6 months period. (c) A group of peaks presumably caused by long-term changes in the parameters of the earth's orbit.
2. Intermediate, between peak groups, parts of the spectrum which are larger than intervals between peaks within specific peak group: (a) The one between synoptic and seasonal frequencies; it plays an important role in Hasselmann's theory of stochastic climate models because the absolute minimum of the energy of climatic variability seems to be reached here (around 1 cycle/month frequency). (b) Another interval with gradual spectral changes is between seasonal and Milankovitch groups of peaks and can be considered as a classical interval of climate changes.

The interval between approximately several hundreds of years and a few dozen ka is relevant: the spectrum here is close to the straight line with tangent equal to - 2. In the next Chapter it will be shown that this feature of climate variability is described by Hasslemann's law of climatic variability without feedbacks.

The schematic spectrum in Fig. 1.15 reflects the local variability of near-surface air temperature. If we consider an analogous spectrum for another climatic parameter, e.g., for local sea surface temperature, there will be some modifications in the spectral pattern. For example, periodical inertial oscillations will appear, and the slope of the "basal" spectrum to the left of approximately 1 month period will be steeper and closer to the -2 degree line.

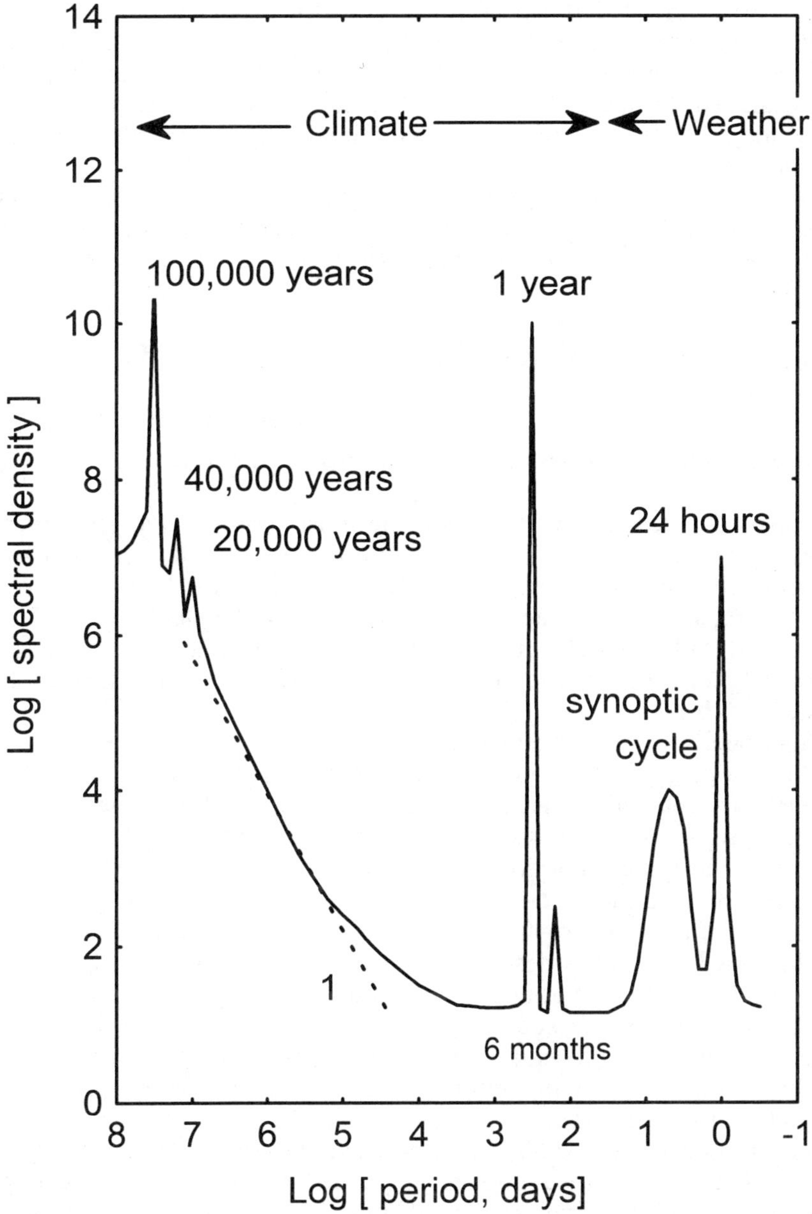

Fig. 1.15. Schematic spectrum of atmospheric variability. *Dashed line 1* corresponds to Hasselmann's -2 degree law of climatic variability without feedbacks

Major modifications of generalized spectrum, in comparison with Fig. 1.15, will be needed if we consider globally averaged parameters, e.g., global temperature. In this case, diurnal, seasonal, and 6-month peaks will be much smaller, the synoptic peak will disappear, and the whole spectrum will be much closer to the -2 degree line. In the next Chapters these and other features of generalized climatic spectrum will be discussed in more detail.

Obviously, there are features of climatic variations, either local or dubious, which are not shown in Fig. 1.15. The 20-ka and 40-ka cycles mentioned above are also subject to discussion.

1.4 Deterministic Climate Models

1.4.1 General Circulation Models and Coupled Models

So-called coupled models represent the most sophisticated instrument of climatic research today. They comprise models of the atmosphere (general circulation models or GCMs) as well as models of the ocean, cryosphere, and land surface. The components of coupled models can also be called full models in that they include the three-dimensional representation and interaction of these components on a global time--dependent basis.

Historically, the atmospheric component of coupled models, i.e., GCMs, originated from models for weather forecasting. They had to simulate synoptic eddies and other relatively small-scale phenomena in the atmosphere, so their spatial and temporal resolution had to be rather detailed. The typical horizontal resolution of GCMs is a few degrees of latitude and longitude, and they can comprise several dozens of vertical layers in the troposphere and stratosphere. The main task in constructing the first GCMs was the description and understanding of the present-day climatic distribution of atmospheric parameters over the globe with fixed boundary conditions (fixed sea-surface temperatures and solar forcing). Indeed, already by the mid-1960s GCMs satisfactorily described global fields of atmospheric temperature and air pressure at global, continental, and subcontinental scales.

By the end of the 1960s the first coupled ocean-atmosphere models appeared; they described important global-scale features of the climate system although several serious difficulties appeared that anticipated present-day problems related to climate modeling. Since the 1960s computer speed has increased by millions of times (10^6 to 10^{12} operations per second), but serious problems still remain.

Figure 1.16 and Table 1.2 illustrate the present-day skills of the coupled models. In Fig. 1.16 results of 11 full models of the climate system are

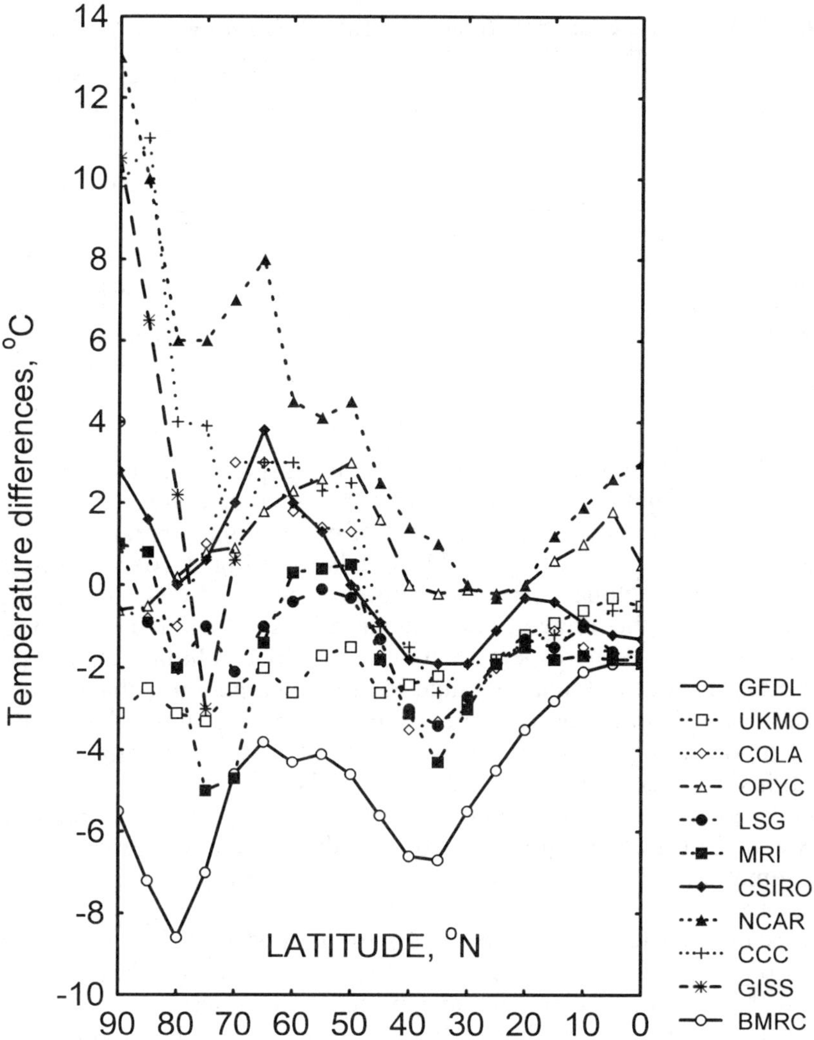

Fig. 1.16. The zonally averaged differences in surface air temperatures of 11 coupled models from observations for December-January-February. (Fragment of Fig. 5.1g from Houghton et al. 1996, p. 241)

presented - zonally averaged differences of surface air temperatures from observations for winter conditions of the Northern Hemisphere (December-January-February). It is evident that the errors in model temperatures are large: the range of estimations reaches about 20 °C in the Arctic, 12 °C in the temperate zone, and 7° C in the subtropics. (Differences for June - July - August are smaller in the Northern

Table 1.2. Mean square root differences in globally averaged parameters of 11 coupled models from observations. Calculated by the author using values from Tables 5.2 - 5.4 (Houghton et al. 1996). Observational data for sea ice are from Ropelewski (1989)

Parameter	Winter (DJF)		Summer (JJA)	
	Absolute error	%	Absolute error	%
Surface air temperature	1.44 °C	-	1.46 °C	-
Precipitation	0.49 mm/day	18%	0.13 mm/day	4.5%
NH snow cover	10.3×10^6 km^2	23%	7.9×10^6 km^2	99%
NH sea ice cover	3.9×10^6 km^2	26%	5.3×10^6 km^2	54%
SH sea ice cover	8.8×10^6 km^2	90%	6.5×10^6 km^2	123%

Hemisphere, but larger in the Southern Hemisphere). It is evident that errors in the estimation of local temperatures are much larger.

In Table 1.2 globally averaged errors of main climatic parameters (mean square root estimations using the 11 models from Fig. 1.16) are presented. Errors in mean global air temperatures, about 1.5 °C, are small in comparison with the range of seasonal variations in local temperatures, but are considerable with respect to long-term global climate changes. For example, it is believed that the range of global temperature changes during the past century was approximately 0.6 °C, which is much less than the above model error. Therefore, it is very difficult to study global climatic variations using contemporary coupled models.

Model skills to assess such key parameters of the climate as the area of snow and sea ice covers are even worse. Table 1.2 shows that the errors in the estimations of these parameters have an order of respective observed values, i.e., errors are often close to 100%.

The causes of these and other errors of coupled models can be divided into two groups. The first type of problem can be called technical, and are related to an insufficient understanding and description of specific physical, chemical, or biological mechanisms within the components of the climate system. We can also speak of technical problems related to the lack of observational data, which are needed to construct physical models and to validate their results.

Here are some examples of problems of the above kind (Houghton et al. 1996):

- Radiative effects of clouds remain an area of difficulty.
- The large-scale dynamics of current ocean models are not completely validated, in part because of a dearth of appropriate observations.
- Fluxes at the ocean-atmosphere interface have not been yet fully examined; there may be serious errors.

- Existing oceanic models use very crude parametrizations of subgrid scale processes for near-surface and interior mixing and for deep convection.
- The role of sea ice in climate change is especially uncertain because of poorly known interface feedbacks.
- Current models used for climate change studies do not satisfactorily simulate ENSO processes.
- Solar fluxes at the land surface are currently highlighted as being significantly in error compared to observations due to the inadequate treatment of clouds.
- There are intrinsic difficulties in parametrizing slope effects to determine runoff in a climate model.
- Biases and uncertainties in the surface energy balance and radiation and water budgets are a significant source of error in simulations of climate.
- Models of runoff, in general show great uncertainty in global models, there are no convincing treatments of the scaling of the responsible processes over the many orders of magnitude involved, and in high latitudes of the effects of frozen soils.
- How the integrated stomatal resistance for global vegetation might change with changing climate and CO_2 concentrations is largely unknown.

The above list can be easily continued: see recent works of Connolley and King (1996), Hecht (1996), Pate-Cornell (1996), Cane et al. (1997), Gualdi et al. (1997), Kachiola and Reck (1997), Kerr (1997), Ramanathan and Vogelmann (1997), etc. At the same time, it seems that problems like these, in principle, can be solved. At least, essential improvements might be expected in the future.

Unfortunately, there are much more serious problems with coupled models, related to the fundamental properties and the very nature of the climate system.

At first, the problems of spin up, flux adjustment and model drift represent major difficulties today as well as 30 years ago despite the much better understanding of specific climate mechanisms and the enormous increase in computer speed since the 1960s. These problems are related, on the one hand, to the very strategy of climatic research and, on the other, to fundamental features of nature: the intrinsic differences between climatic subsystems, and between respective timescales.

It must be recalled that the aim of modelers, while studying contemporary climate, is to integrate the coupled model from arbitrary initial conditions to a climatic statistical equilibrium under fixed (but, evidently, seasonally varying) solar forcing. The dearth of such experiments and long oceanic time scales make it impossible to reach this aim directly, by brute force, so the strategy is to spin-up atmospheric and oceanic models separately before coupling them. The spin-up procedure leads to the artificial absence of atmosphere-ocean feedbacks between the moments of adjustment. This, as well as possible errors in model ocean--atmosphere fluxes, "result in a drift away from a realistic current climate when the models are coupled" (Houghton et al. 1996). Several procedures are used in order to improve this situation: acceleration techniques for reaching ocean model equilibrium under surface forcing, initializing the ocean model with the aid of three-dimensional ocean data, flux adjustment techniques - see Manabe and Stouffer (1988), Sausen et al. (1988), Gates et al. (1993), Gleckler et al. (1994), Kerr (1994), Bye (1996), etc.

The above approaches to climate studies rise the following doubts:

- Spin-up, acceleration, initializing, and other similar techniques are artificial. They are not related to natural features of climate but are caused by computational necessities and the shortcomings of existing climatic subsystem models.
- Moreover, these procedures lead to distortions in representing key climatic physical processes - ocean/atmosphere feedbacks and processes within climatic subsystems.
- Some of the above procedures use explicitly (initializing), and others use implicitly data on the present-day state of the climate system in order to describe the same, contemporary climate.
- The very notion climatic equilibrium seems vague. It is not at all evident that this equilibrium exists under fixed (seasonally varying) boundary conditions. On the contrary, we know from paleoreconstructions that, in the past, abrupt changes in global climate occurred under stable external forcing. Also dramatic changes occurred during ice ages when variations in solar forcing were minimal.
- The current strategy of climatic research supposes implicitly that a climatic equilibrium state (if any) exactly fits the contemporary state of ice sheets, which is not evident. Also ice sheets are fixed in the experiments although models are integrated for thousands of years, and there must be changes in ice sheets during these periods of time.
- Even if an equilibrium state of the climate system were real and were satisfactorily described by coupled models, their ability to describe long-range climatic changes would not be evident.

Moreover, the overall deterministic character of the present-day concept of coupled models seems to be a source of major mistakes in formulating a strategy of research and analysis of its results. Coupled models presume that initial and boundary conditions exactly determine all parameters of the climate system at any moment in the future. However, we know that nonlinearities in the system, as well as other circumstances (e.g., the vague character of the very notions boundary, initial conditions at small spatial and timescales, multiplicity of climatic mechanisms, etc.), lead to intrinsic uncertainties in the behavior of the system at time scales of more than several weeks or months. That is, the nature of the climate system allows an infinite number of variants of the future evolution for the same initial and boundary conditions.

Therefore, the contemporary approach to climate variations might lead to a misunderstanding of the nature of global climatic evolution. Coupled GCMs/ocean models inevitably represent global climatic variations as a result of external forcing and (to a lesser extent) internal deterministic processes, whereas real global climatic changes seem to be more a result of accumulation of random impulses of heat and water fluxes between main climatic subsystems (primarily, between the ice sheets and the ocean).

So coupled GCMs/ocean models seem to be unable to describe the exact phase and magnitude of natural global climatic changes in the past and in the future, and they are not focused on the description of statistics of these changes. From our point of view, in many cases important features of month-to-month, inerannual, decadal, secular, and paleo natural climatic changes are better described with the aid of simpler but better structured models explicitly taking into consideration the random nature of climate. At the same time, it is evident that "full" models are incomparable for studying specific mechanisms within climatic subsystems, obtaining three-dimensional fields of atmospheric and oceanic parameters for given boundary conditions, examining results of external climate forcing, and in many other important cases. (The discussion on full climate models will be continued in Chapter 2).

1.4.2 Other Types of Climate Models

Besides coupled full climate models, the following earlier or alternative types of deterministic models are used in climate change studies:

- Historically, so-called energy balance models (EBMs) were the first mathematical tools for the description and understanding of climate change. Budyko-Sellers models are perhaps the best-known among them. These are zero-dimensional, box or latitudinally averaged models with a simplified description of climatic processes: usually albedo--temperature feedback is a key mechanism. EBMs were used with considerable success for studying possible man-made climate changes and paleoclimates. EBMs were also used as a deterministic block in stochastic climate models (see following chapters).
- Later, a more sophisticated variant of EBMs appeared - statistical dynamical models (SDMs). In these models more or less exact equations of atmospheric hydrothermodynamics were first averaged over intermediate periods of time (several weeks or 1 month), and then

often averaged globally over latitudinal belts. So SDMs operated not with actual values of atmospheric variables but with statistics of these variables, although SDMs remained deterministic models. In this way, SDMs helped to understand zonally averaged features of atmospheric parameters and circulation.

- Somewhat intermediate, between EBMs, SDMs, and GCMs, is the so-called thermodynamic model of the climate system by Julian Adem (AM). Seemingly, this is the only climate model, apart from GCMs, which has such a long history (more than 30 years) and such a large number of various applications (see Adem 1991a; Dobrovolski 1994). Like the SDMs, the AM operates with time-averaged variables, but it is fitted to the real global geography and takes into consideration a larger number of atmospheric and oceanic processes. The main difference between the ideology of AM and GCMs is that the latter are based on classical hydrothermodynamics (i.e., on the laws initially obtained from observation of fluids and gases on small spatial and temporal scales), whereas the former tries to construct a blended mathematical description of climate based on both classical thermodynamics and a study of climate-scale processes themselves. This approach has its own advantages as well as restrictions; it is successful in describing monthly anomalies of air and sea temperatures (Adem et al. 1994), possible man-made changes of climate (Garduño and Adem 1995), and historical (Garduño et al. 1997) and paleoclimate changes (Adem 1991b). Although AM is a deterministic model, it appeared to be very useful as a component (deterministic block) of stochastic climate models. In the next chapters we will discuss this and other applications of the Adem model.
- One more, important type of climate model is the family of radiative-convective models (RCMs). These are usually one-dimensional models with good vertical resolution and more or less accurate description of vertical convection, radiation (and later some chemical) processes. Many results were obtained, and important hypotheses were formulated using RCMs. For example, a positive water vapor feedback was hypothesized in the earliest simulations of global warming with simple radiative-convective models.

In Chapter 2 a more detailed classification of models will be given.

1.5 Observational Basis for Stochastic Climate Theory

Observational data play a key role in climate theories. Climatic archives are the basis for both construction and validation of climate models. Strictly speaking, climate models and theories are a tool for interpolation and extrapolation of climatic data. Three main problems related to climatic data are: coverage of the climatic subsystems by the observations; means of obtaining indirect information where direct observations are not possible; errors in the data.

1.5.1 Data on Variables of "Thick" Climatic Subsystems

1.5.1.1 Near-Surface Air Temperature

There are two most important climatic variables: near-surface air temperatures (NSAT) and sea surface temperatures (SST). The first is critical with respect to the life on the earth's surface, whereas the second is, perhaps, the most important climatic state parameter that determines, to a large extent, local air temperature, air-sea interactions, and globally averaged temperatures.

Historically, NSAT was one of the first climatic variables to be measured directly, and NSAT data are available from many data bases. For instance, monthly NSAT time series from several thousands meteorological stations as well as precipitation and atmospheric pressure series (also 16 other archives) together with the appropriate software are available from the CD-ROM World Weather Disk (1994). Frequently, NSAT, averaged over the globe, was used as global temperature. Combinations of NSATs over the continents and SSTs for oceans also served as estimates of the global temperature (see, e.g., Folland et al. 1984).

However, even air temperature series are rather short: the length of rare longest monthly NSAT records is a few hundreds of years, and the typical length is several dozens of years. Thus, it is very difficult, using the available data, to analyze features of climatic variations with characteristic scales larger than few dozens of years. Also the coverage of the earth's surface by meteorological stations was uneven: the territories of the North America, Europe, and the former USSR were better represented, whereas meteorological stations in other regions were sparsely settled. Finally, the problem of man-made impact on the estimations of NSATs exists.

The situation with observations over the oceans (i.e., over 71% of the earth's surface) was dramatic: for a long time only observations at island stations, scientific ship stations, and commercial ships were available.

Complete data, even for the period since 1949, are available only for areas southward of 75° N and northward of 50°S (Chen 1985a,b). With respect to NSAT direct observations over the ocean, also a problem of ship coordinates and positioning arose. Calculations by Larin (1982) showed that irregular errors of monthly NSATs over oceans for 5° trapezia could be of an order of monthly anomalies' values themselves. Other works (Larin 1986a,b; Esbensen and Reynolds 1981; Oort and Maher 1985) demonstrated that errors in directly measured monthly NSATs over the ocean are so important that at present it is possible to reveal only the most general features of NSATs variabilities.

Recently, satellite-based NSAT measurements have become available. Global coverage is a great advantage of these data, although the problem of errors in remotely sensed NSATs remains. Timofeev et al. (1996), Konda et al. (1996), and other authors suggest that these errors are of an order of 1 °C, i.e., of an order of monthly NSAT anomalies and more than secular changes in global temperatures.

1.5.1.2
Other Atmospheric Variables

Note that time series of aerological and remotely sensed data on atmospheric variables, in general, are much shorter than series of near-surface variables - a few dozens of years. For instance, there is a global set of satellite-based measurements of temperature in the low 6-km layer (2.5° x 2.5° grid) of the atmosphere since the late 1970s. By the way, an analogous archive of NSATs based on observations at meteorological stations since 1979 (5° x 5° grid) demonstrates that long-term trends in NSATs and 6-km layer temperatures do not coincide (Balling 1996).

Important information on the errors in satellite estimations of long-term changes in air temperatures is given in Susskind (1993). It is shown that errors in June 1988 - June 1980 differences of the 500-1,000 mb layer mean that remotely sensed temperatures have a variance of 0.97 °C (global mean value), i.e., are of the order of climatic thermal changes themselves.

Remotely sensed data are an important source of information on cloud coverage and precipitation on the global scale. Strictly speaking, rainfall itself is not detected directly by satellites; existing precipitation products are scaled cold cloud frequencies which are, at least for certain areas (especially for tropics and subtropics) and times of year, highly correlated with spatially averaged accumulated rainfall. The spatial precision of estimates derived from cloudiness, however, is certainly suspect, since the expanse of cloud is often substantially greater than that of the associated precipitation (Arkin and Janowiak 1993). Also it is well known that traditional estimations of precipitation (at meteorological stations) have large errors: errors in monthly liquid precipitation can reach 10%, and errors in solid precipitation can have an order of several dozens of percent.

Moreover, in polar, desert, and mountainous regions, such errors can easily reach 100% (Kuznetsova 1998, pers. comm.). Therefore, it is very difficult to validate satellite-based estimations of precipitation, and to estimate climatic changes of precipitation.

Large errors are also characteristic for the estimations of wind stress over the sea surface: about 20% (Janssen 1997), or 2.0-6.5 ms^{-1} (Kishtawal et al. 1996).

1.5.1.3
Sea Surface Temperature

There are two main ways of estimating SST: using routine in situ observations (data from ships and buoys) and remotely sensed data. Strictly speaking, these two methods give estimates of two different parameters: the temperature of the near-surface water layer (from several dozens centimeters to several meters) in the first case, and the temperature of a thin surface film in the second. The respective difference of estimations can be 1 - 2°C (Birman and Balashova 1986).

Traditional in situ data cover a time period since the end of the 1940s; they are usually averaged over 5° x 5° or 2.5° x 2.5° squares and over a time period of 1 month. The coverage of the ocean by observations of this type is, however, limited; The North Atlantic and some other regions are much better represented than vast parts of the ocean in the Southern Hemisphere and central parts of the Pacific and Indian Ocean. Maps illustrating the spatial distribution of routine SST data can be found in Oort and Maher (1985) and Reynolds (1988).

However, even SST data in squares frequently crossed by ships contain large errors (Esbensen and Reynolds 1981; Larin 1982, 1986a,b; Barnett 1984; Folland et al. 1984; Oort and Maher 1985, etc.). Larin (1986a,b) shows that for various 5° trapezia in the North Atlantic (far from the subarctic hydrological front) average errors of monthly SST estimations are 0.1–0.7 °C, i.e., 15–76% of the respective SST standard deviations. In Dobrovolski (1992) it is demonstrated that appropriate errors in the zone of the subarctic hydrological front can reach 1.5–2.0 °C, i.e., up to 100% of monthly SST standards. Note that these errors cannot be diminished appreciably by the increase in the amount of in situ data because the spatial distribution of ship observations remains uneven.

Finally, note that monthly SST series even in the North Atlantic contain many blanks, especially in equatorial, tropical, and polar latitudes. On the whole, it is clear that at present only the most general conclusions on the features of in situ SST climatic variations can be made.

Many papers are devoted to the analysis of SST remote sensing techniques and errors. It is interesting that during the past 10 years (since the 1980s) progress in acquiring satellite-based SST estimations has not been very impressive. In the 1980s average errors of individual SST measurements using satellite data varied from 0.5 °C (Measuring ... 1987; Yokoyama and Tanba 1988; Gorodetski 1989) to 1 °C (Ahmad et al. 1985; Milman and Wilheit 1985; Malkevich and Gorodetsky 1988). In the 1990s analogous estimations of errors were 2–3 °C (Kerenyi 1993), 1 °C (Mutlow et al. 1994), and 1.54–1.83 °C (Trishchenko and Khokhlova 1996). Respective errors in monthly SST can exceed 1 °C. Thus, we see that our ability to suggest sophisticated models of SST long-term variations is restricted by uncertainties in observational data. Seemingly, only the most important features of SST climate changes have a chance of being detected.

1.5.1.4
Sea Level

Local sea level height at a coastal station is the easiest measuring parameter of the climate system; data of this kind are available (for some tide gauges) from the beginning of 19th century. Many authors of global sea level change reviews, based on tide gauge observations, use the Permanent Service for Mean Sea Level (PSMSL) data set (Spencer and Woodworth 1993). This data set is also available on INTERNET.

Now global sea level monitoring is based on a network of modern tide gauges. The Global Sea Level Observing System (GLOSS) of the Intergovernmental Oceanographic Commission (IOC) is a coordinated project for the monitoring of long-term global sea level change and is intended to serve the various purposes of oceanographic and climate change research (IOC 1990; Woodworth 1991). GLOSS consists of a network of approximately 300 tide gauges worldwide, of which over two-thirds are now operational. Technical developments in recent years have seen many of the traditional float and stilling well tide gauges replaced by modern systems based on pneumatic and acoustic principles (Spencer 1992). Many of these have satellite or telephone data transmission equipment, enabling real-time data access and fault checking. Bottom pressure recorders (Spencer and Vassie 1985), inverted echo sounders (Wimbush 1990), and thermistor chain moorings for dynamic height (McPhaden 1993) now provide quasi-sea level measurements in several areas of the deep ocean, which will provide information on the ocean circulation (Warrick et al. 1996; see also earlier works by Baker et al. 1989; Roach et al. 1989; Wyrtki 1989).

However, estimation of global mean sea level (GMSL), its interpretation and attribution, is still a very difficult task. Recently (for the past few years) a time series of satellite-based GMSL–TOPEX/POSEIDON data (Hendricks et al. 1996) has become available, but assessment of possible errors within these data is related to many uncertainties. In Chapter 3 we will discuss these problems, using results of stochastic analysis of existing GMSL registrations.

Observational data on other than SST and sea level variables - deep sea variables, currents, fluxes, and others - usually do not provide reliable long time series with global coverage although such information can be expected in the future (data on sea ice cover will be discussed below).

1.5.1.5 Ice Sheets

At this time, despite more than four decades of sustained international field investigations costing several hundreds of millions of dollars, we still know very little about either of the polar ice sheets. It is not even known whether they are growing larger or smaller, and it certainly cannot be predicted what they will do in response to a warmer climate. Research in the past has been hampered by the vast size of the ice sheets, and the harshness of the polar environment. The advent of satellite remote sensing has allowed making measurements over very large areas of ice sheets, but "there is still a long way to go" (Thomas 1993).

However, the uncertainty in estimates of ice-sheet mass balance is equivalent to an uncertainty of ± 3 mm per year in the rate of change of global sea level (ibid.), i.e., about $\pm\ 10^3\ km^3$ in the mass balance of ice sheets. This means that even the high estimates of more than 2 mm per year sea level increase during the recent past (Peltier and Tushingham 1989) could be entirely caused by wastage from the polar ice sheets.

There are also some indirect data on recent changes in mass balance of ice sheets (information about icebergs) as well as paleodata on ice sheets. These time series will be analyzed in Chapter 3.

1.5.2 Data on Variables of Thin Earth Covers

1.5.2.1 Snow Cover

Unlike atmosphere and ocean, quantitative information on climatic changes in thin covers on a global scale can be obtained only by using satellite-borne data. In turn, remote sensing data are available, at best, for the past 20-30 years (snow cover, sea ice cover). Global-scale information on vegetation cover and soil moisture became available only 3-4 years ago or still remains at an initial, experimental stage. So only month-to-month or year-to-year variability can be investigated using such data.

Since 1966, continental snow cover extent has been monitored weekly using NOAA satellites in the visible and near infrared wavelengths, thus providing a data base with spatial and temporal continuity (Matson et al. 1986). To obtain quantitative information concerning the volume of snow, snow depth as well as snow extent must be known. The scanning Multi-

Channel Microwave Radiometer (SMMR) on board the Nimbus-7 satellite has been acquiring passive microwave data since 1978. This instrument has the capability of providing global estimates of snow depth and area by using an algorithm derived from microwave radiometric observations of snow (Foster and Chang 1993). The error bands are lacking for both products. As a reference, it can be noted that the difference between SMMR and NOAA snow area estimates is, on average, about 10%. Also, for the month of March, for example, the Northern Hemisphere "climatological" snow mass is 290 x 10^{16} g and 364 x 10^{16} g as determined from the SMMR and RAND snow depth data (based on meteorological stations archives), i.e., the difference is about 23%. The error bands are unknown and may be large; however, this temporal data set is the only source of monthly snow depth currently available (Chang et al. 1990).

1.5.2.2 Sea Ice

Remotely sensed data on sea ice cover have been available since 1972 (Electrically Scanning Microwave Radiometer, ESMR - Nimbus-5 satellite). From 1978, data on sea ice were produced by SMMR (Nimbus-7) (Parkinson and Gloersen 1993). The spatial resolution varies from approximately 30 to 150 km (Gloersen and Barath 1977). Later a special sensor microwave imager (SSMI) (Cavalieri et al. 1991) and Synthetic Aperture Radar (SAR) (Carsey and Pihos 1989, etc.) were used to provide sea ice data.

Taking all the various uncertainties into consideration, accuracy estimates for the calculated sea ice concentrations are 15-25% for the ESMR data (Zwally et al. 1983; Parkinson et al. 1987) and 5-9% for the SMMR data (Cavalieri et al. 1984; Steffen and Maslanik 1988). The highest uncertainties (about 25%) for the ESMR data occur in situations with an unknown mixture of first-year and multiyear ice. In the case of multiyear ice concentrations calculated from the SMMR data, the precision is estimated to be 13-25% (Cavalieri et al. 1984). The SSMI algorithm appears to overestimate multiyear ice concentrations by about 9%, while the mean differences between total ice concentrations obtained from the SSMI and those obtained from the aircraft range from -3 to +5% (Cavalieri et al. 1991).

1.5.2.3 Vegetation Cover

Detailed description of methods for obtaining remotely sensed parameters of global vegetation cover is given in Choudhury (1989, 1993) and Choudhury and DiGirolamo (1994), etc. The spectral, spatial, and temporal characteristics of some of the instruments useful for monitoring changes in the vegetation cover are given in Table 1 from Choudhury (1993, p.135). All in all, more than seven types of sensors can be used for this purpose, information from more than nine satellites is available, spatial resolution can vary from 30 m to 145 km, time of data acquisition from every 30 min to every 3 h (or fixed), and repeat cycle 12 hours to 18 days (or stationary). Seemingly, polarization difference in the spectral range of 37 GHz is the most convenient tool for acquisition data on vegetation cover. The parameter of vegetation cover which can be obtained from satellite observations can be called "spatial density". Nowadays, the length of time series of vegetation cover parameters does not exceed several years.

Remotely sensed data on vegetation cover and precipitation as well as in situ data are being used to obtain information on soil moisture, although there are still many problems in acquiring such information. These problems will be discussed in more detail in Chapter 3.

1.5.3 Data on Discrete and Local Climatic Objects

1.5.3.1 River Runoff

Because of their relatively small size or discrete character (rivers), objects of this type are easy to observe using routine, in situ methods, although many problems arise with respect to acquiring data for remote territories. Monthly and annual time series of the most important hydrological variable, river runoff, start at best at the beginning of the 19th century (data from Global Runoff Data Centre, Koblenz, FRG). Errors in these data vary with specific river; within economically developed territories it is very difficult to restore "natural runoff", taking into consideration water losses for irrigation and industry. Moreover, the natural variability of the runoff in many cases is a source of larger errors. For instance, the error in annual Volga runoff is estimated to be approximately 7%, and the error in the estimation of Volga mean climatological runoff is thought to be ± 19 km^3/ya, i.e., it exceeds possible man-induced runoff changes because of irrigation, additional evaporation from the surface of reservoirs, etc. Global coverage of runoff observations also represents a major problem. For example, reliable data on Amazon runoff has become available only recently, and information on many rivers in tropical regions is still poor.

1.5.3.2 Lakes

Water level height in lakes is, perhaps, the most accurately measured parameter of the climate system. Possibly, here we have the only case where errors in estimating an important climatic parameter are close to zero. That is why lake water heights were often considered as important indicators of climate change (even as indicators of global climate change, which was not always true). In fact, the technical accuracy of water level height measurements in lakes is not worse than that of sea level height, but unlike for the ocean, lake water level height measured at a specific station (gauge) is usually representative for the whole lake. Also, changes in lake water level are usually larger than changes in global mean sea level. For example, characteristic changes in the Caspian Sea level during the past century were several meters, i.e., an order of magnitude larger than changes in global mean sea level. Finally, it is much easier to attribute data on lake water level changes because geological, geographical, and other natural conditions around a specific lake are more homogeneous than a global environment affecting the behavior of larger climatic subsystems and objects.

The above circumstances make it possible to successfully use data on lake levels for many purposes: calculating lake water balances using traditional methods, application of aerological method for lake water bodies balance and for lake watershed balance, etc.

1.5.3.3 Mountain Glaciers

The international database on mountain glaciers contains two different kinds of information about glaciers of the 20th century : (1) glacier inventory data describing spatial variability and (2) glacier fluctuation data documented changes over time. The World Glacier Monitoring Service (WGMS) of the International Commission on Snow and Ice (ICSI/IAHS) collects and publishes worldwide standardized data as a contribution to the Global Environment Monitoring System (GEMS) of the United Nations Environment Programme (UNEP) and to the International Hydrological Programme (IHP) of the United Nations Educational, Scientific and Cultural Organization (UNESCO) (Hoelzle and Trindler 1998).

These data are being made accessible in different ways, traditionally through publications, and today also through electronic data access. The data appear in the following three publications of the WGMS:

- *The World Glacier Inventory.* This is a worldwide glacier inventory of existing perennial ice- and snowmasses (see Haeberli and Hoelzle 1995).
- *The Fluctuations of Glaciers* series. In this database, long-term measurements of glaciers, mainly as time series, are stored in a standardized way. Data can presently be found for 1,440 glaciers relating to mass balance, length variations, changes in volume, area, and thickness

[IAHS(ICSI)/UNESCO 1967, 1973, 1977, 1985; IAHS(ICSI)/UNEP/UNESCO 1988, 1993a].

• *The Glacier Mass Balance Bulletin.* In this bulletin summary information is presented for about 50 glaciers, and extensive information for 10 glaciers [IAHS(ICSI)/UNEP/UNESCO 1991, 1993b, 1994].

All in all, not less than 30 time series of relatively long annual mass balances of mountain glaciers can be found in the above sources, although errors in these time series are largely unknown (Reynaud and Dobrovolski 1998). Results of stochastic analysis of these series will be discussed in Chapter 3.

1.5.4 Conclusions on Observational Data

Finally, in Table 1.3 expert estimations of possible errors and other parameters of climatic time series are shown. The most important result of the analysis in this table is that time series are usually short, and errors are often comparable with climatic-scale changes in the respective variables themselves. This means that, in principle, our ability to model and analyze climatic variations is strongly restricted today. It is evident that only the most general features of climatic variations can be detected on the basis of such an observational background. On the contrary, artificially sophisticated models that need many parameters to be assessed, are risky.

Table 1.3. Observational data on main variables of the climate system (expert estimations)

Variable	Trad.[1]/ Sat.[2]	Coverage	Characteristic series lengths	Characteristic errors
Near surface T_a^m	Trad.	Land parts	Dozens of years	10^{-1} °C
Low 6-km T_a^m	Sat.	Global	~20 years	10^0 °C
Precipitation^m	Trad.	Land parts	Dozens of years	5% - 100%
Wind stress^m	Sat.	Ocean	~20 years	~20%
SST^m	Trad.	Oc.parts	40-50 years	10^0 °C
SST^m	Sat.	Global	~20 years	10^0 °C
Sea level^a	Trad.	Coastal	Dozens of years	10^0 cm
Sea level^m	Sat.	Global	A few years	10^0 cm
Ice sheets^a	Sat.	Global	A few years	10^3 km^3
Snow cover area^m	Sat.	Global	~20 years	~20%
Sea ice area^m	Sat.	Global	~20 years	5 - 25%
Vegetation density^m	Sat.	Global	~20 years	?
River runoff^m	Trad.	Selected	Dozens of years	~10% and more
Lakes level heights^m	Trad.	Selected	Dozens of years	Small errors
Mountain glac. bal.^{a}	Trad.	Selected	Dozens of years	?

Trad.–traditional data; Sat.–satellite-based data; T_a–air temperature; ^{a} - annual; ^{m} - monthly

2 Theoretical Foundations of the Stochastic Approach to Climate Variability Studies

2.1 Basic Ideas and Principles of the Stochastic Climate Theory

2.1.1 Mathematical Models and Natural Processes

Investigations in the research area of climate change are characterized by the intensive use of sophisticated mathematical models (most of them deterministic): general circulation models and coupled models, deterministic chaos and neural networks, etc. However, many questions remain unanswered. Computer speed has increased by millions of times since the 1960s, information (e.g., remotely sensed) about the climate system has increased by zillions of times, detailed models of hundreds of specific mechanisms within the system have been proposed, but our skill forecasting the behavior of the climate system with a lead time of more than several days is still limited. Researchers still wonder why recent global "heating" is accompanied by a negative thermal trend in the lower troposphere, and why, 17,000 years ago, under approximately same solar conditions, there was a 1-km-thick ice sheet over Chicago.

The author of the present book does not claim to find solutions to all these problems. He would simply like to discuss these matters and try to understand, together with his colleagues - authors of publications in different research areas - something common and fundamental in the main features of climatic changes. For this purpose, it is necessary to return to the very beginnings of the stochastic (random) processes theory and to discover the possible climatic meaning and analogies of this theory. First of all, it is useful to recall the main problem of mathematical modeling: relationships between the natural process and its model (Fig. 2.1). Here, area *A* denotes natural process(es) under investigation, and *B* denotes a class of mathematical models we want to use in our research. Mathematical models *B* are a simplified description of natural objects or processes. The structure and principles of behavior of *B* are known to the researcher. Because of the relative simplicity of the mathematical model, knowing only part of information on the natural process *A* (e.g., from observational data), we can artificially "reconstruct" *A* obtaining an "identified" variant of mathematical model, *C*.

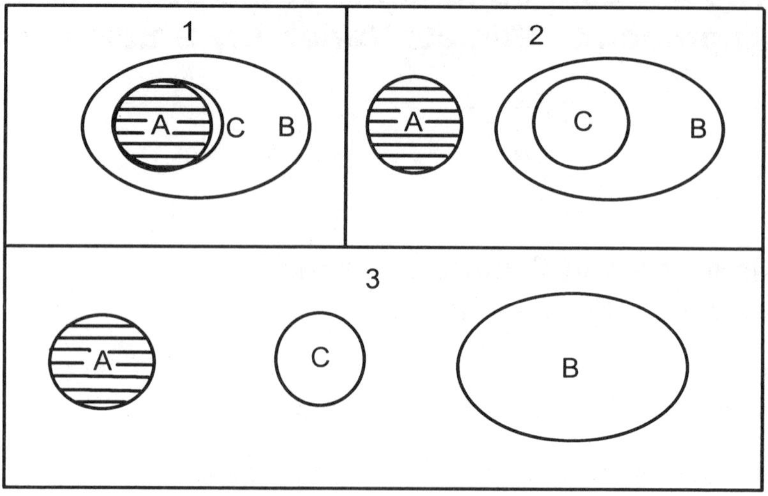

Fig. 2.1. Relationships between natural process (*A*), a class of mathematical models for the description of this process (*B*), identified specific variant of model with parameters estimated using observational data (*C*). *1* "ideal" situation; *2,3* methodological errors

In Fig. 2.1.1 an "ideal" variant is presented when the class of mathematical models involves the process under consideration, and an identified variant of the model *C* adequately describes *A*.

Figure 2.1.2 represents a case when the choice of model class is an error. Consequently, the identified model *C* is an error, too, and has nothing in common with the process *A* , so the whole investigation is a failure. Anticipating the next sections of this chapter, we can give the following example. If the process *A* is close to the realization of a nonstationary random sequence, and we use models of stationary processes for its analysis, the results will be absolutely wrong.

Moreover, the results can be even more confused if, in addition to a wrong choice of model class, we use incorrect methods of model identification and model parameter assessment (Fig. 2.1.3). In this case, the three areas, *A*, *B*, *C*, have nothing in common with each other at all. For instance, this case can occur if the estimated parameters of the model in Fig. 2.1.2 contradict the supposition of stationarity of the process. (It will be shown in the next sections that this can occur if one or several roots of the characteristic equation of autoregression are on or within the unit circle on the complex plane).

Thus, the task of a researcher studying climate changes is rather difficult and complex. He must: (a) determine the processes he wants to investigate, their spatial and temporal scales, discreteness intervals, and the length he is interested in; (b) collect observational data on the processes, fill in the blanks (if needed), and assess possible errors in the data; (c) choose the appropriate class of mathematical models for the description of the processes under consideration; (d) identify the specific type of the model(s) and assess, using observational data and apriori information, model parameters; (e) conduct experiments with the model, compare their results with observational data of the processes and with initial suppositions made while constructing the model, etc. Stages a and b were discussed in Chapter 1; in Section 2.1 we will deal with point c; Section 2.2 will be devoted to stage d of climatic change analysis; other chapters and sections of the book will be about modeling real climatic processes and applications of stochastic climate theory.

In Sections 2.1 and 2.2, we will use and discuss fundamental papers and monographs by Yaglom (1952, 1962, 1981, 1987), Kolmogorov (1956), Box and Jenkins (1970), Van den Boss (1971), Privalsky (1985), Privalsky et al.(1992), Privalski and Jensen(1993), Box, Jenkins, and Reinsel (1994).

2.1.2 A Climatic Variable as a Random Variable

There are two possible approaches to the study of climatic variability (as well as to the study of many other natural and man-made processes). The first is deterministic, i.e., we postulate that the value of some variable of the climate system, ξ, is unequivocally determined by the value of this and other variables of the climate system, also by relevant parameter outside the climate system, at previous moments of time. From this point of view, the task of the climatologist is to obtain as much information on climatic variables and parameters and to study physically as many relevant processes as possible. The ideal of such a strategy is to obtain enough detailed information to be able to give an exact forecast of the variable under consideration. In accordance with the deterministic approach to the climate, ξ is an exact value, and if it is not so with respect to present, past, and future moments of time, the reason is errors and blanks in observational data, and shortcomings of deterministic, hydrothermodynamical, and other models.

This point of view on climatic (atmospheric) variables and parameters is, to a large extent, adequate when we deal with synoptic temporal scales. Indeed, if we have detailed information on the state of the atmosphere and oceanic surface for the past few weeks, we can deterministically describe and explain the present and future (for the next few days) behavior of the atmosphere.

However, if we consider climatic time scales (approximately 1 month and more), the deterministic approach demonstrates its limits. In practice, forecasts of major atmospheric variables - air temperature, atmospheric pressure, wind speed, especially cloudiness and precipitation, etc. - are very difficult if the lead time is more than several days. Perspectives of such forecasts are not promising either: the history of climate forecasts shows that increasing computer speed by millions of times and the improvement in observational techniques since the 1960s have enabled us to increase the lead time of successful forecasts by only a few days.

There are also the following explanations to such a situation. When we try to forecast and analyze processes with time scales of several weeks and more, we inevitably deal with the accumulation of a myriad tiny effects (related to nonlinear processes within the atmosphere and the ocean as well as to the very notions of initial and boundary conditions), which destroy the foundations of the causal approach. In other words, uncertainties which were only small annoying obstacles while forecasting weather become the very nature and the most powerful mechanism of the climate changes. (Recall that many outstanding contemporary physicists think that probabilistic features are the basis of all the laws of nature - see, e.g., Yaglom 1981, p. 7).

In the next sections we will continue the discussion on these extremely important matters; here, we would like to introduce some initial notions of the random process theory in order to be able to start and continue, step by step, a quantitative analysis of the probabilistic and stochastic properties of the climate. As is well known, a variable such as mean monthly air temperature (and many other temporally averaged climatic variables), both local and spatially averaged, is not determined unequivocally by the boundary conditions (sea surface temperature, etc.) during this month and by the evolution of the climate system during previous months.

In other words, the value of a climatic variable can be different for the same boundary and initial conditions. So it might be considered not as a deterministic variable, but as a *random variable*, ξ . Other examples of random climatic variables are the following: monthly anomalies (with previously removed seasonal dynamics) of spatially or globally averaged atmospheric and oceanic variables, annual values of variables, linear and nonlinear combinations of the above variables, etc.

A random variable ξ is regarded as specified if one knows its *distribution function F*

$$F(x) = \mathbf{P}\{\xi < x\}, \tag{2.1}$$

where the symbol $\mathbf{P}\{\ \}$ will be used to denote the probability occurrence of the relation which appears between the braces; x is a value of the random variable. In addition to the distribution function, one often uses its derivative, which is called *probability density*:

$$f(x) = \frac{d}{dx} F(x) . \tag{2.2}$$

In Fig. 2.2 an example of probability density of air temperatures is shown.

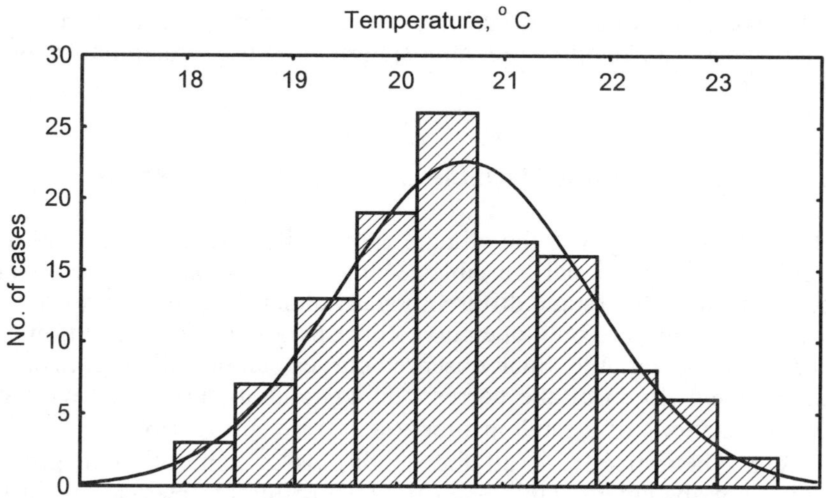

Fig. 2.2. Theoretical, normal probability distribution (*solid curve*) fitted to sample probability distribution (*histogram*). Mean monthly April air temperatures in New Orleans, 1874-1990

The histogram in Fig. 2.2 demonstrates the estimation of probability density using observational data, and the solid curve the theoretical normal distribution. The data are from World Weather Disk (1994). Evidently, in reality we always have only one observed, deterministic value of climatic variable (in our case, air temperature) for a specific moment of time (here, for a specific month). It can be considered as a *sample* value of the random variable. However, if the probability distribution is stable enough, we can use other April temperatures for the estimation of the distribution function or probability density (see below on stationarity and ergodicity of random functions).

We see from the definitions [Eqs. (2.1), (2.2)] that, roughly speaking, probability density indicates relative "chances" of finding a random variable within various intervals of values. It is evident that in Fig. 2.2 simple and fundamental normal distribution almost perfectly fits observational data. Here, we will not discuss the problem of probability distribution types. However, it is important to note that probability distributions of many variables of thick climatic subsystems are close to normal, whereas distributions of thin and local subsystem variables are often asymmetric and truncated.

Finally, if we consider n different climatic variables $\xi_1, \xi_2, \dots, \xi_n$, they can be regarded as a single *multidimensional random variable* which is specified by a *multidimensional distribution function*

$$F(x_1, x_2, \dots, x_n) = \mathbf{P}\{\xi_1 < x_1, \xi_2 < x_2, \dots, \xi_n < x_n\}, \tag{2.3}$$

which is a function of several variables. Such a representation can be useful if we consider several variables of the climate system in the same month (or year): air temperature, atmospheric pressure, precipitation, wind velocity, etc.

2.1.3 Evolution of a Climatic Variable as a Random Function

While studying changes in the climate system, it is necessary to consider not only one-dimensional variables, but also *random functions*, i.e., processes which can assume different values under the same conditions. In reality, we can observe only one curve corresponding to the evolution of a specific climatic variable. Such observed curves can be called *realizations* (or *sample functions*) of the corresponding random functions. We shall denote realizations of a random function $\xi(t)$ by superscripts, e.g., $\xi^{(1)}(t)$, $\xi^{(2)}(t)$, etc. Theoretically, if all the conditions of some climate process evolution are absolutely stable, we can make different observations of the same random climatic function and obtain different realizations of the random function in question.

In Fig. 2.3 a scheme of a random function, sample function (realization), and random variable is given. Here, really observed variations of annual temperatures in New Orleans are plotted as one of the sample functions (dashed line). It is evident that this curve is not the only one possible, and it is easy to imagine other realizations of temperature variations which are plotted using solid lines. It is important to stress that, contrary to a widespread and mistaken opinion, the realization of a random function $\xi^{(n)}(t)$ is, in turn, not a random but an absolutely deterministic (real) function of time. It is also important to note that the number of realizations $\xi^{(1)}(t)$, $\xi^{(2)}(t)$, etc. is generally supposed to be infinite.

Let us introduce some formal description of a random function. Let T be a set of moments of time t_1, t_2, ... By a random function of an argument t in T we can mean a function $\xi(t)$ whose values are random variables. Thus, a random function on T is a family of random variables $\xi(t_1)$, $\xi(t_2)$, ... , corresponding to all moments t_1, t_2, ... of time T. If the set T contains only a finite number of elements (moments of time), then the random function $\xi(t)$ is a finite family of random variables. In this case we can regard $\xi(t)$ as a single multidimensional random variable. Such random variables are studied in classical probability theory, and are specified by using multidimensional distribution functions [Eq. (2.3)]. Formally, a time series of climatic variables can be regarded as multidimensional variables because the number of points within the time series is always finite. However, we are usually interested not only in the behavior of a climatic variable within the time interval of available observations, but in its behavior in the future and in the past as well as in general features of climatic variabilities. So it is more convenient to consider an infinite set

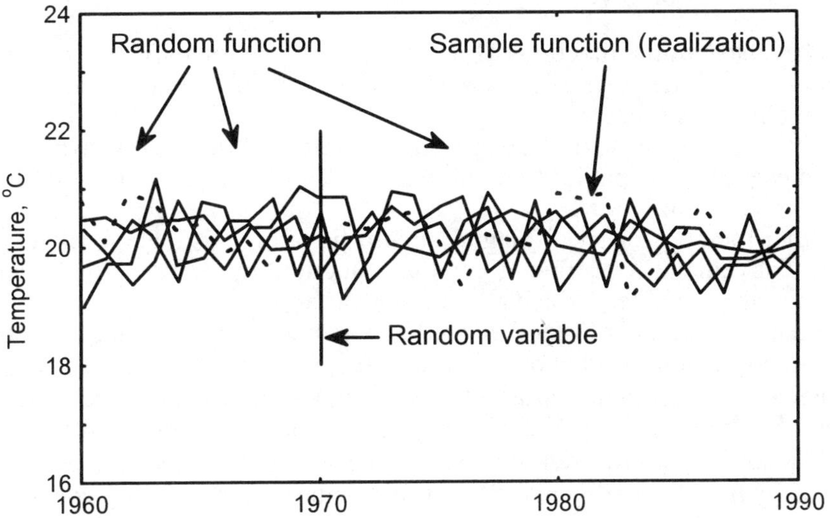

Fig. 2.3. Scheme of a random function $\xi(t)$, sample functions (realizations) $\xi^{(1)}(t)$, $\xi^{(2)}(t)$, ... , random variable $\xi(t_1)$. One of the sample functions (*dashed line*) corresponds to the registration of New Orleans annual temperatures (see Fig. 1.12)

T, then $\xi(t)$ is an infinite family of random variables. Such families of random variables are not studied in the classical probability theory, and hence the specification of $\xi(t)$ requires a special definition.

Following Slutski, we can regard *the random function* $\xi(t)$ *as being specified if for each element (moment of time) t in the set T we are given the distribution function*

$$F_t(x) = \mathbf{P}\ \{\xi(t) < x\} \tag{2.4}$$

of the quantity $\xi(t)$, *if for each pair of elements* t_1, t_2 *in the set T, we are given the distribution function*

$$F_{t1,t2}(x_1, x_2) = \mathbf{P}\ \{\xi(t_1) < x_1, \xi(t_2) < x_2\} \tag{2.5}$$

of the two-dimensional random variable $\xi\ (t_1, t_2) = [\xi(t_1), \xi(t_2)]$, *and so on, and in general if for any n elements* t_1, t_2, ... , t_n *in the set T we are given the distribution function*

$$F_{t1,t2,\dots,tn}(x_1, x_2, \dots, x_n) = \mathbf{P}\ \{\xi(t_1) < x_1, \dots, \xi(t_n) < x_n\} \tag{2.6}$$

of the n-dimensional random variable

$$\xi(t_1, t_2, \ldots, t_n) = [\xi(t_1), \xi(t_2), \ldots, \xi(t_n)].$$

The distribution functions [Eq. (2.6)] must also satisfy the *symmetry condition* and the *compatibility condition*, which are seemingly natural for a real climatic processes.

Note that any system of distribution functions [Eq. (2.6)] satisfying the symmetry and compatibility conditions can be regarded as defining some random function.

Other methods of specifying a random function exist. Thus, sometimes it is convenient to define a random function by an analytic formula containing parameters which are random variables.

Finally, there is still another method of specifying a random function. As already emphasized above, in order to apply probabilistic methods, we can regard the climate variability as an experiment which, theoretically, could be "repeated" by Nature many times under similar conditions, and which can lead to different outcomes. The set Ω of all possible outcomes, ω, of such an "experiment" (the set of "elementary events") plays a basic role in the contemporary axiomatic formulation of probability theory (Kolmogorov 1956). In fact, by a random variable ξ is meant a quantity which takes different numerical values for different outcomes of a natural experiment with the earth's climate. In other words, a random variable is a numerical function of the point ω of the set Ω. Therefore, with respect to a random variable, it would be more accurate to write $\xi(\omega)$ instead of ξ. Since we have defined a random function $\xi(t)$ on T to be a family of random variables corresponding to all the elements t of the set T, from this point of view a random function is a function $\xi(\omega, t)$ of the two variables ω and t.

If we fix the value of the argument ω in $\xi(\omega, t)$, we have a definite (deterministic) function $\xi(\omega, t) = \xi^{(\omega)}(t)$ of the time variable t, depending on the parameter ω. Thus, a random function $\xi(\omega, t)$ can be regarded either as a family of random variables $\xi_t(\omega)$, depending on time, t, or as a family of realizations, $\xi^{(\omega)}(t)$, depending on the parameter ω. In the latter case, to specify the random function we have to give the probability (or probability density) of occurrence of the various realizations. This leads to the definition of a particular measure P (the *probability measure*) on the functional space of realizations, and by specifying this measure, we specify the random function.

This approach to the concept of a random function, originating in Kolmogorov's work (Kolmogorov 1956) on the basis of the probability theory, has been very fruitful for the further development of the theory. It can be shown that specifying a random function by defining a probability measure on the set of realizations of the function includes as special cases all the other ways of specifying a random function discussed above (see

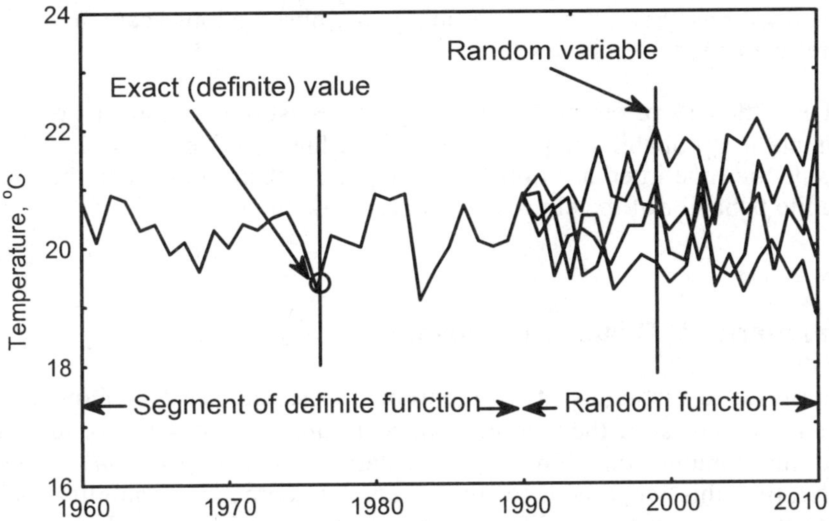

Fig. 2.4. Scheme of a forecast taking into consideration the random nature of climatic variability. *Curve 1960-1990* corresponds to the existing registration of annual temperatures in New Orleans

the fundamental theorem in Kolmogorov 1956, p.29), and moreover has many advantages over the other approaches.

The above considerations are based on a supposition that the random function is infinite. However, in reality, we always possess only one realization of climatic random function, and this realization has a limited length. So it is important to know how to treat the missing parts of random function and the lacking realizations from the point of view of the theory of random functions.

Figure 2.4 illustrates this situation. We can consider an existing segment of observed realization (before 1990, in our case) together with unknown parts of climatic random function (after 1990) as a specific random function, $\xi^{(\omega 1)}(t)$. As well as the general random function of this climatic variable $[\xi^{(\omega)}(t)]$, function $\xi^{(\omega 1)}(t)$ has an infinite number of realizations (or values of random parameter ω). Specific properties of $\xi^{(\omega 1)}(t)$ are the following:

1. The number of realizations of the function $\xi^{(\omega 1)}(t)$ is infinitely less than the number of realizations of $\xi^{(\omega)}(t)$.
2. Within the time interval of observations, all realizations of the function $\xi^{(\omega 1)}(t)$ coincide with each other and with the observed registration of our climatic variable.
3. Outside the time interval of observations, realizations of $\xi^{(\omega 1)}(t)$ differ from each other. The family of these realizations after the end of the

observations can be called forecast whereas the family of respective realizations before the beginning of observations can be called retrospection.

Thus, the task of researchers making a forecast of climatic changes is to estimate the probabilistic parameters of the family of realizations, $\xi^{(\omega 1)}(t)$, outside the time interval of observations, rather than to estimate the exact, definite behavior of the climatic variable in the future.

2.1.4 Stationarity of Climatic Processes

It is useful to distinguish two important cases. The set T of moments of time t can consist of the whole set of real numbers $-\infty < t < \infty$, in which case the random function $\xi(t)$ is usually called a (*random*) *process*. Evidently, this approach to the study of climatic variability is an abstraction, because in reality climatic registrations are not only limited in time, but also consist of discrete quantities. At the same time, many important theoretical results have been obtained for this case, and sometimes climatic records, with some reserve, can be regarded as realizations of a random process.

On other occasions, T can consist only of the set of all integers: ... , -2, -1, 0, 1, 2, ... , in which case it is convenient to call $\xi(t)$ a *random sequence*. The term random function is often reserved for situations in which we wish to consider random processes and random sequences simultaneously.

The random function $\xi(t)$ *is called stationary if all the finite--dimensional distribution functions* [Eq. (2.6)] *defining* $\xi(t)$ *remain the same if the whole group of points* $t_1, t_2, \ldots, t_n$ *is shifted along the time axis, i.e., if*

$$F_{t_1+\tau, t_2+\tau, \ldots, t_n+\tau}(x_1, x_2, \ldots, x_n) = F_{t_1, t_2, \ldots, t_n}(x_1, x_2, \ldots, x_n) \tag{2.7}$$

for any $n, t_1, t_2, \ldots, t_n$ *and* τ. In particular, this implies that for a stationary random function, all the one-dimensional distribution functions [Eq. (2.1)] and probability densities [Eq. (2.2)] have to be identical, i.e., $F_t(x)$ cannot depend on t . Also all the two-dimensional distribution functions [Eq. (2.3)] can only depend on the difference $t_2 - t_1$, and so on.

From the purely physical point of view, the meaning of the concept of stationarity can be the following (Yaglom 1952). "It means that $\xi(t)$ describes the time variation of a numerical characteristic ξ of an event such that none of the observed macroscopic factors influencing the occurrence of the event changes in time".

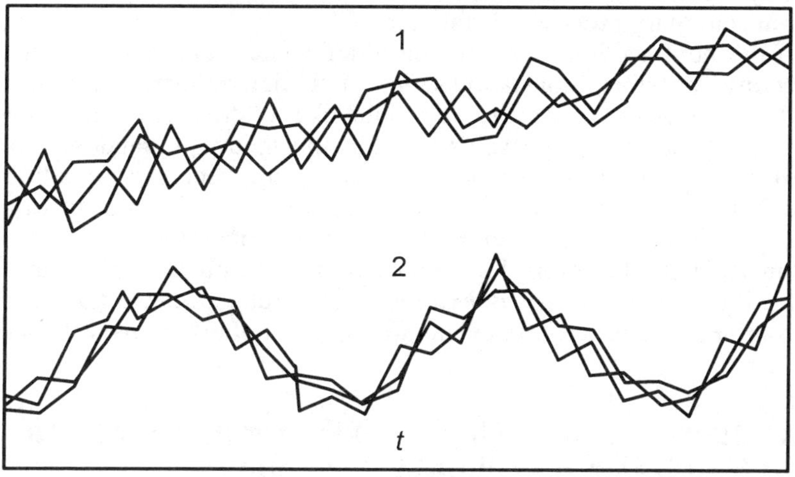

Fig. 2.5. Examples of climatic processes which are not stationary: *1* process with linear deterministic component ("trend"); *2* process with periodic deterministic component

The climatic meaning of the concept of stationarity deserves additional comment. Of course, macroscopic factors influencing climatic processes are never absolutely stable. At the same time, we can speak about external conditions of climatic events, especially on local scale and within thin climatic subsystems, as relatively stable during a certain period of time. Thus, processes in Figures 1.3 and 1.12 resemble realizations of stationary sequences. At the same time, curves in Figures 1.2, 1.9, 1.10, 1.11, 1.13, 1.14 do not give an impression of being stationary sample functions.

In Fig. 2.5 we give, schematically, examples of two random functions, which are not stationary. Function 2.5.1 is a process with deterministic linear trend. Processes of this type, for instance, can be detected in the registrations of sea level heights in the regions with vertical coastal movements. In Fig. 2.5.2 a periodic deterministic function is a component of the random process. In both cases, distribution functions and their parameters are not constant in time.

The latter case represents a special interest because, visually, nonstationary processes with periodic deterministic components (e.g., seasonal or diurnal dynamics) can resemble stationary processes with spectral peaks (atmospheric synoptic variability or inertial oscillations in the ocean). The difference between these two variants is that the phase of periodic (quasiperiodic) component of seasonal or diurnal dynamics is stable and definite (i.e., summer is always warmer than winter, day is usually warmer than night) whereas the phase of quasiperiodic oscillations within stationary processes is random.

In the next sections we will introduce some quantitative methods for detecting possible nonstationarities and deterministic trends within random processes. However, we would like to stress that the problem of stationarity can never be solved within the theory of random processes itself (which, in principle, can indicate only necessary conditions of stationarity). Other considerations - comparisons with processes of other types, intuition, wide analogies, etc. - are very important.

Finally, note that even if absolutely stationary processes are rare in the real climate, their theory is extremely important because the models of nonstationary processes are eventually based on the stationary theory.

2.2 Introduction to the Theory of Random Functions with Emphasis on Climate Variability

2.2.1 Moments, Mean Value, Correlation Function

To give a complete description of a stationary random function, we have to know all its finite-dimensional distribution functions [Eq. (2.6)], which, according to Eq. (2.7), depend only on differences $t_j - t_1$ ($j = 2, \dots, n$). However, in climatic problems it is very difficult or impossible to estimate these distribution functions. Therefore, in developing a theory of stationary random functions, it was natural to restrict oneself (at least, initially) to a study of properties determined by only the simplest numerical characteristics of the multidimensional distributions [Eq. (2.6)]. The *moments* of a probability distribution serve as simple numerical characteristics of the distribution, and are widely used in the probability theory as well as in the stochastic climate theory. In the case of the distribution [Eq. (2.6)], the moments take the form

$$\mu_{m_1,m_2,\dots,m_n}(t_1,t_2,\dots,t_n) = \mathbf{E}\,\xi^{m_1}(t_1)\xi^{m_2}(t_2)\dots\xi^{m_n}(t_n)$$

$$= \int_{-\infty}^{\infty}\int_{-\infty}^{\infty}\dots\int_{-\infty}^{\infty} x_1^{m_1}x_2^{m_2}\dots x_n^{m_n}\,dF_{t_1,t_2,\dots t_n}(x_1,x_2,\dots,x_n) \tag{2.8}$$

where the symbol $\mathbf{E}$ denotes the averaging operator, i.e., $\mathbf{E}\eta$ denotes the mathematical expectation of the random variable η.

The integral in the right-hand side of Eq. (2.8) is an n-fold Stieltjes integral. Recall that the ordinary Stieltjes integral is defined as the limit

$$\int_a^b f(x)dF(x) = \lim_{\max|x_k - x_{k-1}| \to 0} \sum_{k=1}^{n} f(x_k')[F(x_k) - F(x_{k-1})], \tag{2.9}$$

where $a = x_0 < x_1 < ... < x_{n-1} < x_n = b,\ \ x_{k-1} \le x'_k \le x_k$.

The improper Stieltjes integral is defined in the usual way as the limit

$$\int_{-\infty}^{\infty} f(x)dF(x) = \lim_{a\to-\infty, b\to\infty} \int_a^b f(x)dF(x). \tag{2.10}$$

Multiple Stieltjes integrals are defined similarly. The simplest of the moments [Eq. (2.8)] is the first moment

$$\mu_1(t) = m(t) = \mathbf{E}\xi(t) = \int_{-\infty}^{\infty} x dF_t(x), \tag{2.11}$$

the *mean value* of the random function $\xi(t)$.

We would like to stress that there is an essential difference between the above definition of the mean value, and the term "mean value" often used from the point of view of the common sense. In the latter case one frequently means a value obtained by averaging over some interval of time, whereas in the first case averaging over realizations of the random function is meant. Possibilities of replacing averaging over realizations by averaging over time intervals will be discussed in Section 2.2.2

An illustration of a possible confusion related to the two types of notion mean value is given in Fig. 2.6. Here the changes of level of the Caspian Sea level since 1830 are shown. The mean value of the sea level, calculated using the traditional formula for sample estimation, is shown by the solid line, and formally assessed 95% confidence limits of this estimation (using Statistica-5 software) are shown by dashed lines. The mean value in Fig. 2.6 is usually considered as a true average, stable level of the Caspian Sea.

In fact, arithmetical mean value and corresponding confidence intervals are usually (and in this example) estimated for variables which are not correlated with each other. However, annual sea levels are strongly correlated in time, and sea level changes are close to a nonstationary process. In Fig. 2.6 we see only one realization of the random function describing the behavior of the sea level. Other realizations can strongly differ from the curve in Fig. 2.6, and the mean value (mathematical expectation) defined by Eq. (2.11) can differ greatly from -26.7 m. Our estimations show that the error in the estimation of the mean value has an order of the sea level variations themselves, so it is not even possible to indicate $m(t)$ in Fig. 2.6.

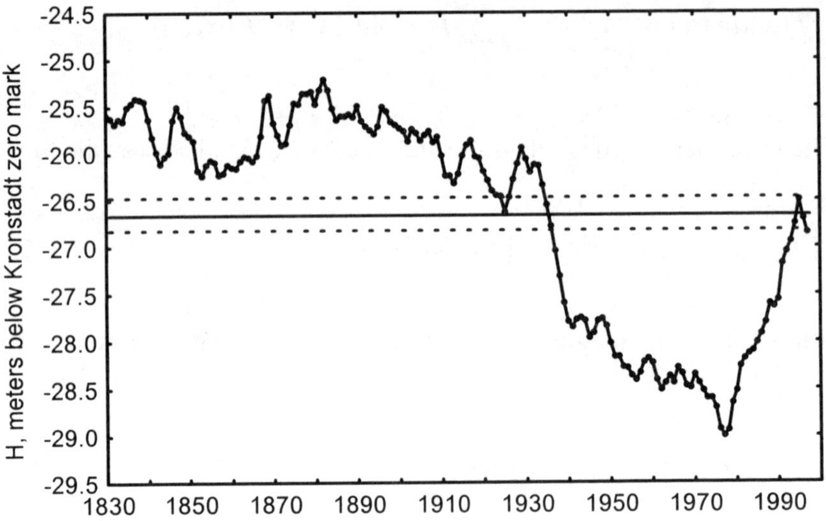

Fig. 2.6. Caspian Sea level changes, 1830-1997 (estimation). (Data from R.K.Kliege 1997, pers. comm.). *Solid line* denotes arithmetical mean value; *dashed lines* 95% confidence intervals calculated using standard STATISTICA-5 procedures

Continuing the discussion on mean values, note that if $\xi(t)$ is stationary, then the distribution function $F_t(x)$ does not depend on t, and then the mean value of $\xi(t)$ is obviously just a numerical constant:

$$\mathbf{E}\xi(t) = m \, . \tag{2.12}$$

The mean value is a very important characteristic of a random function $\xi(t)$, but it can describe only the coarsest properties of the climatic process. More detailed features of climatic variations $\xi(t)$ are given by the second moment

$$\mathbf{E}\xi(t)\,\xi(s) = \mu_{1,1}(t, s) = B(t, s) \, . \tag{2.13}$$

The function $B(t, s)$ is called the *correlation function* of $\xi(t)$, and the theory which takes into account only those properties of random functions which are determined by their first and second moments is called the *correlation theory* of random functions. The centered function

$$B_c(t, s) = B(t, s) - m(t)\, m(s) \tag{2.14}$$

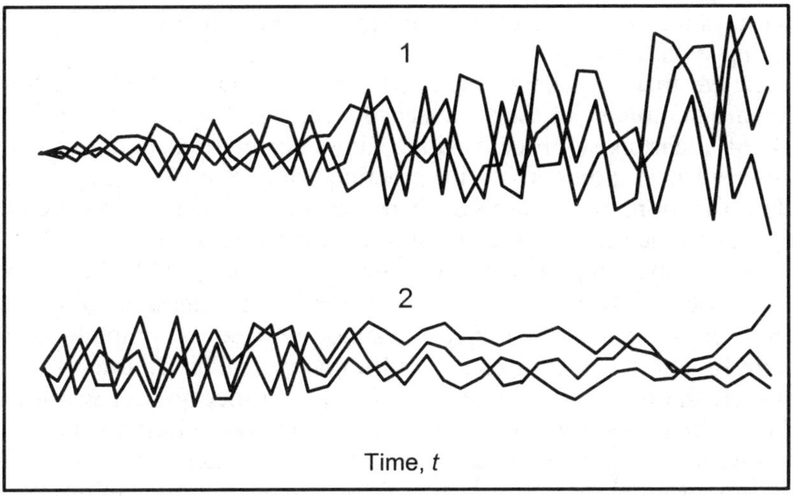

Fig. 2.7. Examples of nonstationary random functions with constant mean value. *1* $B(0) \neq \text{const}\,(t)$; *2* $B(\tau > 0) \neq \text{const}\,(t)$

is often called the *covariance* (function). Also note that the term correlation function, or *autocorrelation function*, is often reserved for a *stationary* covariance.

It is clear that if $\xi(t)$ is stationary, then the correlation function $B(t, s)$ depends only on the difference $t - s$, i.e.,

$$\mathbf{E}\xi(t)\,\xi(s) = B(\,t - s)\,. \tag{2.15}$$

Thus, in the correlation theory, stationary random functions are characterized by one constant m and one function $B(\tau)$ of a single variable τ, which takes integral values for random sequences and arbitrary real values for random processes.

The fact that the multidimensional distribution function [Eq. (2.6)] plays no role in the correlation theory suggests that the definition of stationarity used above can be changed. In the correlation theory, a random function can be called stationary if its mean value [Eq. (2.11)] is a constant and if its correlation function [Eq. (2.13)] depends only on $t - s$. Random functions satisfying these two conditions are said to be *stationary in the wide sense* or *stationary in Khinchin's sense*. [We assume here that the mean value and the correlation function of $\xi(t)$ are finite]. Throughout this book the word stationary will always be meant in this sense.

In Fig. 2.7, two examples of nonstationarities with respect to the second moment are shown (in addition to Fig. 2.5, where nonstationarities were related to the first moment). Random functions in Fig. 2.7 have stable mean values, but their correlation functions change in time. In Fig. 2.7.1 $B(0)$ (variance or standard deviation) grows with time; in Fig. 2.7.2 $B(\tau)$ is changing with time.

It is clear that, in general, the condition (2.7) may not hold for functions which are stationary in the wide sense: see, e.g., the example in Khinchin 1934) of a stationary process with a given correlation function. However, Yaglom (1952) notes that, in practice, one almost never encounters random functions which are stationary in the wide sense but which are not stationary in the sense of the definition given in Eq. (2.7).

It should be emphasized that the mean value and the correlation function obviously do not specify the random function $\xi(t)$ uniquely. Therefore, the correlation theory cannot replace a complete theory of random functions which uses the multidimensional distribution [Eq. (2.6)]. However, the practical view of the correlation theory is growing because of the fact that the random functions encountered in climatology often turn out to be *normal* (or *Gaussian* - see Fig. 2.2), which means that all their finite-dimensional distribution functions [Eq. (2.6)] are multidimensional normal (or Gaussian) (see Cramer 1946; Gnedenko 1962). For normal random functions $\xi(t)$, the mean value and the correlation function completely specify $\xi(t)$, i.e., they determine all the distribution functions [Eq. (2.6)]. Therefore, in principle, the correlation theory can answer any question pertaining to a normal random function $\xi(t)$.

However, even in the case of random functions $\xi(t)$ which are not normal, the correlation theory can answer a whole series of questions which are important in the climatic applications of the random process theory. It should also be noted that for normal random functions, the concepts of stationarity in the wide sense and of stationarity in the *strict sense* [i.e., in the sense of the definition (2.7)] are exactly the same: a normal random function always has finite moments $m(t)$ and $B(t,s)$, and if $m(t)$ is independent of t and $B(t,s)$ depends only on $t - s$, it can be shown that all the multidimensional distribution functions, which in this case are normal, satisfy the condition (2.7).

A further remark concerning normal random functions: in climatological problems, the probability distribution functions [Eq. (2.6)] can be regarded as normal because of the *central limit theorem* of probability theory (Hasselmann 1976) or using analogies with the normality of Maxwell's distribution of molecular velocities - the theory of thermal noise (Middleton 1960).

Also note that there are many variants of criteria which permit one to verify the normality of random functions (see early fundamental work by Bunimovich and Leontovich 1946). Returning to the example in Fig. 2.2 (air temperatures in New Orleans), the normal character of the distribution can be demonstrated even more convincingly if we calculate the cumulative distribution using the observed values, and compare it with the theoretical normal cumulative distribution (Fig. 2.7). In this case,

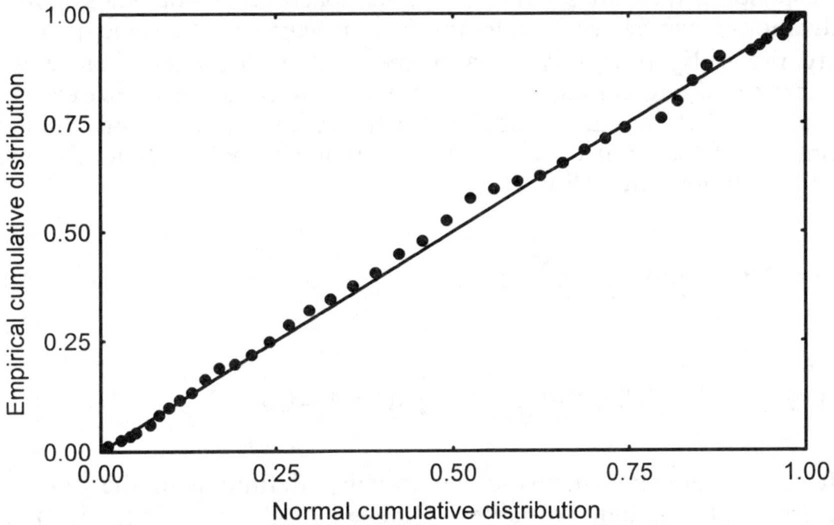

Fig. 2.8. Empirical probability-normal theoretical probability plot of the monthly (April) air temperatures in New Orleans, 1874 - 1990

calculations give values of the most important parameters of empirical distribution equal, if compared with the errors of estimations, to the theoretical (normal) ones: skewness 0.15 (standard error 0.22), kurtosis 0.29 (standard error 0.44).

2.2.2 The Ergodicity of Climate Variability

In practice, we always have only one realization, $\xi^{(j)}(t)$, of the random function $\xi(t)$ which describes the variations of climatic parameter (variable). Thus, it is impossible to use the above formulas to estimate the moments of the random function. Fortunately, in many cases, changes in climatic variable can be considered, at a first approximation, as a segment of realization of a stationary random function. In turn, if $\xi(t)$ is stationary, its mean value m and the correlation function $B(\tau)$ can often be calculated by using just one realization of $\xi(t)$.

The possibility of calculating these characteristics of a stationary random function from a single realization is a consequence of the fact that the so-called *ergodic theorem* (or *law of large numbers*) is often applicable to stationary random functions (or at least to many climatic variations usually encountered in reality).

According to the ergodic theorem, the mathematical expectation of both the quantity $\xi(t)$ and the quantity $\xi(t+\tau)\xi(t)$, obtained by averaging the corresponding quantities over the whole space Ω of outcomes (over all realizations), can be replaced by the *time averages* of the same quantities. More precisely, if $\xi(t)$ is a stationary random sequence (here t is an integer) satisfying certain very general conditions to be indicated below (which are often met in practice in climatic research), then, with a suitable definition of the limit of a sequence of random variables, the following limiting relations are valid:

$$m = \mathbf{E}\xi(t) = \lim_{N\to\infty} \frac{1}{N+1} \sum_{t=0}^{N} \xi(t), \tag{2.16}$$

$$B(\tau) = \mathbf{E}\xi(t+\tau)\xi(t) = \lim_{N\to\infty} \frac{1}{N+1} \sum_{t=0}^{N} \xi(t+\tau)\xi(t). \tag{2.17}$$

It can be shown that, using a reasonable definition of the limit of a sequence of random variables (Loeve 1960, pp. 151, 157) and Chebyshev's inequality, the approximate formula for the mean value (if N is sufficiently large) is

$$m \approx \frac{1}{N+1} \sum_{t=0}^{N} \xi^{(1)}(t), \tag{2.18}$$

where $\approx$ denotes approximate equality, and $\xi^{(1)}(t)$ is a realization of the stationary sequence $\xi(t)$, i.e., a set of observed values of the sequence. Similarly, the relation (2.17) allows us to make an approximate calculation of the correlation function $B(\tau)$ of a stationary sequence, by using the formula

$$B(\tau) \approx \frac{1}{N+1} \sum_{t=0}^{N} \xi^{(1)}(t+\tau)\xi^{(1)}(t). \tag{2.19}$$

An analogous formula can be introduced for the random processes (see, e.g., Yaglom 1952).

Note that several additional variants of the *correlation function* are also used:

1. The normalized correlation function

$$\overline{B}(\tau) = \frac{B(\tau)}{B(0)}. \tag{2.20}$$

2. The *centered correlation function*

$$R(\tau) = \mathbf{E}\,[\,\xi(t+\tau) - m][\,\xi(t) - m] = B(\tau) - m^2\,. \tag{2.21}$$

3. The *correlation coefficient*

$$\overline{R}(\tau) = \frac{R(\tau)}{R(0)}. \tag{2.22}$$

We now discuss conditions which have to be imposed on a stationary random function $\xi(t)$ in order to guarantee the validity of the ergodic theorem. As was shown first by Slutski (1938), formula (2.16) is true if and only if

$$\lim_{N\to\infty} \frac{1}{N+1} \sum_{t=0}^{N} R(\tau) = 0. \tag{2.23}$$

In practice, the function $R(\tau)$ often approaches zero at $\tau\to\infty$ quickly enough, and Eq. (2.23) is true, since the dependence between the random variables $\xi(t+\tau)$ and $\xi(t)$ often becomes progressively weaker as $\tau\to\infty$. A condition analogous to Eq. (2.23) can be formulated in order to guarantee the validity of the ergodic theorem for the second moment, Eq. (2.17).

However, the above theoretical considerations need some comment. First, the expression "correlation function approaches zero quickly enough" must be elucidated. Here quickly means that $R(\tau)$ must decrease sufficiently with time lag, τ, which has to be much less than the length of the existing realization of the random function.

In Fig. 2.9 we demonstrate two different cases related to the behavior of the correlation function and the ergodic theorem. In Fig. 2.9.1 sample correlation coefficients for the Caspian Sea level (1830-1997) are shown. It is evident that correlation coefficients do not decrease quickly, at least in comparison with the length of existing realization. So the conditions of the ergodic theorems are not satisfied, and the mean value of the sequence cannot be accurately estimated (cf. Fig. 2.6 and discussion on pp. 57-58).

In Fig. 2.9.2 correlation coefficients for the Volga runoff (1887-1992) are presented. Evidently, correlation function decreases much more quickly in this case, and there are many more chances to assess the mean value of the process.

Another important remark. As follows from Eq. (2.23) and similar equations, information about the correlation function is needed in order to validate the applicability of the ergodic theorems. At the same time, in order to obtain the estimation of the correlation function, we must use the

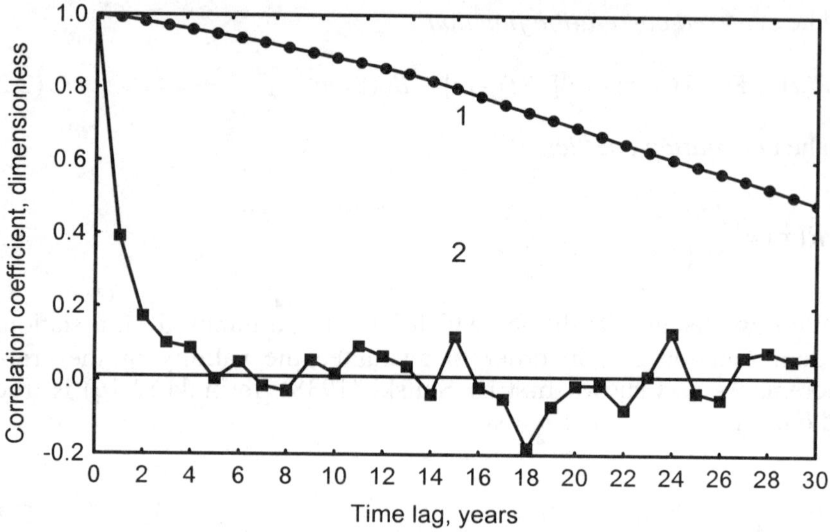

Fig. 2.9. Sample correlation coefficients. *1* Caspian Sea levels (1830 - 1997); *2* Volga runoff (1887 - 1992). (Data from R.K.Kliege 1994, *2*; Kliege 1997, pers. comm., *1*)

ergodic theorems. So, we have a vicious circle here, and cannot obtain strict proofs of the ergodicity (although we can obtain a "negative" result if the correlation function does not decrease "quickly"). In other words, in order to make a decision concerning the ergodicity of a climatic process, we have to use not only mathematical tools, but also to take into consideration empirical analogies with other types of processes, to use experience, intuition, etc.

2.2.3 Examples of Stationary Random Sequences

2.2.3.1 Uncorrelated Random Variables

The simplest example of a stationary random sequence is a sequence $\eta(t)$, $t = \dots, -2, -2, 0, 1, 2, \dots$ of independent random variables with identical probability distributions. Let the mean value $\mathrm{E}\eta(t)$ of the random variables $\eta(t)$ equal 0, and let their variance

$$\mathbf{E}[\eta(t) - \mathbf{E}\eta(t)]^2 = \mathbf{E}\eta^2(t)\,, \tag{2.24}$$

equal 1. Then, the correlation function of the sequence $\eta(t)$ equals

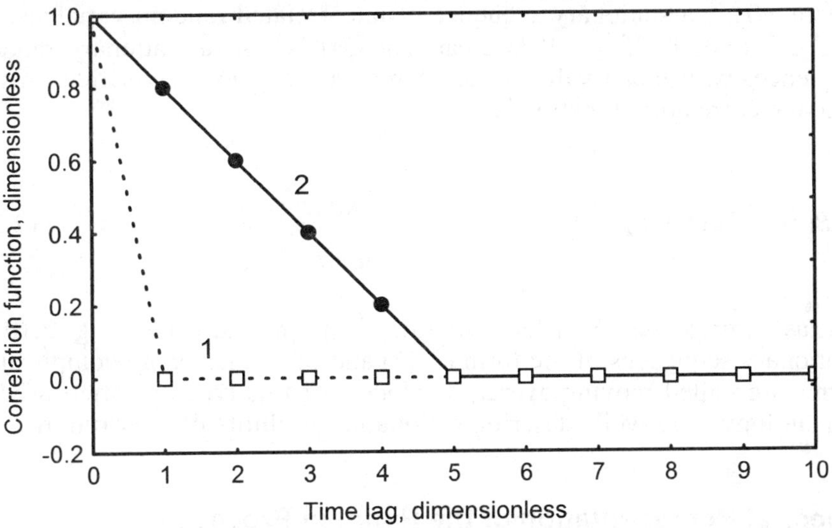

Fig. 2.10. Correlation functions: *1* the sequence of uncorrelated random variables; *2* moving average ($m = 5$)

$$B(\tau) = \mathbf{E}\,\eta(t+\tau)\,\overline{\eta(t)} = \begin{cases} 1 & \text{for} \quad \tau = 0, \\ 0 & \text{for} \quad \tau \neq 0. \end{cases} \tag{2.25}$$

The correlation function (2.25) is shown in Fig. 2.10.1. In the general case, the elements $\eta(t)$ of a stationary sequence with mean value zero and correlation function (2.25) may not be independent. Formula (2.25) only implies that the random variables $\eta(t)$ are *uncorrelated.* For this reason we shall refer to a sequence of the type considered here as a *sequence of uncorrelated random variables*. Note that in the special case of *normal* stationary sequences, lack of correlation is equivalent to independence. (Also note that the consequences of the above type are often called white noise). It will be shown in the next pages that the sequence of uncorrelated random variables is often a good model for the description of sequences of monthly anomalies of wind velocity, fluxes at the atmosphere-ocean interface and other climatic processes.

2.2.3.2
Moving Averages

Let $$\xi(t) = \frac{\eta(t) + \eta(t-1) + \ldots + \eta(t-m+1)}{\sqrt{m}}, \tag{2.26}$$

where $\eta(t)$ is a stationary sequence of uncorrelated random variables, $t =$ ... , -2, -1, 0, 1, 2, ... It is clear that $\xi(t)$ is also a stationary random sequence, with mean value zero. Moreover, as is easily verified, in this case the correlation function is

$$B(\tau) = \mathbf{E}\,\xi(t+\tau)\overline{\xi(t)} = \begin{cases} \dfrac{m-|\tau|}{m} & for|\tau| \le m, \\ 0 & for|\tau| > m, \end{cases} \qquad (2.27)$$

i.e., as τ increases, $B(\tau)$ falls off linearly from 1 to 0 (see Fig. 2.10.2). Stationary sequences of the form (2.26) and of similar, more complicated, forms are called moving averages. Models of this type are often used in climatology, e.g., while describing climatic variability of river runoff.

2.2.4 Spectral Representation of the Random Process

The correlation function of the random process $B(\tau)$ is a function depending on time lags, τ . It shows how the features of the process are related to the time intervals. However, it is often important, for the description of a random process as well as of climatic processes, to introduce a function which would depend on frequency, λ . In fact, the energy of temporal pulsations of many natural (and climatic) processes is more directly related to the frequency of these pulsations than to their periods.

For these purposes the spectral representation of the random process was proposed by Khinchin (1934) in his very important paper. In fact, Khinchin proved that the correlation function of *any* stationary random process can be represented in the form of an integral

$$B(\tau) = \int_{-\infty}^{\infty} e^{i\lambda\tau}\, dF(\lambda), \qquad (2.28)$$

where $F(\lambda)$ is a real nondecreasing bounded function, and λ is an angular frequency. Conversely, every function of the form (2.28), where $F(\lambda)$ has the properties just specified, is a correlation function of a stationary random process. This theorem of Khinchin plays a very important role in the general theory of stationary random functions, and is vital in most applications of the theory.

The function $F(\lambda)$ is called the *spectral distribution function* of the stationary random process $\xi(t)$. Because of Eq. (2.28) we have

$$B(0) = \mathbf{E}\left|\xi(t)\right|^2 = \int_{-\infty}^{\infty} dF(\lambda), \tag{2.29}$$

which means that

$$\int_{-\infty}^{\infty} dF(\lambda) = F(+\infty) - F(-\infty) < \infty. \tag{2.30}$$

Thus, the spectral distribution function is a bounded function.

Theoretically, if we know the correlation function $B(\tau)$ of the process $\xi(t)$, we can calculate its spectral distribution function. To do so, we have to use the formula for inversion of Fourier-Stieltjes integrals. For many applications (including climatic applications) we only need a special case of this inversion formula, pertaining to the most important situation encountered in practice, i.e., where the absolute value of $B(\tau)$ falls off so rapidly as $|\tau| \to \infty$ that

$$\int_{-\infty}^{\infty} |B(\tau)| d\tau < \infty. \tag{2.31}$$

In this case, $B(\tau)$ can be represented as the Fourier integral

$$B(\tau) = \int_{-\infty}^{\infty} e^{i\lambda\tau} f(\lambda) d\lambda. \tag{2.32}$$

Formula (2.32) is a special case of Eq. (2.28) and shows that if Eq. (2.31) holds, then the spectral distribution function $F(\lambda)$ can be written in the form

$$F(\lambda) = \int_{-\infty}^{\lambda} f(\lambda) d\lambda, \tag{2.33}$$

where obviously

$$f(\lambda) = F'(\lambda), \tag{2.34}$$

so that the function $F(\lambda)$ is everywhere differentiable. The function $f(\lambda)$ is called the *spectral density* (function) of the process $\xi(t)$. Note that one often uses simply *spectrum* instead of *spectral density* as well as instead of *spectral distribution*. Since $F(\lambda)$ is a nondecreasing function, the spectral density is non-negative everywhere, i.e., $f(\lambda) \geq 0$.

According to Eq. (2.32), the spectral density is the Fourier transform of $B(\tau)$. Hence, if the correlation function $B(\tau)$ is known, the spectral density $f(\lambda)$, in principle, can be obtained by using the formula for the inversion of a Fourier transformation:

$$f(\lambda) = \frac{1}{2\pi} \int_{-\infty}^{\infty} e^{-i\lambda\tau} B(\tau) d\tau. \tag{2.35}$$

Then, we can use Eq. (2.33) to find the spectral distribution function $F(\lambda)$ as well. Thus, formulas (2.35) and (2.33), taken together, are precisely the special case of the inversion formula for Fourier-Stieltjes integrals which is applicable when the condition (2.31) holds.

In the case of a real process $\xi(t)$, i.e., in most climatic applications, the spectral distribution function $F(\lambda)$ is an even function of λ , to within an additive constant, and then Eq. (2.28) can be written in the form

$$B(\tau) = \int_{0}^{\infty} \cos \lambda\tau dF_1(\lambda), \tag{2.36}$$

where

$$F_1(\lambda) = 2F(\lambda) + const. \tag{2.37}$$

Moreover, if the condition (2.31) holds, the function $B(\tau)$ can be represented in the form (2.32), where in the case of real $\xi(t)$, the function $f(\lambda)$ is even, i.e., $f(-\lambda) = f(\lambda)$. Since $B(\tau)$ is then also an even (real) function, formulas (2.28) and (2.35) can be rewritten in the form

$$B(\tau) = \int_{0}^{\infty} \cos \lambda\tau f_1(\lambda) d\lambda, \tag{2.38}$$

$$f_1(\lambda) = \frac{2}{\pi} \int_{0}^{\infty} \cos \lambda\tau B(\tau) d\tau , \tag{2.39}$$

where

$$f_1(\lambda) = F_1'(\lambda) = 2f(\lambda). \tag{2.40}$$

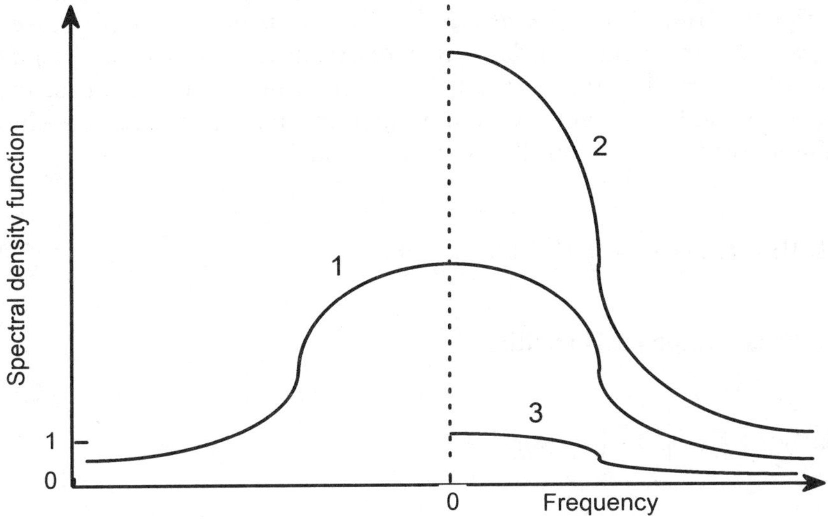

Fig. 2.11. Scheme of different variants of spectral densities:
1: $f(\lambda)$; *2*: $f_1(\lambda)$; *3*: $\tilde{f}_1(\lambda)$

The functions $F_1(\lambda)$ and $f_1(\lambda)$ (Fig. 2.11) are also called the *spectral distribution function* and the *spectral density* (*function*), respectively, of the process $\xi(t)$; they differ from the functions $F(\lambda)$ and $f_1(\lambda)$ introduced earlier only by the factor 2 and by the fact that they are defined only for $\lambda \geq 0$, instead of for $-\infty < \lambda < \infty$. Also so-called *normalized spectral density* is often used in climatic applications of the random function theory:

$$\tilde{f}_1(\lambda) = \frac{f_1(\lambda)}{B(0)}. \tag{2.41}$$

Finally, note that all the above variants of spectral distribution functions and spectral densities are often called spectra or spectrum. Moreover, below we will add some more variants of definitions of similar functions.

2.2.5 Climatic Meanings of the Spectral Distribution Function

It can be shown that if $\xi(t)$ is a real stationary random process, the difference $F_1(\lambda+\Delta\lambda)$ - $F_1(\lambda)$ equals the average power of the spectral (harmonic) components of the process corresponding to the frequency interval $[\lambda, \lambda+\Delta\lambda]$, i.e., to the average energy dissipated per unit time by the aggregate of all oscillations making up $\xi(t)$ which have frequencies between λ and $\lambda+\Delta\lambda$. This makes it clear that the spectral distribution function determines the *frequency distribution of the (average) power of the process*, and hence its derivative $f_1(\lambda)$ can also be called the *power spectral density*. The functions $F(\lambda)$ and $f(\lambda)$, which in the real case differ from $F_1(\lambda)$ and $f_1(\lambda)$ only by unimportant constant factor, have a similar meaning. If we set $\tau = 0$ in (2.38), we find that

$$B(0) = \mathbf{E}\left|\xi(t)\right|^2 = \int_0^\infty dF_1(\lambda), \tag{2.42}$$

or if the condition (2.31) holds,

$$B(0) = \mathbf{E}\left|\xi(t)\right|^2 = \int_0^\infty f_1(\lambda)d\lambda, \tag{2.43}$$

which shows that the total power of the process $\xi(t)$ is obtained by adding up the powers of the separate harmonic components of $\xi(t)$. It also follows from Eq. (2.43) that

$$\int_0^\infty \tilde{f}_1(\lambda)d\lambda = 1. \tag{2.44}$$

It is clear from Eq. (2.44) that the integral $\int_{\lambda_1}^{\lambda_2} \tilde{f}_1(\lambda)d\lambda$ shows what part of the total energy of oscillations [within the process $\xi(t)$] is concentrated between frequencies λ_1 and λ_2 . In other words, the spectral density of the variations of a climatic parameter (and its mathematical prototype–random function) demonstrates the distribution of the energy of the process over frequencies (i.e., over temporal scales).

In Fig. 2.11 an example of spectral density is shown: a fragment of the spectral density function for sunspot numbers (1760-1990) obtained using the traditional method of spectral estimation, "Fast Fourier Transform". The time series itself (divided in two segments) is shown in Fig. 2.12. It is evident that the large part of the energy of the process is concentrated near frequency 0.09 cycles/year (which corresponds to the 11-year period of quasi-oscillations). Also, a large part of the overall energy is concentrated near zero frequency. While considering this registration, researchers usually pay attention to the 11-year quasi-periodicity. However, spectral peak at zero frequency is much more important: it is about five times higher than the 11-year peak.

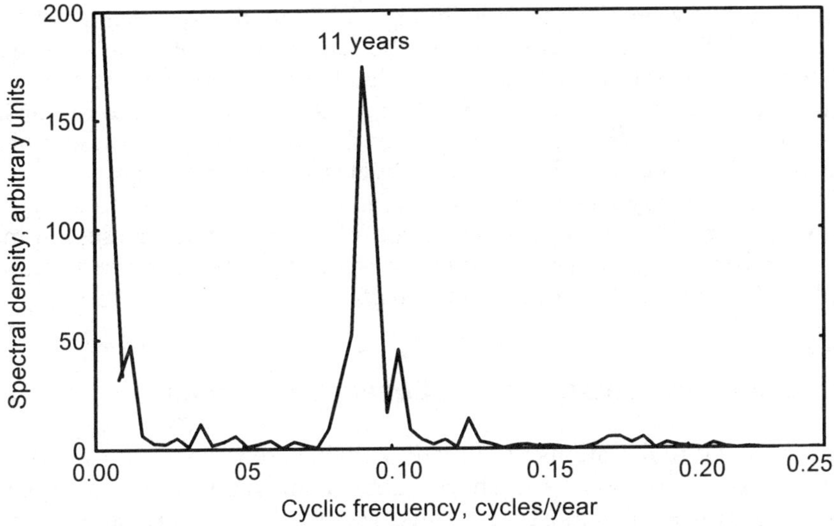

Fig. 2.12. Fragment of the spectral density function estimated using the FFT method. Sunspot annual numbers, 1760-1990 (231 points)

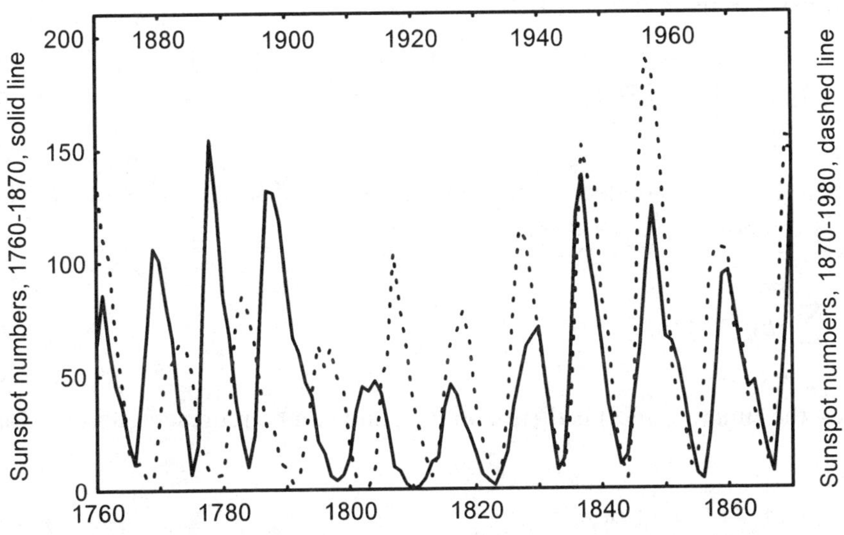

Fig. 2.13. Sunspot number registration, 1760–1980, divided in two segments: 1760–1870 (*solid line*); 1870–1980 (*dashed line*). (Data from R.K.Kliege 1994, pers. comm.)

Also, we would like to note an important feature of the process in Fig. 2.12. At a glance, it seems that the type of this process is the same as the type of processes with seasonal dynamics or diurnal variations: all these registrations have a sharp spectral peak. However, there is an essential difference between the process in Fig. 2.12 on the one hand, and the seasonal and diurnal processes on the other: sunspot numbers really resemble a realization of a stationary random process with stable mean value and random phase of quasi-oscillations. The last feature of sunspot numbers is evident from Fig. 2.12: when the whole registration is divided into two parts, the phases of 11-year cycles are sometimes synchronous, and sometimes asynchronous. Returning to the climate processes with strong deterministic periodic component (e.g., seasonal dynamics), it must be emphasized that a special type of spectra is used in these cases: so--called *discrete spectral distribution functions*.

2.2.6 Spectral Representation of Stationary Sequences

In practice, all registrations of climatic variables represent time series, i.e., realizations of random sequences more than realizations of random functions. However, almost all results described in the previous sections can be carried over to the case of stationary random sequences as well, by making some changes which are hardly of a fundamental nature.

The correlation function $B(\tau)$ of a stationary sequence can always be represented in the form

$$B(\tau) = \int_{-\pi}^{\pi} e^{i\lambda\tau} dF(\lambda), \tag{2.45}$$

where $F(\lambda)$ is a nondecreasing function. If the absolute value of $B(\tau)$ falls off so rapidly as $|\tau| \to \infty$ that

$$\sum_{\tau=-\infty}^{\infty} |B(\tau)| < \infty, \tag{2.46}$$

then the function $F(\lambda)$ is differentiable, and can be represented in the form

$$F(\lambda) = \int_{-\pi}^{\lambda} f(\lambda) d\lambda,$$

where

$$f(\lambda) = F'(\lambda) \geq 0. \tag{2.47}$$

In this case, Eq. (2.45) can be replaced by the formula

$$B(\tau) = \int_{-\pi}^{\pi} e^{i\lambda\tau} f(\lambda) d\lambda. \tag{2.48}$$

It is clear from Eq. (2.48) that the quantities

$$\frac{1}{2\pi} B(-\tau) \qquad (\tau = 0, \pm 1, \pm 2, \ldots)$$

are the Fourier coefficients of the functions $f(\lambda)$, so that the expansion of $B(\tau)$ in Fourier series takes the form

$$f(\lambda) = \frac{1}{2\pi} \sum_{\tau=-\infty}^{\infty} e^{-i\lambda\tau} B(\tau). \tag{2.49}$$

Obviously, Eq. (2.49) is the analog of formula (2.35) in the theory of stationary random processes.

The function $F(\lambda)$ is called the *spectral distribution function of the stationary sequence* $\xi(t)$, and the function $f(\lambda)$, if it exists, is called the *spectral density* (*function*) of $\xi(t)$.

2.2.7 The Markov Sequence

A *Markov sequence* (synonymously, a *Markov process with discrete time*) means a random sequence such that the conditional distribution of the random variable $\xi(t + m)$ if $\xi(t - k_1)$, $\xi(t - k_2)$, ... , $\xi(t - k_n)$, take given values (where $m \geq 0$, and $1 \leq k_1 < k_2 < \ldots < k_n$ are arbitrary), depends only on the value of $\xi(t - k_1)$ and not on the values of the other random variables $\xi(t - k2)$, ... , $\xi(t - k_n)$. This property is sometimes stated as follows (in a way which is not completely accurate): the future of a Markov sequence $\xi(t)$ is determined only by its present, and does not depend on the past behavior of the sequence.

The simplest (and very important) variant of the Markov process with discrete time is the normal stationary Markov sequence:

$$\xi(t) = c\xi(t-1) + a(t), \tag{2.50}$$

where t is an arbitrary moment of time ($t = \ldots, -2, -1, 0, 1, 2, \ldots$), $0 < c < 1$– constant coefficient, and $a(t)$ normal uncorrelated random sequence; it is supposed in Eq.(2.50) that the mean value of $a(t)$ and $\xi(t)$ is zero. The random sequence (2.50) is often called the *first-order Markov process* (although this term sometimes has other meanings). In hydrology similar

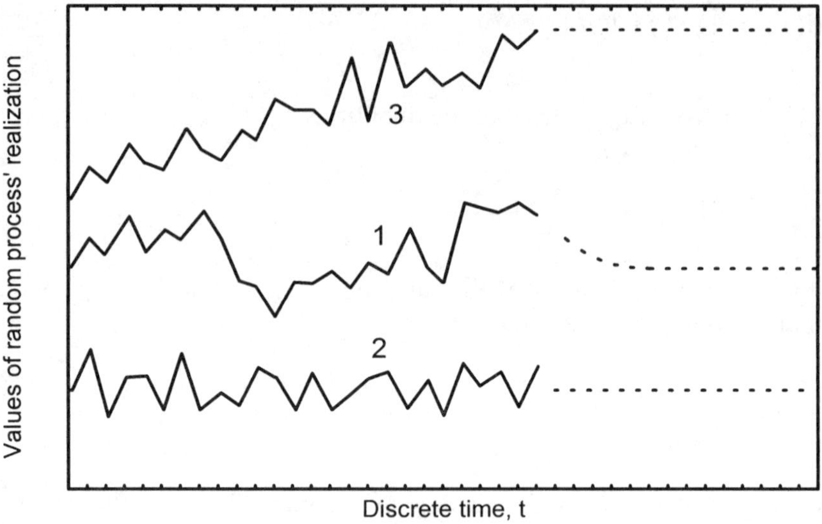

Fig. 2.14. Schemes of realizations of the first-order Markov process (*1*), the sequence of uncorrelated random values (white noise) (*2*), random sequence with uncorrelated increments (*3*) and their respective optimal forecasts

(but not Gaussian) sequences are often called a *simple Markov chain*. Similar sequences are also called the *first order autoregressive processes*. Visually, the first-order Markov process is characterized by deviations longer, in comparison with the white noise, from its mean values (see Fig. 2.14).

It is easy to show that for the first-order Markov process its correlation function (Fig. 2.15) equals

$$B(\tau) = Gc^{|\tau|} \qquad (G > 0;\ \tau = 0, \pm 1, \pm 2, \dots). \tag{2.51}$$

Using formula (2.49) we can at once obtain a formula for the spectral density function (see Fig. 2.15) of the first-order Markov process:

$$f(\lambda) = \frac{G}{2\pi} \frac{1 - c^2}{\left|e^{i\lambda} - c\right|^2}, \tag{2.52}$$

so that $f(\lambda) \geq 0$. Finally, if we consider a realization of the first order Markov process $\xi^{(1)}(t)$, it is easy to show that its optimal forecast will be

$$\xi^{(1)}(t+m) = c^m \xi^{(1)}(t), \tag{2.53}$$

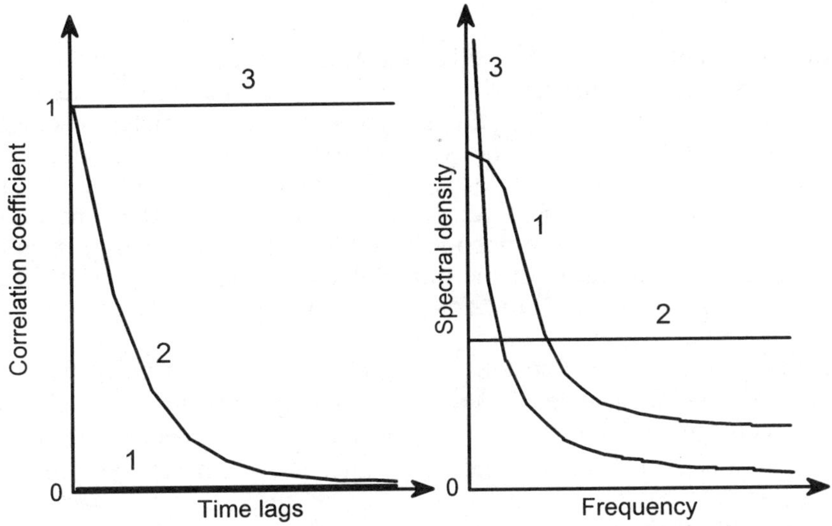

Fig. 2.15. *Left*: Schemes of the correlation coefficients of the first-order Markov process (*1*), the sequence of uncorrelated random variables (white noise) (*2*), the random sequence with uncorrelated increments (*3*). *Right* Schemes of appropriate spectral density patterns

where $\xi^{(1)}(t)$ is a given value of random process realization at moment t , m = 1, 2, ... (see Fig. 2.15). Thus, the future of the first-order Markov process is really determined by its present, and does not depend explicitly on the past behavior of the sequence. We can note, nevertheless, that the very value $\xi^{(1)}(t)$ is evidently determined by the previous behavior of the process.

Note that if c in Eq. (2.50) is close to zero, the first-order Markov process is close to the uncorrelated random sequence (see Sect. 2.2.3.1). On the other hand, if c is close to 1, the first-order Markov process tends to be a process with uncorrelated increments or the discrete Wiener process (see next section).

It will be shown in the next sections and chapters that the Markov process and the first-order Markov process, as well as the uncorrelated random sequence and the discrete Wiener process, play an extremely important role in the study of climatic variability.

2.2.8 The Discrete Wiener Process

If coefficient c in formula (2.50) equals 1, we have a random sequence which is called the *random sequence with uncorrelated increments* or the *discrete Wiener process* (also called *random walk*):

$$\xi(t) = \xi(t-1) + a(t)\,. \tag{2.54}$$

Random function (2.54), as well as the Markov process and white noise, plays a key role in climate variability. It has a very simple mathematical description - like the white noise $a(t)$, it has only one parameter, a standard deviation of $a(t)$ [note than the Markov sequence (2.50) has one more parameter to be assessed - coefficient c]. Moreover, in practice, climatic processes described by Eq. (2.54) are even simpler than processes described by the sequence of uncorrelated random variables. In fact, in the latter case the mean value of the uncorrelated process must be assessed, whereas the random walk [Eq. (2.54)], generally speaking, might not have any mean value.

Sequence (2.54) is a nonstationary function. If we formally fix some moment of time t_0 , also set $\xi(t_0)$ to equal some given value (e.g., $\xi(t_0)=0$), and consider $\xi(t)$, t = 1, 2, ... , we can say that the mean value of this process will be equal to $\xi(t_0)$, and its variance grows linearly with time:

$$B(0) = B_a(0)(t - t_0)\,, \tag{2.55}$$

where $B_a(\tau)$ is the correlation function of $a(t)$.

However, fixation of the random walk at some given moment of time, in a climatological context, seems to be an absolutely artificial assumption. In reality, climatic processes which are close to random walks (global sea and air temperatures, global mean sea level, global ice volumes, etc.) do not manifest, at least at time scales up to thousands of years, any stable mean value. Thus, climatic processes described by the random walk model [Eq. (2.54)] seem to be even simpler structures than the white noise. So it has most chances of being realized, because the simplest structures are the most frequent in nature.

Because of the nonstationarity of the random walk [Eq. (2.54)], the above theory of stationary random processes is not applicable for its analysis (although it is applicable for the analysis of the random walk's increments). Nevertheless, we can avoid some additional mathematical definitions simply by considering the sequence (2.54) a limit case of the first-order Markov process, Eq. (2.50) when $c \to 1$. In this case, roughly speaking, the correlation coefficient will tend to 1 (Fig. 2.15), the optimal forecast will be a so-called inertial forecast (predicted value tends to the last observed value of the random walk), and formally calculated spectral density [Eq. (2.52)] will have a singularity at $\lambda = 0$, i.e., $f(\lambda \to 0) \to \infty$.

Although these features are known in mathematics (the famous mathematician Bernoulli studied a similar *Bernoulli symmetric random walk* 300 years ago), climatologists still mistrust its applications in climate change studies. Indeed, the features of this very simple process are stunning. It represents the temporal summation of an absolutely stationary

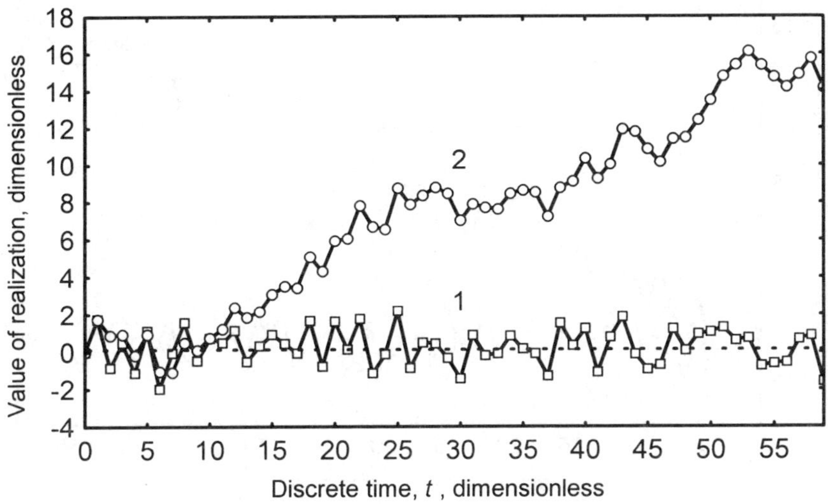

Fig. 2.16. Realizations of the sequence of normal uncorrelated random variables $a(t)$ (*curve 1*), random walk [Eq. (2.54)] (*curve 2*)

process, but is not stationary. It equals the integral of a random process with equal probabilities of positive and negative pulsations, and at the same time constantly moves away from its initial value, etc. (also see Dobrovolski 1991 on this subject).

Figures 2.16 and 2.17 illustrate these paradoxical but very important properties of the random walk [Eq. (2.54)]. Curve 2.16.1 represents a realization of white noise $a(t)$: a stationary sequence of uncorrelated random variables with mean value zero and unite variance. (In fact, these are the first random values generated by the Pentium processor.) Curve 2.16.2 is a temporal summation of values 2.16.1, in accordance with Eq. (2.54). We see that curve 2.16.2 is far from being stationary, it moves far from the initial value of the process. We stress that the realization 2.16.1 was not specially chosen, it is a typical registration of uncorrelated sequence. Later, process 2 in Fig. 2.16 can change its direction, but deviation from the initial (at $t = 0$) value will grow constantly.

Another variant of the nonstationary process can be obtained using the realization of $a(t)$ in the following way. In Fig. 2.17 curve 1 is the same realization of a white noise as in Fig. 2.17.1. The curve 2.17.2 is obtained by adding linear function (linear trend) to curve 1:

$$\xi(t) = a(t) + 0.27t\,. \tag{2.56}$$

Curve 2 in Fig. 2.17 is really a process with linear trend. Moreover, there is something in common between curves 2 in Figs. 2.16 and 2.17

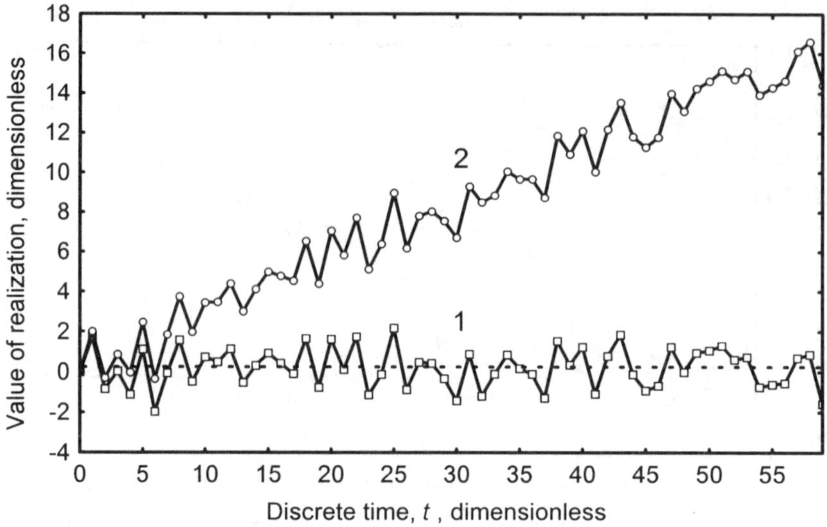

Fig. 2.17. Realization of the sequence of normal random uncorrelated variables (*1*), process with linear trend [Eq. (2.56)] (*2*)

although their structures are absolutely different. Using these examples, we want to say that a researcher cannot restrict his study by simply establishing a positive or negative tendency in a registered climatic variable and searching for the causes of this tendency. The primary task of investigation is to study the structure of this "tendency", because in most cases there may be no deterministic causes of "tendencies" at all. Finally, we can add that there are mathematical tools for distinguishing between the cases in Figs. 2.16.2 and 2.17.2; they will be examined in the next sections and chapters of the book.

2.2.9 Other Types of Random Functions

Only the simplest (but most fundamental) types of random functions (sequences) were considered in the above sections. Below, some other types of random sequences will be used:

2.2.9.1 Autoregressive Models

Consider a stationary normal random sequence $\xi(t)$ with mean value zero:

$$\xi(t) = c_1\xi(t-1) + c_2\xi(t-2) + \ldots + \xi_M(t-M) + a(t), \tag{2.57}$$

where $c_1, c_2, \ldots, c_m$ are the constant (autoregression) coefficients, $t = \ldots, -2, -1, 0, 1, 2, \ldots$, $a(t)$ is the normal uncorrelated random sequence with mean value zero (white noise). Parameter M is called the *order of autoregression*. It is evident that if $M = 1$, sequence (2.57) becomes the first order normal Markov process (2.50). Model (2.57) plays a very important role in the theory of parametric estimations of spectral density (see Sect. 2.3). A detailed description of AR models and their modifications can be found in Box and Jenkins (1970), Privalsky (1985), and Privalsky et al. (1992).

2.2.9.2 Seasonal Models

If random sequence $\xi(t)$ describes a time series of monthly values of some climatic variable, it contains a strong seasonal component. Thus, strictly speaking, $\xi(t)$ cannot be regarded as a stationary random sequence. However, in many cases, it can be transformed into the stationary sequence $\xi_1(t)$ by a simple procedure:

$$\xi_1(t) = \frac{\xi(t) - m_i}{\sigma_i}, \tag{2.58}$$

where m_i and σ_i are the mean value and standard deviation of the respective calendar month, i.

A more sophisticated case of the seasonal model can be considered if we suppose that not only the mean value and variance (standard deviation), but also the correlation coefficients (or autoregression coefficients, also other parameters) of the sequence can have a seasonal dynamics. Models of this type are often used in hydrology for the description of seasonally varying runoff (see, e.g., Vecchia 1985).

2.2.9.3 Threshold Models

In many important cases the random sequence $\xi(t)$, which describes some climatic variable, is strictly limited: $L \le \xi(t) \le H$ where L and H are, respectively, low and high thresholds (screens) of the process under consideration. For example, these cases take place when we consider the volume of local objects or the volume of global-scale objects (lakes, mountain glaciers, ice sheets) which can completely or almost completely disappear ($L \to 0$) and, on the other hand, cannot have an infinite volume, i.e., have some high threshold, H.

The threshold models are often described by random sequences with varying parameters, e.g., by the discrete Wiener process [see Eq. (2.54)] with varying standard deviations of the white noise, σ_a. For instance, it is natural to suppose that the range of random increments, in the first approximation, changes linearly with the value of the process:

$$\begin{aligned}&\xi(t) = \xi(t-1) + a(t),\\&\sigma_a(t) = k[\xi(t-1) - L] + \sigma_{\min},\\&L \le \xi(t) \le H\end{aligned} \tag{2.59}$$

where k is a constant coefficient, and $\sigma_{\min}$ is some fixed minimal value of random increment standard $\sigma_a(t)$.

It will be shown in Chapter 4 that process (2.59) possesses surprising properties: it can give an oscillatory regime or provide the behavior of a strange attractor. Also in Chapter 4 it will be demonstrated that system (2.59) can be used for the description of glaciation-deglaciation processes during the past 600,000 years. Moreover, processes like this can be found in living systems and can describe the behavior of ecological and social systems. Similar types of objects can be called asymmetric random systems and seemingly play a very important role in various processes on the surface of the planet.

Of course, a large number of much more sophisticated random (stochastic) functions is used in the earth sciences. However, in the next chapters we will mostly use the stochastic (random) models described in this section as well as related, rather simple, models. The reasons for using these models are the following:

- As a rule, climatic time series are short. Also the errors in observational data are usually large. However, sophisticated stochastic models need a lot of information for the assessment of their numerous parameters. So, simple models have some advantages.
- The application of even simplest stochastic (random) process models for the description of climatic variabilities is not yet sufficiently clear. Methods of such modeling are far from being perfect, and accurate stochastic models of climatic variations are still rare.
- Fortunately, the simplest stochastic processes seem to be the most fundamental models of climatic variability and variations. At least, our 30-year experience of climatic-scale stochastic modeling shows that a few types of models are the most frequent: the white noise, the first order Markov process, the discrete Wiener process, and some related models.

2.3 Estimation of Model Parameters

2.3.1 Theoretical Models and the Practice of Model Identification

In the above theory of random (stochastic) processes we dealt with abstract objects, mathematical models. A further strategy in constructing probabilistic climate theory is to apply the theory of stochastic processes to the observed climatic variations in order to detect the essential features of the latter. Namely, the estimation of the correlation function and the spectral density $f(\lambda)$, using an observed climatic time series, is of special interest because $f(\lambda)$ demonstrates how the energy of climatic variations is distributed over frequencies (time scales).

However, the task of spectral estimations using observational data is rather difficult. Strictly speaking, real climatic variability differs in many aspects from its theoretical prototypes. Below we list the examples of basic assumptions made in Sections 2.1 and 2.2 (evidently, this list is not exhaustive):

- The distribution function of the variable under consideration must possess some degree of stability. It is preferable to deal with normal processes, at least with processes which can be "normalized".
- Multidimensional distribution functions, describing the random process, must satisfy the symmetry condition and the compatibility condition.
- It is preferable to use stationary process models. At least, the mean value of the process and its correlation function must be stable (and, of course, they must be finite) to give us the possibility of using the stationary random process theory.
- The theory of random functions assumes that the number of realizations of the process is infinite. (In reality, in climatic research we always possess only one realization of observed climatic process).
- Because of the unrealistic character of the above assumption, some additional conditions must be satisfied. First of all, the number of points within an observed time series must be "sufficiently large".
- Also the correlation function must approach zero quickly enough to provide the possibility of estimating the mean value of the process and the applicability of the Fourier integral to describe the correlation function.
- Then, some more sophisticated condition must guarantee the validity of the ergodic theorem for the second moment (the correlation function).

Some of the above assumptions are simple and realistic, and others are more sophisticated and present more difficulties with respect to verification and treatment. In any case, it is clear from the list that the use of even the simplest types of random processes in climate theory is not so easy.

The main problem in applying the theory of random functions to the observed climatic changes is the shortness of existing climatic records. The number of points in typical climatic time series is several dozens (for series of annual values). The first tentatives at estimating the spectral density (which are still widespread: methods of periodogram smoothing, Fast Fourier Transform, etc.) do not take explicitly take this fact into consideration. As a result of this negligence, unrealistic assumptions are usually made with respect to the behavior of the correlation function at large time lags. For instance, if, because of the restricted length of a time series, we can estimate only $(M+1)$ covariances - $B(0)$, $B(1)$, ... , $B(M)$, and cannot estimate other covariances, it is supposed that for $\tau > M$, $B(\tau) = 0$, or the values of $B(\tau)$ are repeated. More realistic procedures for the assessment of spectra and obtaining of reliable stochastic models of observed climatic variability have become possible with the use of the informational approach. In the next section, which deals with this subject, we will generally follow the fundamental work by Ulrych and Bishop (1975), as well as some other works.

2.3.2 Informational Approach to the Identification of Stochastic Models

The relationship between the *information*, I, and the probability, p_i, of the occurrence of an event, m_i, can be written as

$$I = k \ln \frac{1}{p_i}, \tag{2.60}$$

where k is a constant and is equal to 1 when the base of the logarithm is 2. If the observation is made for a long time T, we may expect $p_1 T$ of m_1 events, $p_2 T$ of m_2 events, etc. to have happened. The total information about the system will be

$$I_{total} = k(p_1 T \ln \frac{1}{p_1} + p_2 T \ln \frac{1}{p_2} + \ldots). \tag{2.61}$$

The average information per time interval is represented by H and is referred to by the term *entropy*. Thus (Shannon 1948) suggested that

$$H = \frac{I_{total}}{T} = -k\sum_{i=1}^{M} p_i \ln p_i . \tag{2.62}$$

It can be shown at once that the use of the informational approach eliminates the unreasonable constraints that are demanded by conventional spectral analysis (see Sect. 2.3.1). The method of determining the maximum entropy probability distribution is as follows (Jaynes 1968): the process x_t that is being considered, can take on the values $x_1, x_2, \ldots, x_n$. We assume that the available information about x_t is in the form of average values $\bar{f}_1(x_t), \bar{f}_2(x_t), \ldots, \bar{f}_m(x_t)$, $m < n$. The probability distribution $p_t = p(x_t)$ that is consistent with this information but is maximally free of other constraints is the one that maximizes the entropy

$$H = -\sum_i p_i \log_2 p_i . \tag{2.63}$$

The maximization of H is subject to the constraints

$$\sum_i p_i = 1 \quad \text{and} \tag{2.64}$$

$$\sum_i p_i f_k(x_i) = \bar{f}_k(x_t), \qquad k = 1, 2, \ldots, m . \tag{2.65}$$

The solution to this variational problem is well known and can be found in Ulrych and Bishop (1975).

In applying the concept of maximum entropy to spectral analysis we can establish the relationship between the entropy (the *entropy rate*) for an infinite process and the spectral density $f(\lambda)$ of a stationary Gaussian process:

$$H = \frac{1}{4\lambda_N} \int_{-\lambda_N}^{\lambda_N} \log f(\lambda) d\lambda , \tag{2.66}$$

where λ_N is the Nyquist frequency. The derivation of Eq. (2.66) is shown in detail in Smylie et al. (1973). Rewriting Eq. (2.66) in terms of the covariances B_k of the process gives

$$H = \frac{1}{4\lambda_N} \int_{-\lambda_N}^{\lambda_N} \log[\sum_{-\infty}^{\infty} B_k \exp(-i2\pi\lambda k\Delta t)] d\lambda , \tag{2.67}$$

where Δt is the uniform sampling rate (discreteness interval). Maximizing Eq. (2.67) with respect to the unknown B_k with the constraint that $f(\lambda)$ must also be consistent with the known $B(0), \ldots, B(M\text{-}1)$ results in the *Maximum Entropy Method* (MEM) *spectral estimate.*

2.3.3 Maximum Entropy Method and Autoregressive Models

As shown above, the MEM estimate expresses maximum uncertainty with respect to the unknown information but is consistent with the known information. The variational procedure leads to the expression for MEM spectral density estimation (Edward and Fitelson 1973; Smylie et al. 1973):

$$f(\lambda) = \frac{P_m}{\lambda_N \left| 1 + \sum_{i=1}^{M-1} \gamma_j \exp(-i2\pi\lambda j\Delta t) \right|^2} . \tag{2.68}$$

In Eq. (2.68), P_m is a constant, and the γ_j are prediction error coefficients that are determined from the data.

Below in this section we will follow the development of van den Bos (1971), who established the relationship between MEM spectral analysis and the autoregressive (AR) representation of a random process. He has shown that the MEM spectral density of the real random process is

$$f(\lambda) = \frac{2\sigma_a^2}{\left| 1 - \sum_{j=1}^{M} c_j \exp(-i2\pi\lambda j) \right|^2} , \tag{2.69}$$

which compares with Eq. (2.68). In Eq. (2.69) c_j are the coefficients of the autoregressive random sequence, and $\sigma^2(a)$ is the variance of the white noise in Eq. (2.57).

It is indeed impressive that the representation of a stochastic process by an AR model is that representation that exhibits the maximum entropy. In other words, using AR models for the estimation of spectral density and for the description of the observed (in our case, climatic) process means the following. Instead of estimating the spectral density $f(\lambda)$ at each individual frequency λ, we choose some type of (simple) deterministic function for the description of $f(\lambda)$ and try to estimate a few parameters of this function. In turn, the form of this function is determined, in the case of the MEM approach, using informational considerations.

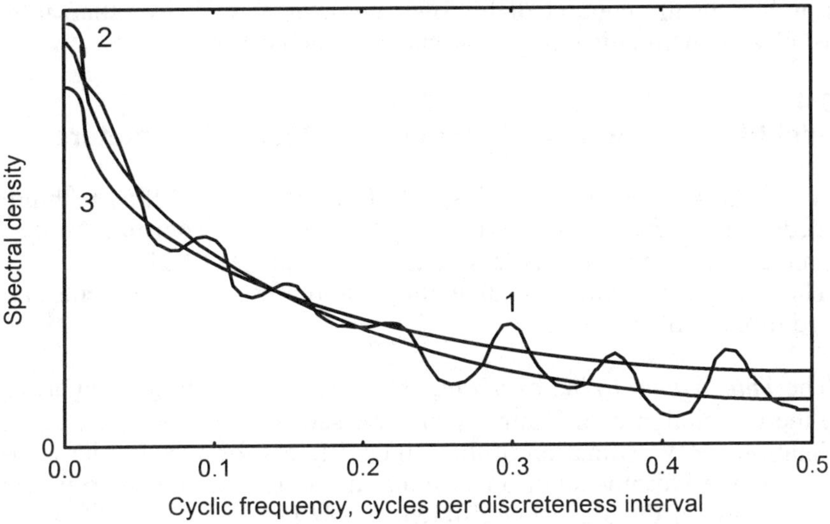

Fig. 2.18. Scheme of spectral density and its estimations. *1* theoretical spectral density; *2* mean estimation of the spectral density with the use of random process segments of a given length; *3* estimation of the spectral density using existing segment of the random process' realization

We can schematically represent the process of estimating spectral density of the random sequence, using a specific method (MEM in our case), in following way (Fig. 2.18). Suppose that the random sequence $\xi(t)$ adequately describes the climatic process under consideration and corresponds to all the conditions described in the above sections. Then the theoretical spectral density $f(\lambda)$ of the sequence $\xi(t)$ exists (curve 1 in Fig. 2.18). This theoretical spectral density is a definite (deterministic) function of frequency λ . In reality, we have only one segment $x(t)$ (of limited length N) of one realization $\xi^{(1)}(t)$ of the random sequence. Estimation of the spectral density using a segment of such a length (and using our specific method of spectral analysis) gives the curve 2 in Fig. 2.18, which is inevitably a random function of frequency λ .

Obviously, we would like to know, the possible errors of such an estimate. We can imagine some "mean" curve (No. 3 in Fig. 2.18) that is an average of all spectral estimations using time series of the sequence $\xi(t)$ with the same length N. Theoretically, this mean spectral density is a definite function of frequency. Thus, the usual way of calculating errors of spectral density estimations is to assess the possible bias and variance of specific spectral estimations 2 with respect to estimation 3. Note that all these variants of spectral density estimations, as well as numerous variants (listed in previous sections) of the very notion *spectral density* and *spectral distribution function* are often called simply spectra. However, a researcher, in the context of his own work or a work by other authors, must clearly distinguish these notions from each other.

2.3.4 Model Identification and Estimation of Model Parameters

Thus, the task of estimating the spectral density, using MEM techniques, is deduced to the determination of the order M, coefficients c_j, and variance $\sigma^2(a)$ of the autoregressive random sequence (2.57).

There are several ways of calculating autoregression coefficients c_j. We tested three of them:

1. The *Yule-Walker* (YW) *estimates*. The main idea of these estimates is to use the correlation coefficients $R(\tau)$, assessed with the help of traditional estimations. YW estimation of the i-th coefficient for a trial order M using the recursive Durbin-Levinson formula are as follows (Box et al. 1994). The formula for the highest coefficient ($i = M$) is:

$$c_{M,M} = \frac{R(M) - \sum_{\tau=1}^{M-1} c_{M-1,\tau} R(M-\tau)}{1 - \sum_{\tau=1}^{M-1} c_{M,\tau} R(\tau)}, \tag{2.70}$$

where $R(\tau)$ are correlation coefficients, calculated according to traditional formulas similar to (2.19). Other trial coefficients ($i = 2, \dots, M - 1$) are:

$$c_{M,i} = c_{M-1,i} - c_{M,M} c_{M-1,M-i} \, . \tag{2.71}$$

The first trial coefficient in this case is adopted as

$$c_{1,1} = R(1) \, . \tag{2.72}$$

Box and Jenkins (1970) point out that the Yule-Walker estimates of the AR coefficients are sensitive to rounding errors, particularly when the AR process is close to the boundaries of stationarity. Another objection to this approach becomes evident when we remember the maximum entropy interpretation of an AR representation. Accordingly, the AR coefficients should be estimated in a manner that is maximally noncommittal with regard to unavailable information. The estimation of the correlation coefficients using traditional formulas of the type (2.19), on the other hand, assumes that $\xi(t) = 0$ for the unknown part of the random sequence, "an assumption that contradicts the principle of maximum entropy" (Ulrych and Bishop 1975).

We would like to make some comments on this suggestion. Our experience of identifying stochastic models for the description of thousands of climatic time series shows that the above doubts are related mostly to autoregressive models with high values of the orders M of autoregression. However, in the practice of climatic stochastic modeling, high orders of autoregression are rare.

2. The *Burg estimates* (BE). Burg (1967, 1968) suggested a method of estimating the AR parameters that does not require prior estimate of the correlation function (correlation coefficients). It differs from the Yule-Walker estimates by the formula for coefficients $c_{M,M}$. In the Burg scheme the equation is solved by minimizing the sum P_M of squares of linear prediction errors in forward and backward directions:

$$P_M = \sum_{t=M+1}^{N} \{x(t) - \sum_{i-1}^{M} c_{M,i} x(t-i)\}^2 +$$
$$\sum_{t=M+1}^{N} \{x(t-M) - \sum_{j=1}^{M} c_{M,j} x(t-M+j)\}^2 , \tag{2.73}$$

where $x(t)$ is the known segment of one of the realizations $\xi^{(1)}(t)$ of the random sequence $\xi(t)$, N is the series length (number of points within the time series). Equalizing partial derivative $\partial P_M / c_{M,M}$ to zero, we obtain

$$c_{M,M} = 2 \frac{\sum_{l=1}^{N-M} [y_{l,M} y'_{l,M}]}{\sum_{l=1}^{N-M} [y_{l,M}^2 + (y'_{l,M})^2]}, \tag{2.74}$$

where

$$y_{l,M} = x(l) - \sum_{j=1}^{M-1} c_{M-1,j} x(l+j), \tag{2.75}$$

$$y'_{l,M} = x(l+M) - \sum_{j=1}^{M-1} c_{M-1,j} x(l+M-j). \tag{2.76}$$

Coefficients $c_{M,i}$ for $2 \leq i \leq M - 1$ are calculated using the Durbin formula (2.71). Estimation of the coefficient $c_{1,1}$, in accordance with the Burg scheme, is

$$c_{1,1} = \frac{\sum_{t=2}^{N} x(t)x(t-1)}{\frac{x^2(1)+x^2(N)}{2} + \sum_{t=2}^{N-1} x^2(t)}. \tag{2.77}$$

Our experience shows that BE estimates, with respect to climatic time series analysis, differ from YW estimates if the order of autoregression M is more than 1 and if the stochastic model is close to the nonstationary one. An example of such a case is the model for the description of terminal lake water height (e.g., Caspian Sea level).

3. It is evident that the BE formula for the first coefficient of autoregression (2.77) does not largely differ from the traditional estimation of the first correlation coefficient $R(1)$. However, the estimation of $c_{1,1}$ is used while calculating higher coefficients. Thus, it is preferable, at least theoretically, to introduce some more convenient estimation of $c_{1,1}$. An example of such an estimation can be as follows. We can choose the value of the first correlation coefficient that provides the smallest possible error of our time series $x(t)$ in the forward and backward directions, similar to Eq. (2.73). In the framework of this approach, other trial coefficients are calculated using the YW scheme.

In the above schemes the estimation of residual white noise variance was

$$\sigma_a^2 = \frac{P_M}{2(N-M)}, \tag{2.78}$$

where the estimation of PM is taken from Eq. (2.73).

A very important problem, with respect to stochastic model identification, is the choice of the order of autoregression M. Following recommendations of Ulrych and Bishop (1975), Privalsky et al. (1992), and Privalsky and Jensen (1993), we tested five criteria for choosing M: the *Akaike criterion* (FPE), the *Akaike informational criterion* (AIC), the *Parzens criterion* (CAT), the *Schwarz-Rissanen criterion* (BIC), and the *Hennan-Quinn criterion* (PSI). All these criteria are some estimation [improved, in comparison with Eq. (2.73)] of the error of the forecast using the identified model: it is supposed that the best variant of the model corresponds to the minimal value of the criterion. Formulas for the criteria are as follows

$$FPE(M) = \frac{N+M+1}{N-M-1} * \frac{P_M}{2(N-M)}, \tag{2.79}$$

$$AIC(M) = \ln \sigma_a^2 + 2\frac{M}{N}, \tag{2.80}$$

$$CAT(M) = \frac{\sum_{i=0}^{M} \sigma_a^{-2}(i)}{N} - \sigma_a^{-2}(M). \tag{2.81}$$

The formulas for BIC and PSI resemble that for AIC, but instead of coefficient 2 we use, respectively, ln N and ln(ln N) (Privalski et al. 1992).

There is also another important problem while estimating stochastic model parameters: the coefficients of autoregression must satisfy the condition of the model's stationarity. To verify this, the *characteristic equation of autoregression* must be solved (Box et al. 1994):

$$1 - \sum_{i=1}^{M} c_{M,i} \varsigma^i = 0. \tag{2.82}$$

The AR model is stationary if all the roots ζ of Eq. (2.82) are outside a unit circle on a complex plane. For instance, for the first-order autoregressive model (the first-order Markov process) the condition of stationarity is satisfied if $-1 < c_1 < 1$. If n roots lie on the unit circle, and all the rest are outside it, the process is "stationarizable", i.e., the n-th order increment of the initial sequence is a stationary process of the order $M - n$. For example, the order of the AR model describing the Caspian Sea level variations (Fig. 2.6) equals 2, and one of the coefficients is close to 1 (actually, it is equal to 0.97). Consequently, the first increments of the Caspian Sea level time series are described by the first-order stationary model with coefficient 0.4 .

Finally, rough values of confidence intervals for spectral density estimations can be determined using the Kromer estimates (Ulrych and Bishop 1975) demonstrating that MEM $f(\lambda)$ estimations are asymptotically unbiased and distributed normally with variance

$$\sigma^2[f(\lambda)] = \frac{2}{DF} f^2(\lambda), \tag{2.83}$$

where DF is the number of degrees of freedom, related to the order of the autoregressive process in the following way: $DF = N/M$.

It follows from Eq. (2.83) that the order of the autoregressive model cannot be too high because the width of confidence intervals of reasonable probability would be comparable with the values of the spectral density itself. With respect to these limitations, and following the recommendation of Haikin (1979), we did not consider orders $M > N/10$.

2.3.5 An Example of Model Identification and Parameter Estimation

We will take as an example the time series of the annual air temperatures in New Orleans, 1874 - 1990 (the largest part of this series was shown in Fig. 1.12). For the estimation of the partial model coefficient, we used the third method described in the previous section. Note that in this case, as well as in many other climatic applications, the Burg scheme gave almost the same results as the Yule-Walker scheme and the modified YW scheme. In Table 2.1 the matrix of model coefficients is presented.

Table 2.1. Matrix of partial coefficients $c_{M,i}$ of the autoregressive models fitted to the New Orleans temperature record (1874 -1990)

Order M	i=1	2	3	4	5	6	7	8	9	10	11	12	13	14
1	0.48													
2	0.38	0.22												
3	0.34	0.15	0.17											
4	0.33	0.14	0.14	0.07										
5	0.32	0.12	0.12	0.03	0.14									
6	0.29	0.12	0.10	0.00	0.08	0.19								
7	0.29	0.12	0.10	0.00	0.07	0.18	0.03							
8	0.28	0.10	0.09	0.00	0.06	0.17	0.01	0.09						
9	0.28	0.10	0.09	0.00	0.06	0.17	0.01	0.09	0.01					
10	0.28	0.09	0.09	0.01	0.06	0.17	0.00	0.08	-0.01	0.07				
11	0.28	0.10	0.08	0.01	0.04	0.16	0.00	0.08	-0.02	0.05	0.08			
12	0.28	0.10	0.08	0.01	0.04	0.17	0.00	0.07	-0.02	0.05	0.10	-0.06		
13	0.27	0.12	0.09	0.01	0.06	0.17	0.04	0.08	-0.02	0.07	0.12	0.00	-0.22	
14	0.28	0.12	0.09	0.02	0.06	0.17	0.04	0.07	-0.02	0.07	0.12	-0.01	-0.23	0.06

In Fig. 2.19 two criteria for the identification of a model are shown: the Akaike criterion and the Schwarz-Rissanen criterion. It is seen from this figure that the values of both criteria fall off from the zero order (model of white noise) to order 1 (first-order autoregressive model or first-order Markov process). Then the Akaike criterion does not change considerably, and the Scharz-Rissanen criterion gradually increases from value 2. In other words, the Akaike criterion shows that the optimal forecast of annual temperatures with unit lead time is improved if we use the first (or

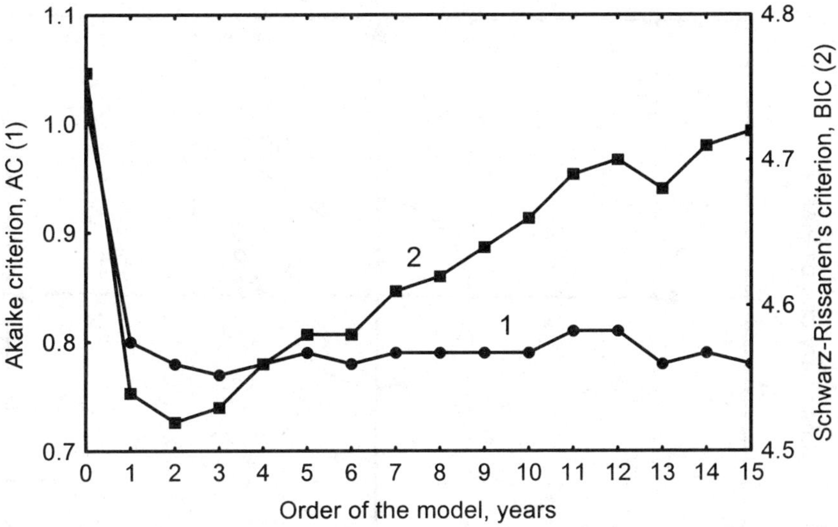

Fig. 2.19. The Akaike criterion (*1*), and the Schwarz-Rissanen criterion (*2*) for stochastic models describing the time series of New Orleans annual near-surface air temperatures (1874 - 1990)

higher) order autoregressive model instead of the white noise model (the sequence of uncorrelated random variables). Formally, the cited criteria give orders of M equal to 2 to 3. Values of other criteria are presented in Table 2.2.

Table 2.2. Criteria for model identification. New Orleans air temperature series

Order M	AC	AIC	CAT	BIC	PSI
0	1.02	4.76	0.000	4.76	4.76
1	0.80	4.51	0.000	4.54	4.51
2	0.78	4.47	-0.001	**4.52**	4.47
3	**0.77**	4.45	-0.003	4.53	4.45
4	0.78	4.47	-0.003	4.56	4.45
5	0.79	4.46	-0.005	4.58	4.44
6	0.78	4.44	-0.010	4.58	4.42
7	0.79	4.44	-0.011	4.61	4.42
8	0.79	4.43	-0.016	4.62	4.40
9	0.79	**4.42**	-0.021	4.64	4.39
10	0.79	4.43	-0.023	4.66	**4.39**
11	0.81	4.43	**-0.024**	4.69	4.39

In accordance with Table 2.2, criteria AIC, CAT, and PSI formally give higher orders: respectively, 9, 11, and 10. Recall that we do not consider

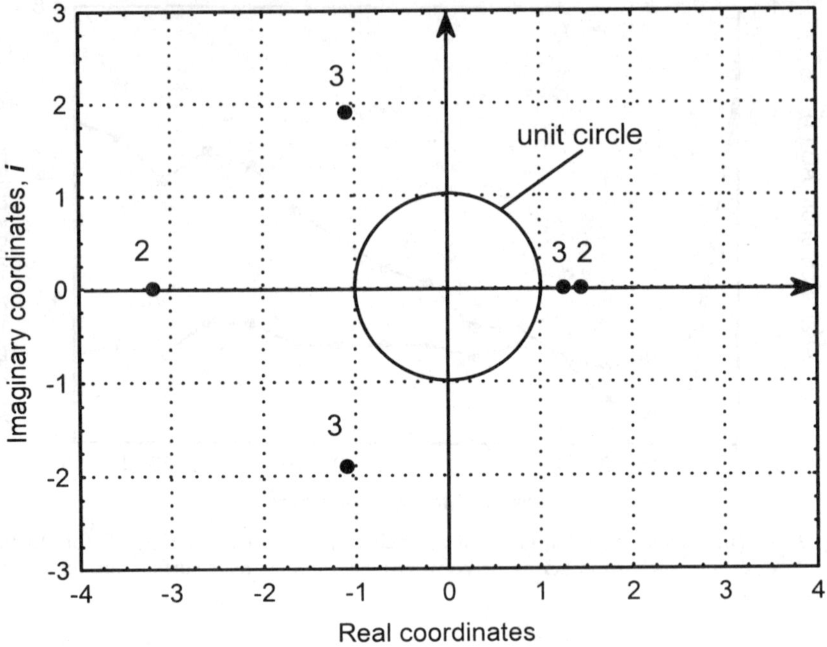

Fig. 2.20. Solutions of the characteristic equation of autoregression for model orders 2 and 3. New Orleans temperature time series

orders higher than N/10 (in our case, approximately 11) because of the Haikin restriction.

However, the solution of the characteristic equation of autoregression shows that model orders 9, 10, 11 provide nonstationary models (not all solutions of Eq. (2.82) lie outside the unite circle on complex plane). On the contrary, models of order 2 and 3 are in accordance with the condition of stationarity (Fig. 2.20).

It is natural to make a final decision, with respect to the order of the model, by comparing different variants of spectral estimations. Six different variants of spectral density estimations (corresponding to orders 1, 2, 3, 9, 10, 11) are shown in Figs. 2.21 and 2.22. In these figures confidence limits for spectral estimations are also shown. (Note that spectral densities are presented in logarithmic scales here: in this case the Kromer confidence limits do not depend on the frequency and can be intercompared). It is evident that, taking into consideration the width of confidence limits, all variants of spectral estimation do not differ largely from each other. (The only exception is the difference between the first--order spectrum and the highest order spectra at low frequencies in Fig. 2.22; however, spectral density at these frequencies, corresponding to the

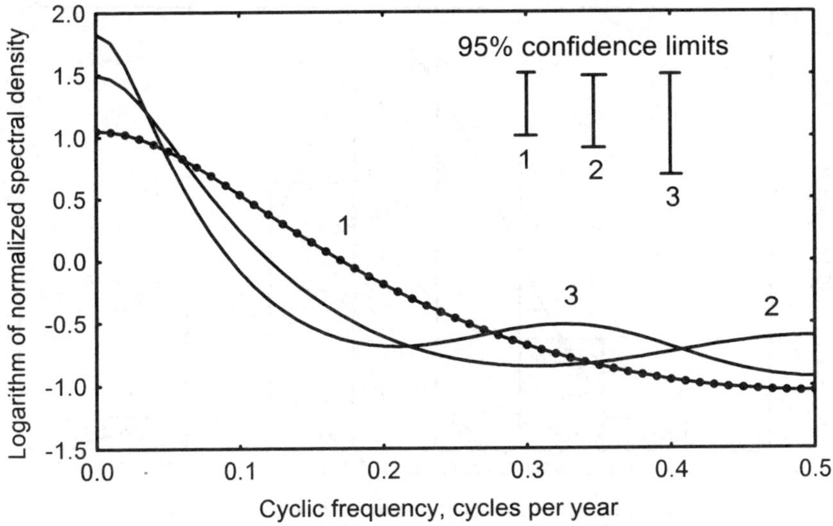

Fig. 2.21. Estimations of the spectral density for the autoregressive model of order 1 (*curve 1*), 2 (*curve 2*), 3 (*curve 3*). *Vertical lines* denote respective 95% confidence limits. New Orleans temperature record, 1874 - 1990

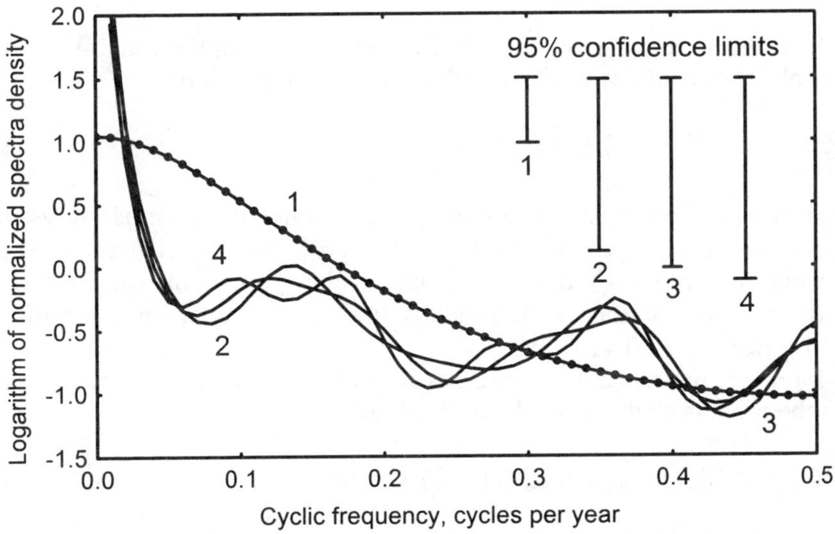

Fig. 2.22. Estimations of the spectral density for the autoregressive model of order 1 (*curve 1*), 9 (*curve 2*), 10 (*curve 3*), 11 (*curve 4*). *Vertical lines* denote respective 95% confidence limits. New Orleans temperature record, 1874 - 1990

periods more than approximately 32 years, cannot be assessed because of the limited length of the time series).

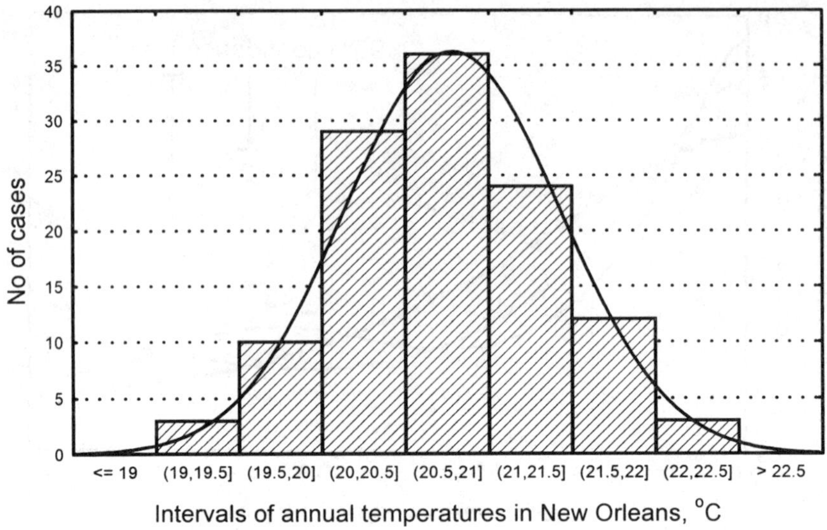

Fig. 2.23. Normal probability distribution (*solid curve*) fitted to the sample probability distribution (*histogram*). Near surface mean annual air temperatures in New Orleans, 1874-1990

Thus, the first-order model AR(1) gives a good approximation of the spectral density (consequently, of the correlation function):

$$\xi(t) = 0.48\xi(t-1) + a_1(t), \tag{2.84}$$

where $\xi(t)$ is taken to be the "anomaly" of the annual air temperatures [the mean value of sequence $\xi(t)$ is taken to be zero], t is time in years, $a_1(t)$ the sequence of normally distributed uncorrelated random values (white noise), and its variance $\sigma^2(a_1)$ equals 0.77 of the total variance of the initial process $\xi(t)$, 0.41 (°C)^2.

Moreover, model (2.84) possesses important advantages in comparison with the models of the second and third order:

$$\xi(t) = 0.38\xi(t-1) + 0.22\xi(t-2) + a_2(t), \tag{2.85}$$

$$\xi(t) = 0.34\xi(t-1) + 0.15\xi(t-2) + 0.17\xi(t-3) + a_3(t), \tag{2.86}$$

where variances $\sigma^2(a_2)$ and $\sigma^2(a_1)$ are respectively 0.74 and 0.72 of the total variance of the process. In fact, models (2.85) and (2.86) contain respectively two and three more coefficients to be assessed than model (2.84). So the errors in estimating these coefficients and other parameters of the model are considerable (compare, for instance, confidence limits in

Fig. 2.21). It is evident from Fig. 2.22 and from the solution of the characteristic equation of autoregression that models of higher orders (9, 10, 11, and others) are even worse.

So the AR(1) - the first-order Markov process - is the best approximation of the initial temperature time series. In Fig. 2.23 the probability distribution of the New Orleans annual temperatures is shown, which fits almost perfectly the normal distribution (cf. Fig. 2.2 for monthly temperatures). This means that in our case, as well as in many other climatic cases, the correlation theory is the optimal one, and the first-order autoregressive model (the first-order Markov process) gives answers to practically all questions related to the behavior of the process.

The above example is characteristic for the study of climatic time series (especially as typical registrations are much shorter). All in all, thousands of time series were processed by the author by "hand", and many others were processed quasi-automatically. Anticipating these results, we would like to discuss the specificity of applying the above methods of time series analysis of the annual values and monthly anomalies of climatic variables:

1. Usually (except for time series related to the local and discrete objects of the climate system) the Yule-Walker method, the Burg scheme, as well as other methods of parametric spectral analysis - Box-Jenkins method, using moving average models and combined models - gave similar results. The relative simplicity of natural climate variation laws could be the cause of this situation. Indeed, we know that the analysis of more deterministic processes - radio signals, changes in electric circuits, etc. - is much more sensitive to the method we use. Perhaps this is because, initially, methods of spectral analysis and the random functions theory were developed for the analysis of man-made processes, the hidden periodicities in them, and so on.
2. As a rule, criteria for the choosing of the model orders provide only reference information. In practice, if we consider all circumstances of the model choice, the first model order turns out to be the best (except for some "exotic" climate processes like terminal lake level heights or behavior of other local objects). This means that confidence intervals of spectral estimations for high model orders, as a rule, overlap the first-order spectral density curve, and the same can be said about other model parameters (correlation function, residual white noise variance, etc.).
3. Thus, the most important task in climate stochastic research is the choice between three fundamental types of models: the white noise, the first order Markov process, the discrete Wiener process (random walk). Obviously, in practice, this choice is related to the estimation of the AR(1) coefficient c_1, i.e., the problem primarily consists in distinguishing three variants: $c_1 = 0$; $0 < c_1 < 1$; $c_1 = 1$. Also, in order

to identify the discrete Wiener process, an analysis of temporal increments of time series is needed. In the next chapters of the book appropriate examples will be given.

2.3.6 Frequency Truncation Method of Normalized Spectral Estimates

In the above sections it was usually supposed that the mean value of the process (m) and the variance of the process $\sigma^2(\xi) = B(0)$ were known. Consequently, formulas for calculating confidence limits of spectral estimations [e.g., Eq. (2.83)] were based on the same assumptions.

However, we always possess only approximate estimations of these parameters. The errors in estimating the variance (standard deviation) can be enormous, especially in the most important case - when the coefficient c_1 is close to 1, and the process is close to the nonstationary one. On the other hand, estimation of $\sigma^2(\xi)$ is included into the formula for the assessment of normalized spectral density. Thus, if we have to compare spectral densities of two or more climatic process, or if we deal with the spectra of multidimensional processes or random fields, some method of diminishing these errors is needed [confidence limit estimations like Eq.(2.83) are not adequate in these cases].

In this section we propose such a method of calculating normalized spectra. Its idea is the following: the largest errors in the estimation of total variance of the process, especially in the case when coefficient c_1 is high, are usually concentrated at low frequencies (near zero frequency). However, estimations of spectral densities at these frequencies are impossible because of the limited length of climatic time series (limited number of points within the series). Thus, it is only natural not to take into consideration the low-frequency segment of the spectral density while calculating normalized spectral density.

Recall that the variance of the random process is equal or proportional - depending on the specific definition of the notion spectral density - to the area under the spectral density curve:

$$B(0) = \sigma^2(\xi) \sim \int_0^{0.5} f(\lambda)d\lambda \ . \tag{2.87}$$

It is easier to use in Eq. (2.87), as in previous figures with spectra, not angular frequency, but "cyclic frequency" λ (2π per discreteness interval). Correspondingly, a part of the variance contained between two frequencies λ_1 and λ_2 can be expressed as

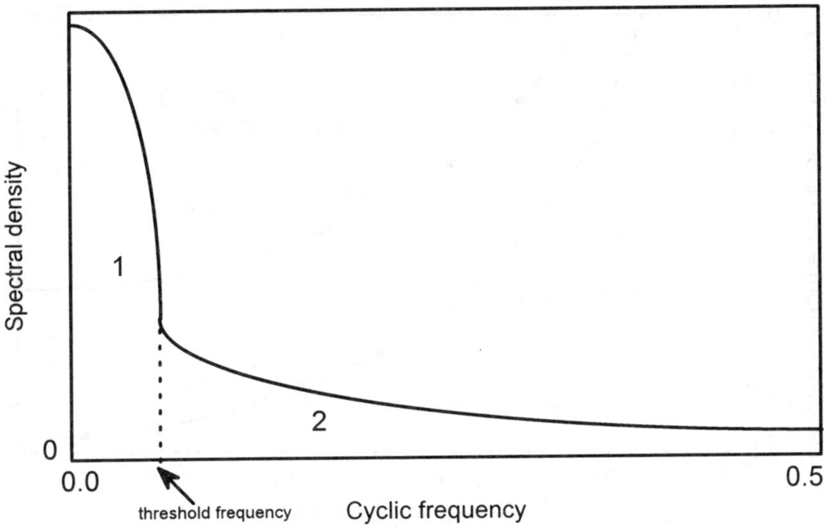

Fig. 2.24. Scheme of the "frequency truncation" method of calculating normalized spectral densities. *Area 1* "truncated" part of variance; *area 1* + *area 2* total variance

$$\sigma^2_{\lambda 1,\lambda 2} \sim \int_{\lambda_1}^{\lambda_2} f(\lambda) d\lambda \,. \tag{2.88}$$

Therefore, while estimating the normalized truncation spectral density $\widetilde{f}_t(\lambda)$, we propose to divide the initial spectral density not by the total variance, but by the part of variance contained between some low "threshold" frequency λ_{th} and 0.5 cyclic frequency (see Fig. 2.24). Thus, calculations are made using the following formula:

$$\widetilde{f}_t(\lambda) = \widetilde{f}(\lambda) \frac{\sigma^2_{0,\lambda_{th}} + \sigma^2_{\lambda_{th},0.5}}{\sigma^2_{\lambda_{th},0.5}} , \tag{2.89}$$

where

$$\sigma^2_{0,\lambda_{th}} = \int_0^{\lambda_{th}} f(\lambda) d\lambda, \qquad \sigma^2_{\lambda_{th},0.5} = \int_{\lambda_{th}}^{0.5} f(\lambda) d\lambda.$$

In Figs. 2.25-2.26 the effect of using the frequency truncation method is demonstrated. In Fig. 2.25 traditional MEM estimations of two time series are shown: the Vostok ice core methane paleorecord, and deuterium paleo-

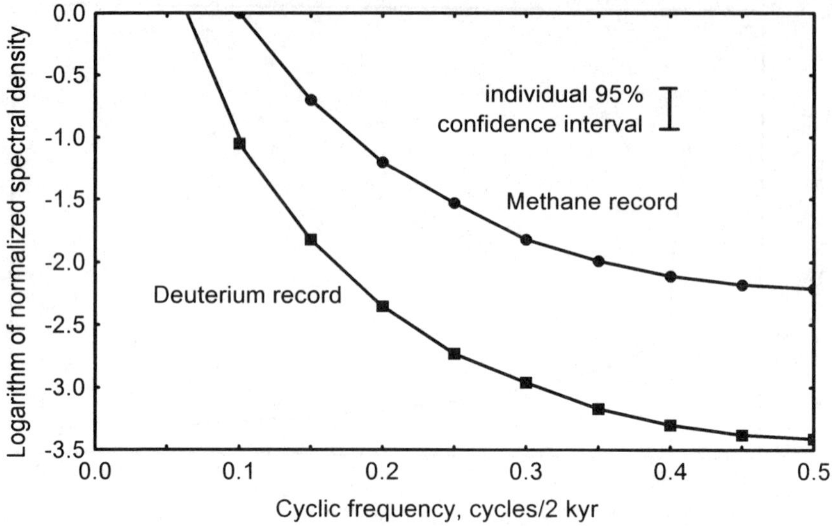

Fig. 2.25. Traditional NEM spectral densities of Vostok methane and deuterium records (Antarctic). *Vertical line* denotes "individual" confidence limits of spectral estimations

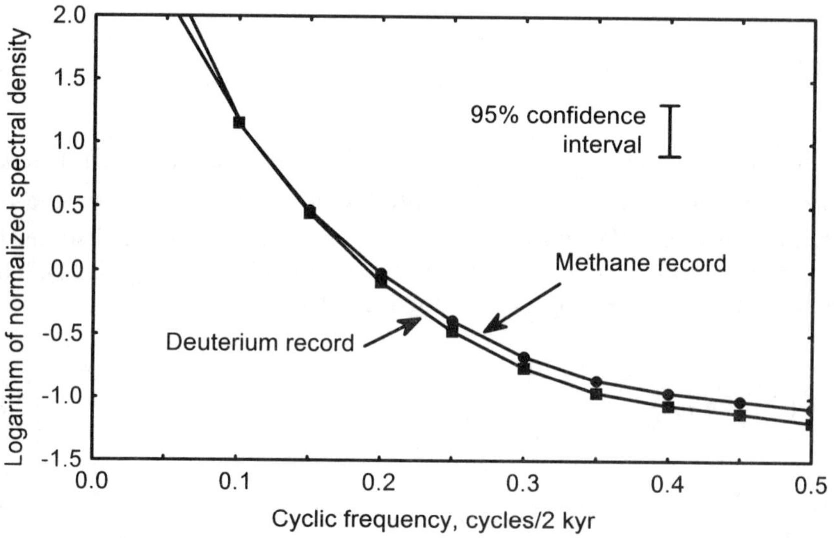

Fig. 2.26. "Frequency truncation" estimations of Vostok methane and deuterium records. *Vertical line* denotes true 95% confidence interval

-record. Traditional, Kromer-like 95% confidence limits are also shown in this figure. Note that these limits are calculated for the case when the variance of time series is presumed to be exactly known. Therefore,

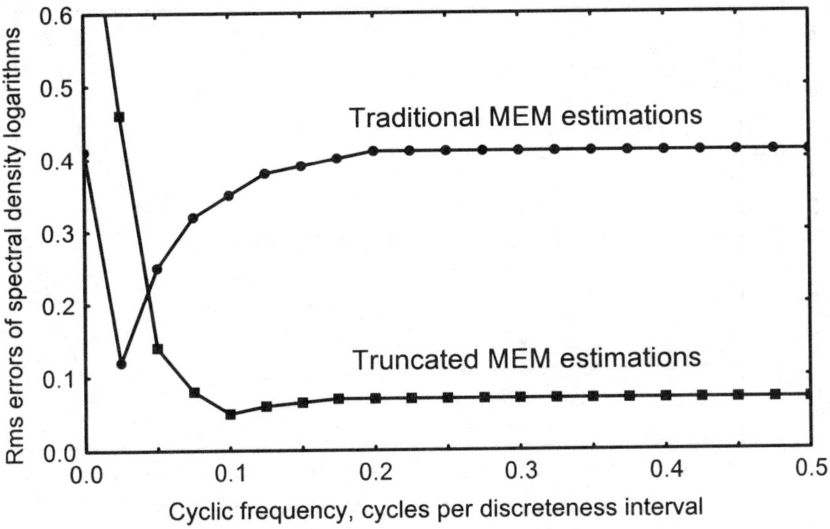

Fig. 2.27. Rms errors of spectral density logarithms estimations for 100-point realizations and coefficient $c_1 = 0.9$

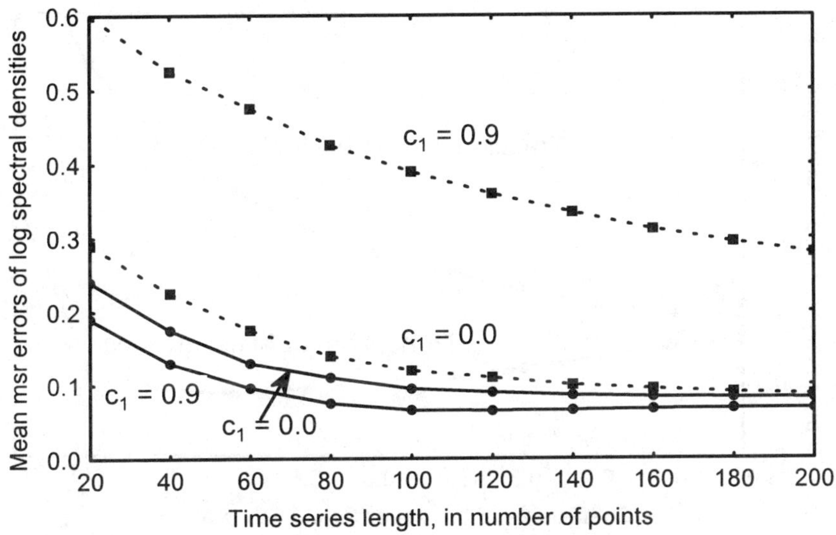

Fig. 2.28. Averaged over frequencies (λ_{th} , 0.5) rms errors of spectral density estimations. Different series lengths (from 20 to 200 points), and different coefficients c_1 (from 0.0 to 0.9). *Dashed lines* traditional MEM estimates; *solid lines* truncation MEM estimates

Kromer confidence limits can be used if we compare different variants of models and spectral densities fitted to the same time series. However, if we want to compare spectral densities of different climatic time series, the

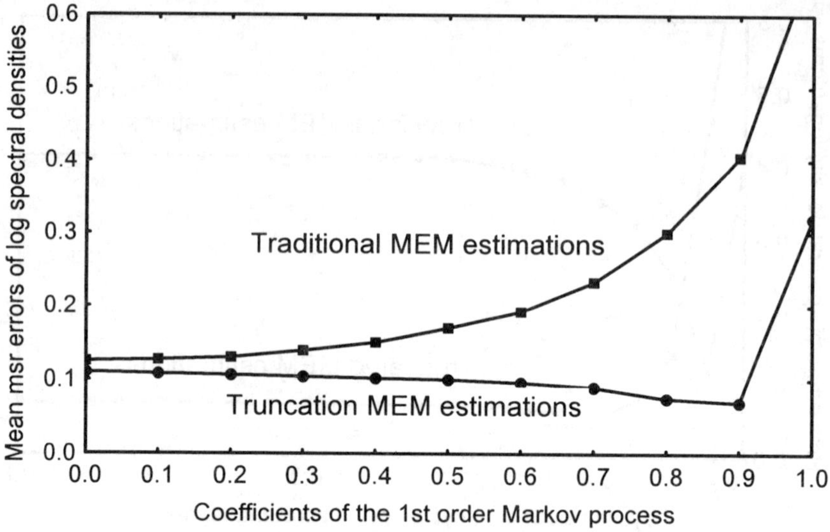

Fig. 2.29. Averaged over frequency interval (λ_{th} , 0.5) rms errors of spectral density estimations for time series length $N = 100$

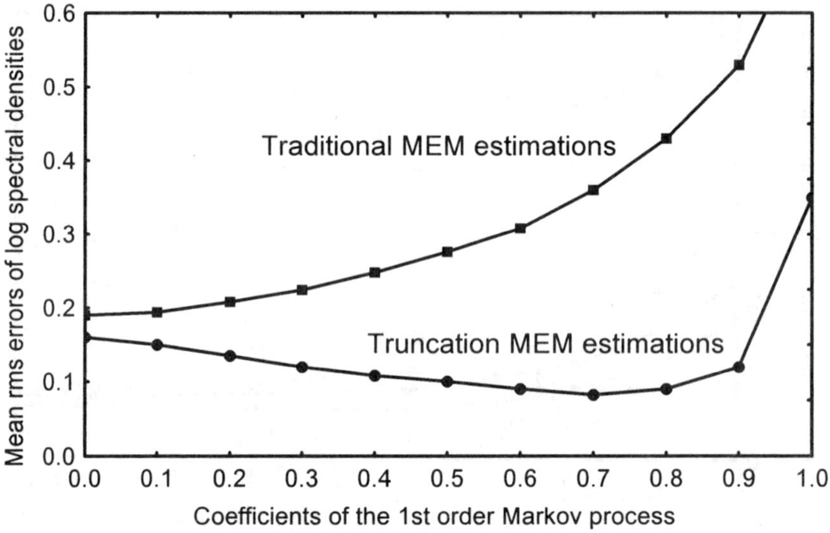

Fig. 2.30. A Fig. 2.29 but for the time series length $N = 50$

Kromer confidence limits do not "work": they do not take into consideration large (and different) errors in the estimations of variances of the two climatic processes. All in all, from the point of view of traditional MEM analysis, the methane spectral density statistically differs from the

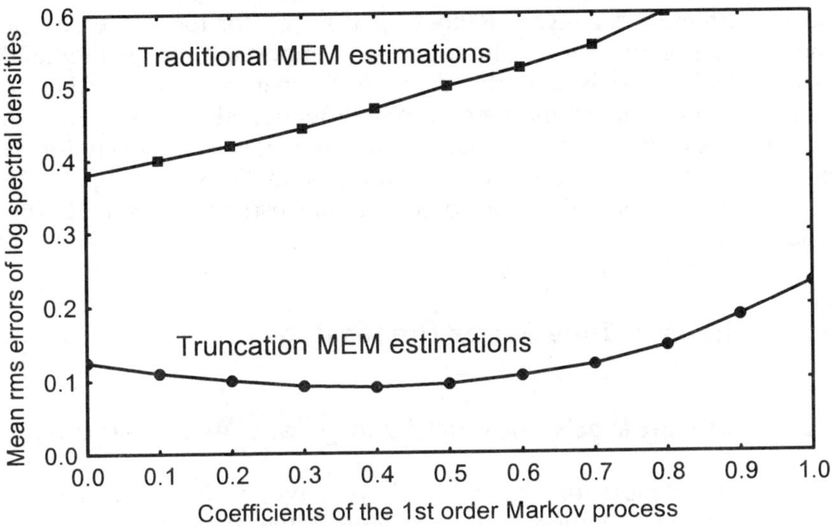

Fig. 2.31. A Fig. 2.29 but for the series length $N = 20$

deuterium spectral density. Therefore, the traditional MEM approach indicates that these are two different processes.

In Fig. 2.26 spectral densities of the same two time series (Vostok methane and deuterium records) are assessed using the frequency truncation method. It is evident that from this point of view, the two spectra densities practically coincide with each other, so the structure of the two processes is identical. The errors of estimation in this case are dozens of times less than the errors in traditional estimations.

Using the Monte Carlo method, a series of experiments, was implemented to assess the effectiveness of the proposed procedure for time series with different lengths and different structure. In Figs. 2.27-- 2.31 the errors of spectral estimations using traditional and frequency--truncated procedures are shown for the most important case: the first--order discrete Markov process (first-order autoregressive model). In Fig. 2.27 mean errors of logarithms of normalized spectral densities are presented for realizations of the AR(1) sequences with coefficient $c_1 = 0.9$ and the length of realizations $N = 100$. In this and other cases, presented in Figs. 2.25 - 2.31, we took the value of threshold frequency λ_{th} equal to $6/N$ cycles/discreteness interval, which appeared to be the most effective λ_{th} estimation for AR(1) sequences.

It is clear that the truncated estimations are much more effective than traditional estimations, save for very low frequencies where spectral estimations are not possible anyway. In Figs. 2.28 - 2.31 mean (between threshold frequency λ_{th} and 0.5 cycles/discreteness interval) errors of spectral estimations are presented for different lengths of realizations and different AR coefficients. It is clear that truncation estimations are especially effective for high AR coefficients, i.e., for the processes close to the nonstationary ones, which is very important for distinguishing between AR(1) models and the discrete Wiener processes. Also it is evident that truncated estimations become effective already for short time series, whereas traditional estimations are not consistent even for very long realizations. Thus, errors of truncated estimations for 20-point time series are much less than errors of traditional estimations for 200-point time series.

2.3.7 Other Methods of Time Series Processing

2.3.7.1 Conventional Methods. Moving Average and ARMA Models

In the previous parts of Section 2.3 we have dealt mostly with the Maximum Entropy Method that is related to the autoregressive models (AR). It is important to compare MEM estimates with those obtained using traditional spectral estimates. On the whole, the comparison of MEM estimates with those using other methods demonstrates considerable advantages of the Maximum Entropy Method (Haykin 1979; Dzhaparidze and Yaglom 1983; Privalsky 1985; Marple 1987; Zhurbenko and Kozhevnikova 1990, etc.). Thus, the resolution properties (the ability to distinguish spectral density peaks of neighboring frequencies) of the MEM exceed that of the method of conventional *periodogram smoothing* and *maximum likelihood spectral estimates*. For the variance and bias of spectral estimates, MEM is not worse than the other two methods (Ulrych and Bishop 1975; Baggeroer 1976; Yaglom 1981; Marple 1987, etc.).

It is also necessary to compare the MEM techniques with other parametric methods of spectral estimations - mainly models of the moving average (AR) and combined *autoregression-moving average models* (ARMA). Our experience of using these types of models (see also Marple 1987) enables us to draw following conclusions:

1. The MA and ARMA processes usually can be satisfactorily described by the autoregressive models.
2. If there are peaks in the spectrum, the resolution capacity of AR estimates is better than that of MA.
3. Statistical characteristics of spectral ARMA estimates, unlike those of the AR estimates, have not yet been satisfactorily studied.
4. In general, algorithms of spectral ARMA estimates (unlike those of the AR estimates) should not be used for the processing of signals in real time.

(At the same time, note that for the processing of the river runoff time series, MA and other, more sophisticated, models can play an important role).

Finally, the problem of types of probability distribution, with respect to time series processing, must be discussed. It was stated above that the correlation theory of random functions can give complete information on the stationary process only if the latter is normally distributed. Below it will be shown that the distribution of most of the variables of thick climatic subsystems can be considered normal in the first approximation. However, distribution of monthly and annual precipitation at specific stations, as well as the distribution of the river runoff and of some other variables, differs from the normal one. Nevertheless, even these variables, being taken at large spatial scales (averaged over large regions), demonstrate probability distributions that are much closer to the Gaussian one. Also Yaglom (1981) and Privalsky (1985) state that the MEM analysis turns out to be useful even when the distribution is not normal: "We do not need an assumption on the Gaussian character of a distribution, since the maximum entropy principle will automatically choose this form as one realized by maximum ways and corresponding to the initial data" (Privalsky 1985). It is emphasized in the same book that the MEM formulas can be deduced without hypothesis on the Gaussian character of the process.

2.3.7.2 "Deterministic Chaos". Other Methods of Nonlinear Analysis

In recent years, the mathematical study of dynamical systems has shown that the solution of simple, nonlinear equations can exhibit complicated temporal and spatial behavior. Many examples now exist of equations showing a variety of behavior ranging from simply periodic to chaotic. An underlying concept that has emerged from such studies is that of a *strange attractor*. Simple attractors such as a system of thermodynamic equilibrium (*point attractor*) or those which decay into periodic states (*limit cycles*) have been known for some time. A strange attractor has the property that the system decays to ("is attracted to") a final state, but this state is not periodic and is extremely complex. In fact, it is chaotic - pseudorandom but arising as the solution of a deterministic set of equations. It is also highly sensitive to initial conditions.

During the past 20 years there have been many attempts to use the above concepts for the description and forecasting of climatic processes. Examples of recent works of this kind can be found in: Carl et al. (1995), Krupchatnikov (1995), Sonechkin and Ivashchenko (1996), Vasechkina et al. (1996), Wang and Fang (1996), Shrirer et al. (1997), Sonechkin et al. (1997), etc. Also packs of computer programs are available for these purposes. For example, the software package Chaos data analyzer

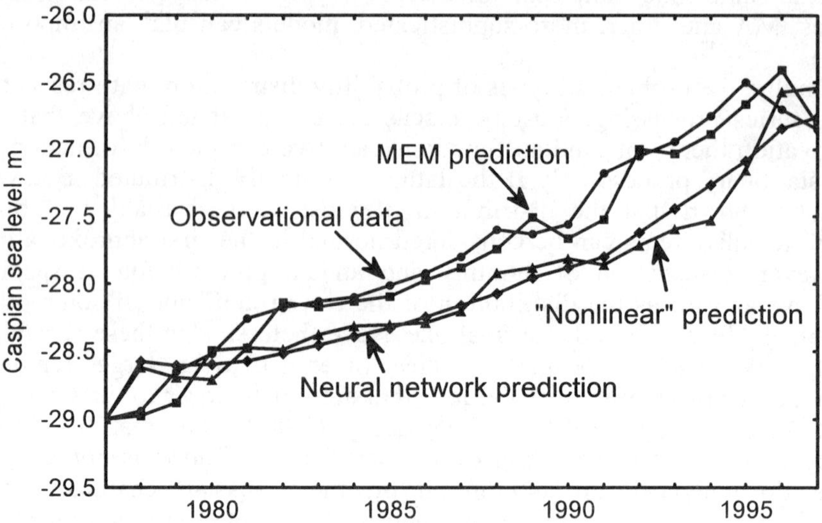

Fig. 2.32. Three variants of the Caspian Sea level prediction and observed curve

(a project of the American Institute of Physics) (Sprott and Rowlands 1995) contains about 20 programs, most of them representing a specific method of nonlinear or deterministic analysis of time series.

However, serious doubts arise with respect to the possibility of describing the climate variations by a system of simple deterministic equations: see, e.g., Gotz (1995) and Sahay and Sreenivsan (1996). On the other hand, strange attractors sometimes can be regarded as interesting prototypes of real climatic processes (see Sect. 2.4 about this).

We tested several methods of fitting chaotic and other nonlinear and deterministic models to different observed climatic time series. For these purposes we used: estimation of the *Lyapunov exponent* (Sprott and Rowlands 1995), calculation of the *capacity dimension* (Mandelbrot 1982; Barnsley et al. 1988), assessment of the *correlation dimension* and "BDS statistics" (Grassberger and Procaccia 1983), the *singular value decomposition* (Sprott and Rowlands 1995), the *wavelet transform* (Chui 1992), the *neural network* method (Tsonis 1992), and the "nonlinear prediction" method and removing noise from the data (Farmer and Sidorowich 1988), and some others.

Unfortunately, we could not obtain positive decisive results using the above methods for the description of variability at climatic temporal scales, and could find no strange attractors or other deterministic structures within the time series of climatic variables. Seemingly, changes in the climatic thick and thin subsystems on a global scale are too random to be satisfactorily described by low-dimensional nonlinear deterministic structures. Moreover, our attempts to use the above methods for the for the analysis of local climatic object variations (which seem to be more promising with respect to nonlinear methods) were unsuccessful.

As a typical example, three variants of forecasts of the Caspian Sea level series are presented in Fig. 2.32. A recent and most intriguing segment of the registration is taken: the period from 1977 that is characterized by intensive sea level rise. Three methods of prediction were tested: prediction using the MEM method, the neural network method, and nonlinear prediction (searching the data record for the nearest points in a D-dimensional embedding). Because of the limited time series length (147 - 166 points), or because of the structure of this climatic registration, the method of singular decomposition did not work at all in this case (for the last three types of prediction we used computer programs from the Chaos data analyzer (1995) pack. However, the problem of limited time series length is typical for climatic registrations; the Caspian series is one of the longest ones.

In Fig. 2.32 predictions with 1-year lead time (prediction time) are intercompared. It is clear that nonlinear methods of prediction give much worse forecasts than the MEM method. Moreover, a simple inertial forecast gives much better results than the neural network and nonlinear techniques.

2.4 Physical Basis of the Stochastic Climate Theory

2.4.1 Atmospheric Forcing of the Climate System

2.4.1.1 Observational Evidence

A key problem of all climate theories is the source of energy for climatic changes. Strictly speaking, the difference between various climate theories is the difference in energy sources indicated and postulated by these theories.

Let us consider a typical spectrum of atmospheric variability (Fig. 1.15). It is unanimously admitted that there is a general decrease in spectrum (energy of variations) with frequency within the range of climatic frequencies: approximately up to 1 cycle/month (except 1-year oscillation and its harmonics). Common sense tells us that energy flows from the frequencies with high energies (spectral density) towards frequencies with low energy, rather than backward. Thus, in the case of climatic variability and variations, energy seems to originate somewhere within the range of low frequencies and large temporal scales.

In turn, it is evident that mechanisms generating slow and powerful variations cannot be found within the atmosphere because the

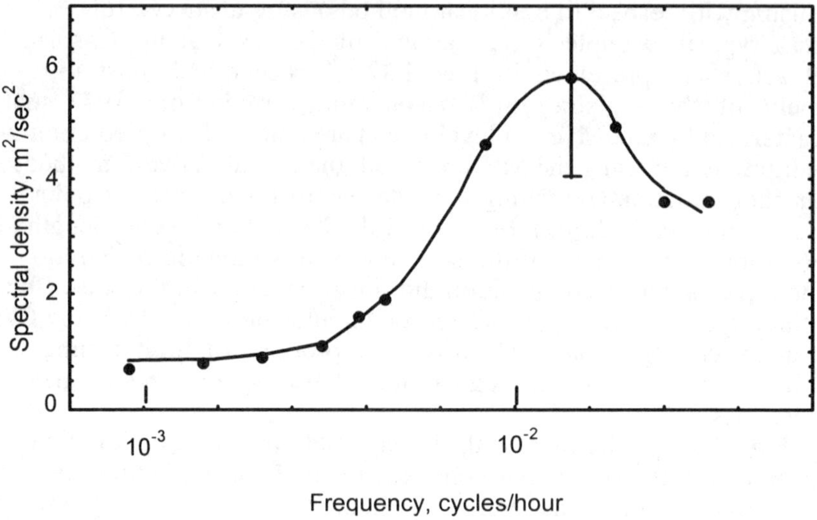

Fig. 2.33. Monin's interpretation of van der Hoven wind velocity spectrum. (After Monin 1969). Some esperimental estimates and a confidence interval are shown

characteristic time of atmospheric variability is only several days, and the temporal inertia of the atmosphere is insignificant: atmospheric heat capacity equals only about 1/400 of the heat capacity of the ocean.

It follows from the above considerations that: (a) we seemingly have to look for the causes of the climate change somewhere outside the atmosphere - in human activities, in the Solar system, in the volcanoes, in sophisticated mechanisms of the oceanic circulation (ENSO and other events), etc.; (b) characteristic temporal scales of climate forcing are likely to be similar to characteristic scales of climate changes, i.e., years, dozens, and hundreds of years; (c) in order to provide a considerable response of the climate system to external forcing (the latter being not great), models with strong (including positive) feedbacks were needed.

Indeed, most scholars in the research area of climate change follow (and followed in the past) the above scheme. As we know, such branches of climatic science as investigations of the greenhouse effect, ozone depletion, ENSO events, and explosive volcanic eruptions absolutely dominate climate change studies.

However, the above approaches still do not give acceptable climate forecasts and do not explain the most important features of climatic variations and variability: the continuous character of climatic spectra and their form. Even the most sophisticated tools of contemporary

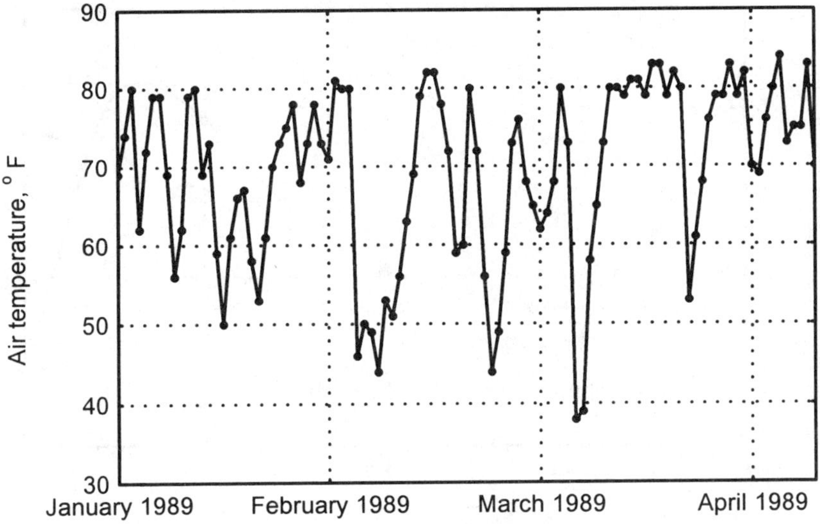

Fig. 2.34. Fragment of New Orleans daily air temperature registration. (Data from World Weather Disc 1994)

climatology, GCMs and coupled models, do not reproduce a realistic form of climatic spectra (see Chapters 1 and 3).

A breakthrough in climate theory was made by K.Hasselmann when he published his stochastic climate models theory (1976, 1981, 1982). His hypotheses were quite opposite to the above, common sense assumptions: (a) the main source of the changes in the climate system is within the atmosphere; (b) the main processes generating climatic variabilities are related to high-frequency fluctuations. In turn, slow changes in the inertial components of the climate system - ocean, cryosphere, land surface (and slow changes in the climate system in general) - represent a response to the atmospheric forcing; (c) hypotheses on the strong positive feedbacks are not needed within this approach.

Thus, the most important notion of the stochastic climate theory is that of *atmospheric random forcing.* It is well known that atmospheric processes have a spectral peak at a period which is equal to several days and is caused by synoptic and related processes. In order to estimate the climatic significance of atmospheric processes, it is important to study the vicinities of the synoptic peak, especially the low-frequency «"wing" of atmospheric spectra.

Back in the 1950s it was shown in a classical work by van der Hoven (Fig. 2.33) that typical wind spectrum becomes flat (close to the horizontal

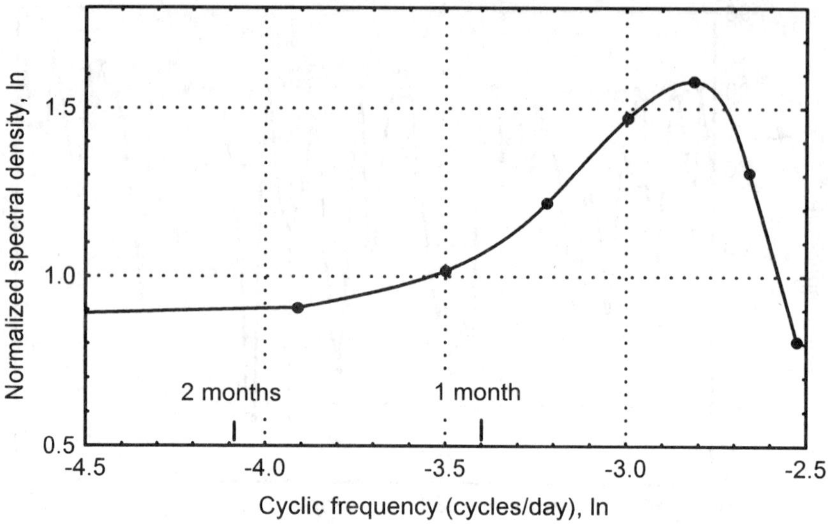

Fig. 2.35. Logarithms of normalized spectral density, air temperature fluctuations in New Orleans

line) beginning with the period of approximately 1 month (i.e., to the left of frequency 1 cycle/month). Recall that horizontal spectral density corresponds to an uncorrelated random process (white noise).

Later, many researchers observed the same pattern of spectra of atmospheric variables to the left of synoptic frequencies: see works by Byshev and Ivanov (1969), Wunsch (1972), Frost (1975), etc. As an example, we present an estimation of spectral density of air temperatures in New Orleans. We have taken a time series of daily temperatures with previously removed seasonal dynamics (Fig. 2.34). The spectral density of this registration is shown in Fig. 2.35. It resembles qualitatively the van der Hoven spectrum: to the left of approximately 1 cycle/month the spectral density is close to the horizontal line, which means that at these frequencies the atmospheric variability can be described by an uncorrelated random process.

2.4.1.2
Atmospheric Model Results

Similar general features of "pure" atmospheric variability are also demonstrated by numerical atmospheric models (Frankignoul and Hasselmann 1977; Kruze and Hasselmann 1986, etc.). Here we would like to present, as an example, results obtained using the simplest possible but

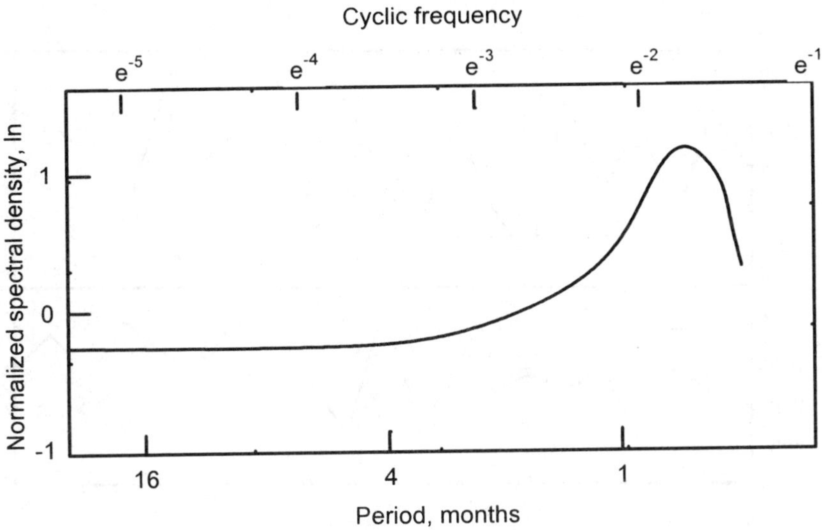

Fig.2.36. Logarithms of normalized spectral densities of the strength of a large-scale westerly-wind current X in the Lorenz model. With a = 0.25, b = 4.0, F = 7.0+2.0 $\cos(2\pi t/\tau)$, G = 1.0, τ = 12 months

most fundamental general circulation model - the low-order geostrophic baroclinic model by Lorenz (1984, 1990). The very well-known equations of the model are:

$$\frac{dX}{dt} = -Y^2 - Z^2 - aX + aF, \qquad (2.90)$$

$$\frac{dY}{dt} = XY - bXZ - Y + G, \qquad (2.91)$$

$$\frac{dZ}{dt} = bXY + XZ - Z, \qquad (2.92)$$

where X represents the strength of a large-scale westerly-wind current, Y and Z are the strengths of the cosine and sine phases of a chain of superposed waves.

The term containing b represents the translation of the waves by the westerly current. The remaining quadratic terms represent a continual transfer of energy, except when X becomes negative, from the westerly flow to the waves, which are assumed to tilt westward with height (Lorenz

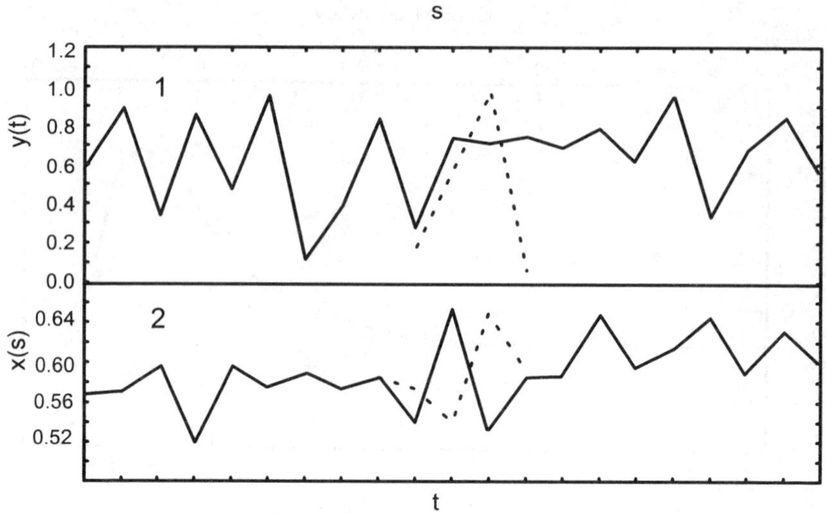

Fig. 2.37. *1* trajectory of the Feigenbaum square one-dimensional mapping x ; *2* trajectory of the smoothed process y

1990). Thermal and mechanic damping are represented by linear terms; constant terms describe thermal forcing.

We have analyzed an X series simulated for the first 10 model years with the help of a Taylor-series scheme with $a = 0.25$; $b = 4.0$; $F = 7.0+2.0\cos(2\pi t/\tau)$, $G = 1.0$; $\tau = 12$ months. These values of parameters were used by Lorenz in the main variant of his experiments (see Fig. 6 from the above paper). Variations of dimensionless X were examined using the MEM technique described in Section 2.3.

The resulting spectral density of X is shown in Fig. 2.36. As in other experiments with atmospheric models, the spectral density of atmospheric variable becomes flat at a frequency approximately equal to 1 cycle/month.

2.4.1.3
Simple Nonlinear Model as Analog of Atmosphere Forcing

Hasselmann (1976) mentioned that the flatness of weather spectrum far from the characteristic frequency of atmospheric processes is a feature, analogous to other nonlinear systems, which possess a characteristic time scale. We can check this supposition using the simplest possible nonlinear model. Consider the following elementary equations for time-dependent weather and climate sequences:

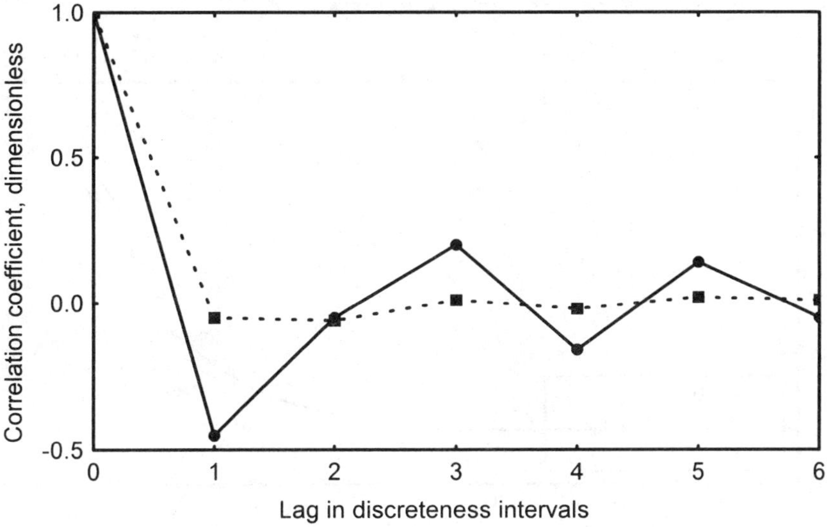

Fig. 2.38. Correlation coefficients (sample estimations) of process *x* (*solid line*), *y* (*dashed line*)

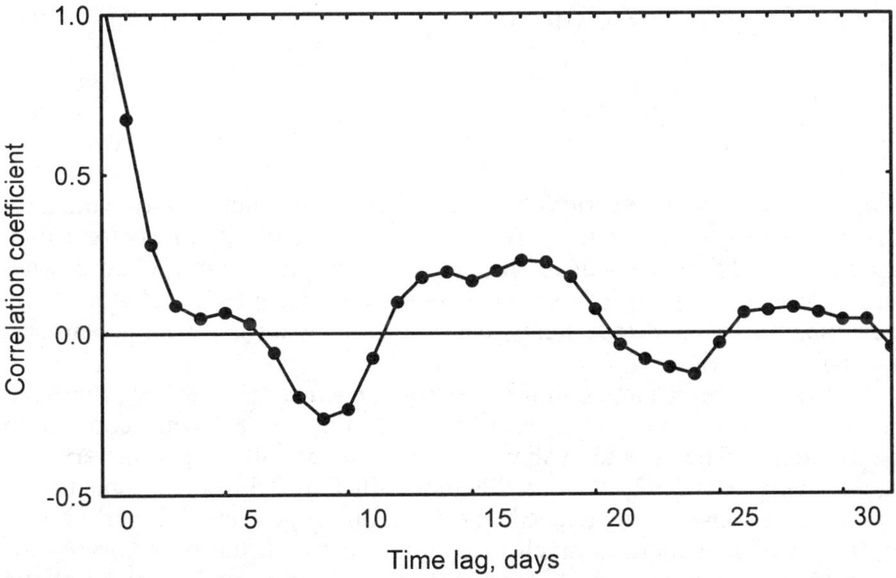

Fig. 2.39. New Orleans daily air temperatures. Sample estimations of correlation coefficients

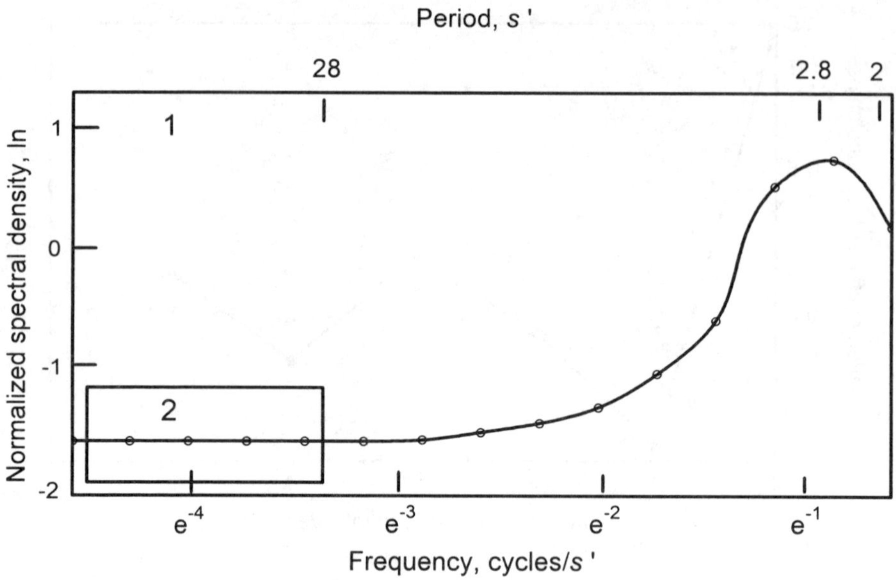

Fig. 2.40. Logarithms of normalized spectral densities of processes x (Fig. 2.37a) and y (Fig. 2.37b); s' discreteness interval of the scale s

$$x(s+1) = rx(s)[1 - x(s)], \tag{2.93}$$

$$y(t) = \frac{1}{N}\sum_{i=1}^{N} x(tN+i), \tag{2.94}$$

where x and y are time series of real numbers, and s and t are two uniform discrete time scales. The time step of the sequence y is N times larger than the time step of the sequence x. In fact, Eq. (2.93) is a simplest square one--dimensional mapping, whose properties were discovered and studied by Feigenbaum in 1978 (see Feigenbaum 1984). In turn, $y(t)$ is a temporal averaging of series x .

Figure 2.37 represents sample segments of sequences x and y , Fig. 2.40 represents their respective spectra, and Fig. 2.38 their correlation coefficients. The initial value of x is 0.75; other parameters are respectively: r = 3.92, N = 14. Segments in Fig. 2.37 correspond to $s >$ 1000, i.e., s considerably exceeds the duration of the transition period.

It is clear that there is much in common between the trajectory of the Feigenbaum square one-dimensional mapping x and real synoptic processes in the atmosphere (Figs. 2.34, 2.35, 2.39). Realizations of both processes resemble the realization of a stochastic (random) sequence. Also there is a qualitative resemblance between their correlation functions, which represent a fading oscillation. Finally, the spectral density of the Feigenbaum mapping trajectory x has the same form as the other spectra shown above in this section: sharp maximum at some characteristic frequency and flat spectrum in the range of relatively low frequencies. Thus, the Feigenbaum trajectory x can be regarded as a registration of weather process, and the averaged curve y can be regarded as an analog of averaged, climatic process.

Of course, there is a difference between the Feigenbaum trajectories on one hand and real atmospheric processes on the other. Here, we deal with a very important problem of stochasticity in the climate system. Indeed, the Feigenbaum mapping is deterministic if its parameters are fixed values (i.e., it has a zero forecast error), whereas real processes are poorly predicted and rather represent random processes. However, this difference, from our point of view, is not so dramatic. Thus, if we consider the specific realization of the atmospheric process, it can be regarded as a deterministic sequence (see Sect. 2.1) as well as the Feigenbaum trajectory.

Feigenbaum mapping is only a prototype (a very important prototype, however) of real processes. In fact, it is unlikely to meet in nature isolated nonlinear structures with a finite number of parameters and absolutely stable values of parameters. In the case of an error in the estimate of initial value x or minor changes in the model parameters, sequences x and y will possess only limited predictability and very quickly converge to a stochastic process. In Fig. 2.37.1. a dashed line shows the forecast corresponding to an error of 0.1 in the initial x value. It is clear that even at the first step the actual and predicted values are situated at the opposite sides from the mean value (0.59) of the process. At the third step the divergence between the forecast and the real curve becomes catastrophic. In turn, process y is much less predictable (see Fig. 2.37.2, dashed line).

We may approximate deterministic Eqs. (2.93) and (2.94) by the following stochastic analogs:

$$x(s) = -0.60[x(s-1) - 0.59] - 0.31[x(s-2) - 0.59] + 0.59 + a_x(s), \tag{2.95}$$

$$y(t) = 0.59 + a_y(t), \tag{2.96}$$

where $a_x(s)$ and $a_y(t)$ are white noises with zero value mean.

It is interesting that stochastic descriptions of x and y [Eqs. (2.95-2.96)] become better than their deterministic initial descriptions [Eqs. (2.93--2.94)], from the point of view of one-step forecast quality, if the error of initial x value is 0.11 and 0.000004 (!) of σ^2_x, respectively. It is clear that such an incredible exactitude in estimation of model parameters is absolutely impossible. Moreover, we cannot even imagine initial and boundary conditions meeting such a requirement.

Thus, simple nonlinear models represent abstract but important prototypes of real atmospheric and climatic processes, which help us to understand the nature of climatic stochasticity. Like data processing and atmospheric numerical models, simple nonlinear models demonstrate a very important feature of atmospheric variability: the flat form of spectral density in the range of low frequencies, far from the characteristic frequency of main (synoptic) oscillation. Perhaps this is one of the fundamental properties of a nonlinear system with maximum (together with other fundamental properties of the simple square mapping described by Feigenbaum).

2.4.2 Hasselmann's Stochastic Climate Models

The first paper explaining the fundamental features of climatic variabilities - the continuous and red character of their spectra - was published by Hasselmann (1976). (Some of the concepts underlying Hasselmann's theory had been expressed previously by Mitchell 1966). In this section we will generally follow the above and some other works by Hasselmann (1981, 1982), adding some comments related to the present state of deterministic and stochastic climate models.

2.4.2.1 Hypothesis on Weather-Climate Two-Scale Separation

A very important notion of weather-climate separation is explicitly introduced by Hasselmann. He describes the state of the atmosphere--ocean-land-cryosphere system by a finite set of discrete variables $\mathbf{z} = (z_1, z_2, \dots)$. Then he represents the evolution of the system as a series of (prognostic) equations

$$\frac{dz_i}{dt} = w_i(\mathbf{z}), \tag{2.97}$$

where w_i is a known function of $\mathbf{z}$. It is supposed that Eq. (2.97) accurately describes the evolution of the system for all times of interest.

A basic implicit assumption of most deterministic climate models is that the system $\mathbf{z}$ can be divided into two subsystems, $\mathbf{z} = (\mathbf{x}, \mathbf{y})$, which are characterized by strongly different response times τ_x, τ_y. Equation (2.17) now can be written in terms of the two subsystems

$$\frac{dx_i}{dt} = u_i(\mathbf{x}, \mathbf{y}), \tag{2.98}$$

$$\frac{dy_i}{dt} = v_i(\mathbf{x}, \mathbf{y}), \tag{2.99}$$

where

$$O\left[x_i\left(\frac{dx_i}{dt}\right)^{-1}\right] = \tau_x << \tau_y = O\left[y_i\left(\frac{dy_i}{dt}\right)^{-1}\right], \tag{2.100}$$

The fast components x_i are the prognostic weather variables used in deterministic GCMs. In turn, slow climate variables, y_i , are the sea surface temperature, ice coverage, land foliage, etc. which are set constant in weather prediction models but represent essential prognostic variables on climatic time scales. τ_x is typically of the order of a few days, whereas most climate variables have response scales τ_y of the order of several months, years, or longer. Thus the inequality (2.100) is generally satisfied. We can add to the above considerations of Hasselmann that the time scale of some climate variables, e.g., ice sheet heights as well as some globally averaged important variables - air and sea temperatures, global mean sea level heights - can be very large and exceed dozens of thousands of years.

2.4.2.2 Classification of Climate Models

Hasselmann stated that with computers of the 1970s it was not possible to integrate Eqs. (2.98) and (2.99) over periods of climatic time scale $O(\tau_y)$. High-resolution GCMs of that time were normally used to integrate the subset of equations (2.98) over an intermediate period τ_i in the range $\tau_x << \tau_i << \tau_y$ for which the "climatic variables" can be regarded as constant, but which is still long to define the statistics of the weather variables $\mathbf{x}$ for a given climatic state $\mathbf{y}$. Thus, *although GCMs provide important information for climate studies, they are not suitable for the simulation of climate variability as such* (emphasized by us - S.D.).

The above statement of Hasselmann is very important. Despite the further development of GCMs and coupled models since the 1970s (compromised by artificial spin-up and acceleration procedures, however), they are still not suitable for the simulation of climate variability. In fact, they are not aimed at the description of the main climatic process: accumulation of random impulses of water and heat within climatic subsystems (see Chap. 1). This is why Hasselmann's stochastic climate model theory is important today, and its further applications are urgent.

Statistical Dynamical Models (SDMs) as well as the Adem thermodynamic model address the subset of equations (2.99). It is usually argued that for the time scales τ_y , the fast terms in the prognostic equation can be ignored, so that Eq. (2.109) can be averaged over the period τ_i, thereby removing the weather fluctuations while still regarding **y** in the right hand side of Eq. (2.109) as a constant:

$$\frac{dy_i}{dt} = < v_i(\mathbf{x}, \mathbf{y}) > , \tag{2.101}$$

where the average < ... > is an ensemble average over a set of realizations **x** for given **y**. If we assume that ergodicity holds (see Sect. 2.2), ensemble averaging and time averaging are equivalent.

Hasselmann (1976) stresses that, although this class of climatic models may be called statistical in the sense that an averaging operation is involved, the reduced Eq. (2.101) is, in fact, deterministic. Actually, SDMs predict a unique, time-dependent asymptotic state for any given initial state. These models cannot generate internally time variable solutions with continuous spectral densities, as required by observation.

By a natural extension of SDMs one can obtain an alternative model that yields continuous "red" spectral density through internal interactions. We can return to Eq. (2.99) and let $\delta\mathbf{y} = \mathbf{y}(t) - \mathbf{y}_0$ denote the change of the climate state relative to a given state $\mathbf{y}(t = 0) = \mathbf{y}_0$ in a time $t < \tau_y$ sufficiently small that **y** can still be regarded as constant in the forcing term on the right hand side of the equation. The term $\delta\mathbf{y}$, in turn, may be divided into mean and fluctuating terms: $\delta\mathbf{y} = <\delta\mathbf{y}> + \mathbf{y}'$ where the ensemble average is taken over all **x** states for fixed $\mathbf{y}_0$. The mean value $<\delta y>$ can be obtained from Eq. (2.101):

$$<\delta y_i> = <v_i> t . \tag{2.102}$$

The rate of change of the fluctuating term is given by the following equation

$$\frac{dy_i'}{dt} = v_i(\mathbf{x}, \mathbf{y}) - <v_i> = v_i' , \tag{2.103}$$

where $<v_i'> = 0$ and $y_i' = 0$ for $t = 0$.

The statistics of $v_i'(t)$ are defined through the statistics of variables $\mathbf{x}(t)$ for given $\mathbf{y}_0$. Hasselmann assumes here that $\mathbf{x}(t)$ and $v(t)$ represent a stationary random process.

2.4.2.3
Analogies with Turbulent Fluid, Brownian Motion, and Other Physical Processes. The Central Limit Theorem

Equation (2.103) is identical to the equation describing the diffusion of a fluid particle in a turbulent fluid, where y_i' represents the coordinate vector of the particle, and v_i' the turbulent (Lagrangian) velocity. It is known that for stationary v_i', the integration of (2.103) gives a non-stationary process y_i' with covariances growing linearly in time. Hasselmann recalls the early work by Taylor (1921), who pointed out that this result could be interpreted physically as the continuum-mechanical analogy to normal molecular diffusion or to Brownian motion.

It is important to stress that for $t >> \tau_x$ it is immaterial for the statistical properties of y_i', involving time scales $>>\tau_x$, whether the forcing is continuous or discontinuous. This means that the behavior of the climate variable y_i' can be described by a random sequences with discrete temporal increments. Therefore, the linear growth of y_i' covariances is analogous to the linear growth of the second moments of the discrete Wiener process given by Eq. (2.55).

Another important consequence of Hasselmann's theory is that the nonstationary response y_i' to stationary random forcing v_i' implies that climate variations would continue to grow indefinitely if feedback effects are not considerable. These will become effective as soon as the integration is carried into the region $t = O\ (\tau_y)$. At the same time, we know (and it will be shown again below in this chapter and in Chap. 4) that for such local variables as ice sheet heights and global sea level and for such globally averaged variables as air and sea temperature, characteristic time can be very large: up to thousands of years. Consequently, the model of the discrete Wiener process (random walk) could be applied for the description of such climatic parameters within a very large range of temporal scales.

Hasselmann (1976) also invokes an important analogy between climate variations and Brownian motion. He states that the climate variables **y** and weather variables **x** may be interpreted in the analogous particle picture as the coordinates of large and small particles, respectively:

> Numerical experiments with GCMs correspond in this picture to the explicit computation of all paths of the small particles for fixed positions of the large particles. Even if the large particles were allowed to vary during the computation, it would normally not be feasible to carry the integrations sufficiently far to consider appreciable deviations of the large particles from their initial position. Finally, the approach used in the stochastic forcing model corresponds to the classical statistical treatment of the

Brownian motion problem, in which the large-particle dispersion is inferred from the statistics of the small particles with which they interact (Ibid, p. 476).

From our point of view, the above statements are still true despite the increased computer speed and improvements in GCMs and coupled models since the 1970s, although Hasselmann's argumentation can be partly modified. Perhaps the spin-up problem will be solved with time, and it will be technically possible to integrate the full system of Eqs. (2.108) and (2.109) over large (climatic) periods of time. However, it seems that initial and boundary conditions for coupled models will always be imperfect, and an infinite number of nonlinearities in the climate system will always make it impossible to describe deterministically the climatic result of fast weather processes (just as it is impossible and senseless to try to describe the behavior of each small particle of Brownian motion).

Hasselmann's analogy between climatic variations and Brownian motion also helps us to answer the criticism of many scholars towards stochastic climate models. Researchers who deal with coupled models and large archives of atmospheric data often say that stochastic (stochastic dynamical) climate models "are not physical" because they do not describe "physically" synoptic and other fast processes in the atmosphere. The answer to these reproaches can be thefollowing: the classical description of Brownian motion is a "physical description", although it is statistical and does not trace the behavior of each small particle.

Analogies with other physical processes also help to understand important features of Hasselmann's stochastic climate models. For instance, the weak-turbulence theories for plasmas or theories of weakly interacting random wave fields in solid state physics, high energy physics, and in various geophysical applications (cf. Hasselmann 1966, 1967) show that conditions similar to time-scale separation [Eq. (2.100)] allow a statistical reduction of the complete climate-weather system without arbitrary closure hypothesis.

Generally speaking, the property (2.110) enables statistical closure through the application of the Central Limit Theorem, whereby "the response of the system is completely determined statistically by the second moments of the input if the forcing consists of a superposition of small, statistically independent pulses of time scale short compared with the response time of the system" (Hasselmann 1976, p. 477). Also statistical reduction of the weather-climate system is possible using the application of the Lyapunov theorem; it was published first in 1901 and can be found in Lyapunov (1954).

2.4.2.4
Spectra and Correlation Functions of the Stochastic Climate Models. Models Without Feedback

Finally, taking into consideration Eqs. (2.98), (2.101), (2.102), and (2.103), the main equation of the stochastic climate model can be introduced in the following way

$$\frac{dy_i}{dt} = < v_i(\mathbf{x}, \mathbf{y}, \mathbf{e}) > + \, v_i^{'}, \tag{2.104}$$

where $\frac{dy_i}{dt}$ describes the behavior of the climate parameter, term $<v_i(\mathbf{x},\mathbf{y},\mathbf{e})>$ corresponds to the relatively slow interactions and processes within the climate system, and $v_i^{'}$ denotes atmospheric random forcing of the system. In Eq. (2.104) we slightly modify Hasselmann's equation, adding a term **e** which expresses external (including man-made) influence on the climate system.

It was shown in the previous sections that within the range of climatic frequencies, the atmospheric stochastic forcing $v_i^{'}$ can be considered, at the first approximation, as white noise - a sequence of uncorrelated random variables. Consider two variants: the first corresponds to the case when the term $< v_i(\mathbf{x}, \mathbf{y}, \mathbf{e}) >$ in Eq. (2.104) equals zero, i.e., there are no feedbacks in the system (in other words, characteristic climatic time τ_y is very large). It was stated above that in this case continuous atmospheric forcing $v_i^{'}$ can be replaced by a sequence of uncorrelated random variables with a time step (discreteness interval) larger than the characteristic time scale of weather variables, $\mathbf{x}$. Analysis of observational data (Sect. 2.4.1) as well as model results shows that time step Δt can be about 1 month; it can be called intermediate time scale. The resulting behavior of climatic variable y_i in this case is similar to the motion of large Brownian particles, and can be described by the random walk model (the discrete Wiener process).

In accordance with Sect. 2.2 - cf. Eq. (2.52) - the spectral density of the climatic variable y_i can be approximately expressed as

$$f_y(\lambda) \approx \frac{const}{\lambda^2} \qquad (\tau_y^{-1} << \lambda << \tau_x^{-1}). \tag{2.105}$$

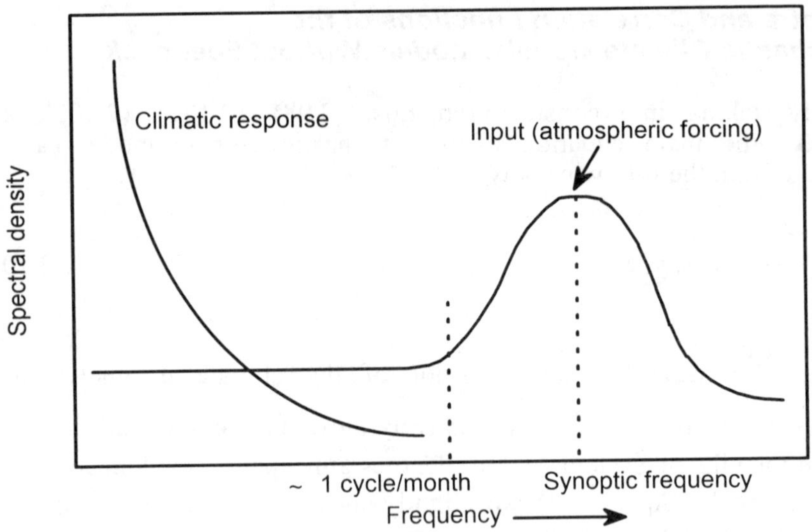

Fig. 2.41. Scheme of the input and response spectral densities of stochastically forced climate model without feedback. (After Hasselmann 1976, p.478, Fig. 1)

The main features of the climatic random walk y_i in the time and frequency domain are presented in Fig. 2.41. It is important to note for further applications of the stochastic climate theory that the constant level of the input spectrum at low frequencies can be determined from relatively short time series of the input, the record length required being governed by the time scale of the input, rather than by the time scale of the response. This is extremely important in the cases when we try to investigate relatively large time scales given rather short observational time series. For instance, the typical length of many important time series is only a few years (several dozens of months); however, even this poor information is often sufficient for estimating, at the first approximation, the first moments of a climatic input series.

Note the existence of a singularity in the response spectral density in Fig. 2.41 at zero frequency. It is consistent with the nonstationarity of the climate term y_i' , i.e., with the linear growth of the variance of climatic variable with time. Evidently, in reality, this growth cannot be unlimited. Thus, a model without feedback works for time scales which are larger than the intermediate time scale but less than the time scale τ_{max} involving the feedback mechanisms. Also note that τ_{max} depends on the specific parameter and process under consideration.

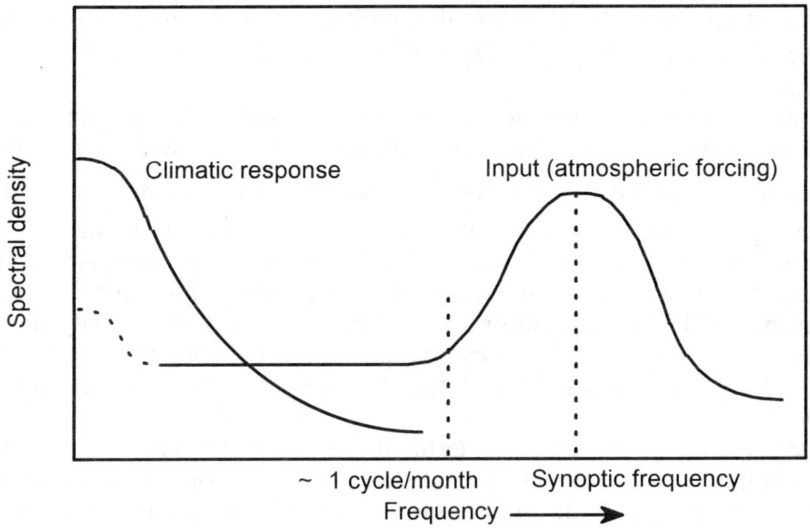

Fig. 2.42. Input and response spectral densities of stochastically forced (single component) climate model with linear feedback. (After Hasselmann 1976, p. 480, Fig. 2)

2.4.2.5
Models with Feedback

Hasselmann (1976) also proposed a general stochastic model (taking into account possible feedbacks) as a Fokker-Planck equation governing the evolution of the probability distribution $p(\mathbf{y}, t)$:

$$\frac{\partial p}{\partial t} = \frac{\partial}{\partial y_i}(\hat{v}_i p) - \frac{\partial}{\partial y_i}(D_{ij}\frac{\partial p}{\partial y_j}) = 0\,, \tag{2.106}$$

where

$$D_{ij} = \frac{\langle y_i' y_j' \rangle}{2\delta t} \tag{2.107}$$

with $y_i' = \delta y_i - \langle \delta y_i \rangle$ as before, and $\hat{v}_i = \langle \delta y_i \rangle / \delta t - \partial D_{ij} / \partial y_j$.

Equation (2.106) describes the evolution of an ensemble of climatic states with an arbitrary initial distribution for arbitrary large times. The propagation and diffusion coefficients $\hat{v}_i$ and D_{ij} are functions of climatic properties of the weather variables **x**. Thus Eq. (2.106) takes into consideration both direct internal coupling through the term $\hat{v}_i$ and indirect feedback through the dependence of the diffusion coefficients on the climatic state.

In accordance with the above general stochastic climate theory, the spectral density of v_i for lower frequencies is related to the slow variations of the mean variables $\langle v_i \rangle$ on the climatic time scales. The increase in the spectral density of the climatic variables y_i towards lower frequencies will be associated with a corresponding increase of the spectral densities of v_i in this range. It is important to note that this is not in conflict with the assumption of a white input spectral density at low frequencies. Essential is that a spectral gap exists between the weather and climate frequency ranges in which the input spectral densities are flat (Fig. 2.42).

The presence of the diffusion terms in Eq. (2.106) is related to the fact that climatic evolution is a statistical rather than a deterministic phenomenon. Even if the initial state of the climate system is perfectly known and prescribed in the form of a δ-function distribution for p, the diffusion term leads to a finite spread of the probability distribution at later times.

The analytical integration of Eq. (2.116) will normally not be possible. However, it is essential for the application of the stochastic climate theory that solutions can be constructed by the Monte-Carlo method, in which Eqs. (2.101) and (2.103) are integrated numerically for an ensemble of realizations using a statistical simulation of v_i'. Within the approximation of the two-timing theory, this term can be represented as a zero'th order Markov process.

In practice, this means that stochastic (stochastic dynamical) climate models can be constructed by adding to existing deterministic dynamic--statistical (or somewhat more sophisticated models) a weather forcing term in the form of a sequence of uncorrelated random values. Because of the two-scale properties of the climate system, the time step of this forcing can equal approximately 1 month. Statistical properties of the weather forcing can be determined, at the first approximation, using observational data.

Thus, in order to check the applicability of the stochastic climate theory to the description of climatic variations, it is necessary:

1. To analyze statistically and stochastically the observational data on climatic parameters (variables). If the theory is true, we can expect that time series of climatic parameters are described by fundamental stochastic models: white noise, the first-order Markov process, random walk (the discrete Wiener process).

2. To try to construct, using the above method, dynamic-stochastic models of climatic processes and to compare their results with observational data.

In the next chapters we will try to use both approaches and to demonstrate that the stochastic climate theory is a useful and powerful tool in climatic investigations. Furthermore, it describes, from our point of view, the essential features and the very nature of the climate system and climate changes.

3 Stochastic Models of Recent Climatic Changes

3.1 Changes in Thick Climatic Subsystems

It was demonstrated in Chapter 1 that there are substantial differences between characteristic dimensions and sizes of thick climatic subsystems on one hand (atmosphere, ocean, ice sheets) and thin subsystems (snow and sea ice cover, precipitation, vegetation cover, soil moisture), on the other. Resulting differences in the temporal behavior of parameters of these subsystems and differences in respective stochastic patterns will be discussed in this section.

3.1.1 Local Changes

There is a striking contrast between stochastic models of local (i.e., at specific stations or within relatively small areas) and global (globally averaged) parameters of thick subsystems. Seemingly, this contrast and resulting patterns of global climatic variability represent the most significant and most interesting feature of recent climate changes. The stochastic structure of local temporal climatic variabilities can be studied using two approaches: (a) stochastic analysis of observed time series and (b) construction of stochastic dynamical zero- or one-dimensional models. We will start with the first approach.

3.1.1.1 Analysis of Observational Data

Applicability of the Correlation Theory. The crucial problem of climate change analysis is to answer the question whether the correlation theory, i.e., the theory of stationary random processes, can adequately describe climatic variations, and if so, what are the spatial and temporal limitations of such a description? Although formal spectral and statistical analysis is widespread nowadays, there is no unanimously acknowledged answer to this question in the literature on stochastic and statistical climate studies. From our point of view, this situation can be explained by the fact that

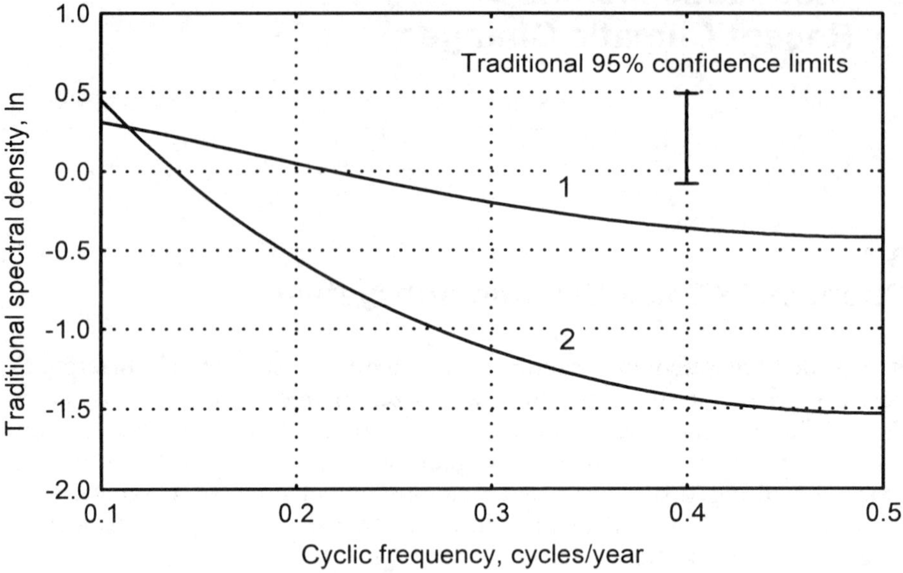

Fig. 3.1. Traditional Maximum Entropy Method estimations of spectral densities. New Orleans near-surface air temperatures: *1* 1974-1932; *2* 1932-1990. *Vertical line* denotes mean Kromer-type estimations of 95% confidence intervals

methods of spectral and stochastic analysis of time series are still imperfect. In Sect. 2.3 we analyzed the errors of existing spectral estimates and tried to demonstrate that these estimates could be improved. Processing of a large number of climatic time series (see Table 3.1) shows that the truncation algorithm of calculating normalized spectra can give decisive results with respect to the applicability of the correlation theory to local climatic variability.

This problem can be illustrated by Figs. 3.1 and 3.2. In Fig. 3.1 traditional MEM estimations of New Orleans temperature spectra, together with Kromer-type confidence limits, are shown for two halves of the series: from 1874 to 1932 (first curve) and from 1932 to 1990 (second curve). (Frequencies 0.0 - 0.1 cycles/year are not shown because it is not possible to assess spectra near the zero frequency). It is evident that the difference between spectra is larger than the confidence interval. Thus, at first sight, the two halves of the series are not identical statistically, and the stationarity condition does not hold. Analysis of most of the annual time series from Table 3.1 gives the same result. In other words, the very concept of the correlation theory of random processes seems to be not applicable for the description of real climatic (local and regional) variability.

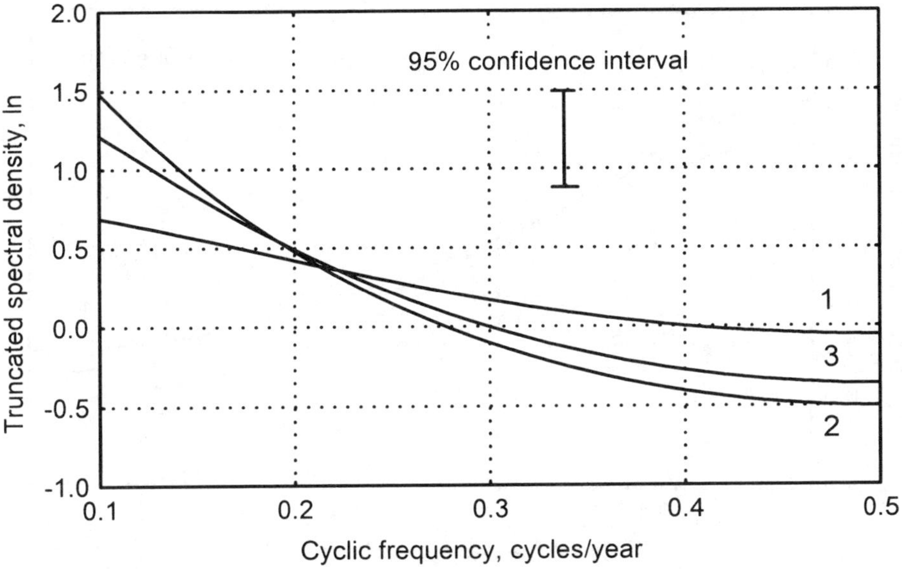

Fig. 3.2. Frequency-truncated MEM estimations of normalized spectral densities. New Orleans near-surface air temperatures. *1* 1974-1932; *2* 1932-1990; 3 1874-1990. *Vertical line* denotes mean 95% confidence intervals. Segments of spectral densities between frequencies 0.0 and 0.1 cycles/year are not shown

However, if we use the truncation algorithm of estimating normalized spectral density (see Sect. 2.3), quite different results can be obtained. In Fig. 3.2 spectra for the same two halves of the New Orleans time series are shown together with the spectrum calculated for the whole series (1874--1990). It is clear that, this time, the three spectra coincide with each other to within the width of the confidence interval. Spectra for other time series from Table 3.1 were also recalculated using the truncation method, and results for practically all registrations were analogous to the above results: improved spectral estimations demonstrated the stationarity of time series with respect to the spectral density and the covariances, i.e., with respect to the second moments.

We consider this result as very important, because it gives us a possibility of applying a fundamental correlation theory of random processes to the study of local climatic change, as well as a chance to use existing, relatively short, registrations of climatic variables for the identification of the main types of stochastic climate models.

The same calculations were made for normalized monthly anomalies (see below) of time series from Table 3.1 with monthly resolution. In this case stationarity was even more evident: very often traditional spectral estimates have already demonstrated the stability of spectra (see, for

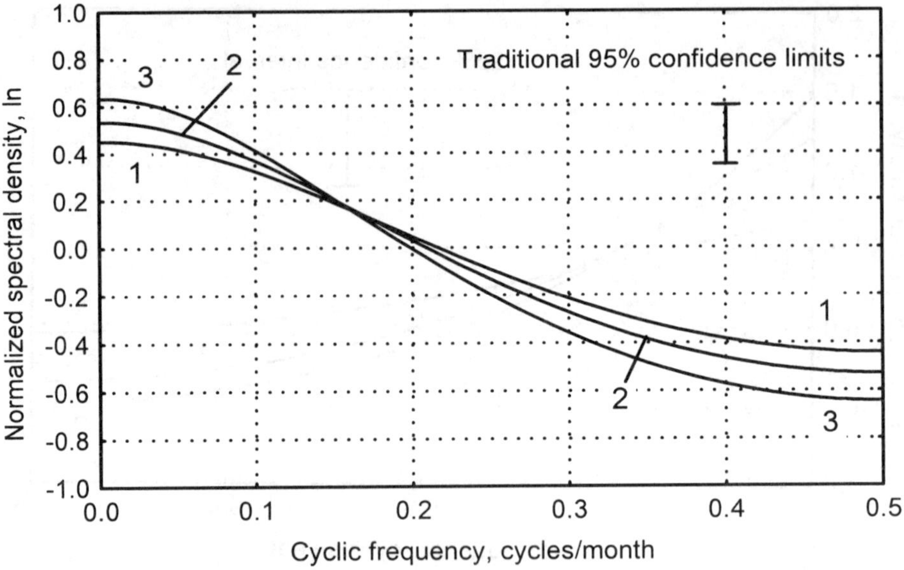

Fig. 3.3. Traditional MEM estimations of spectral densities. New Orleans monthly near--surface air temperature anomalies: *1* 1874-1912; *2* 1913-1951; *3* 1952-1990

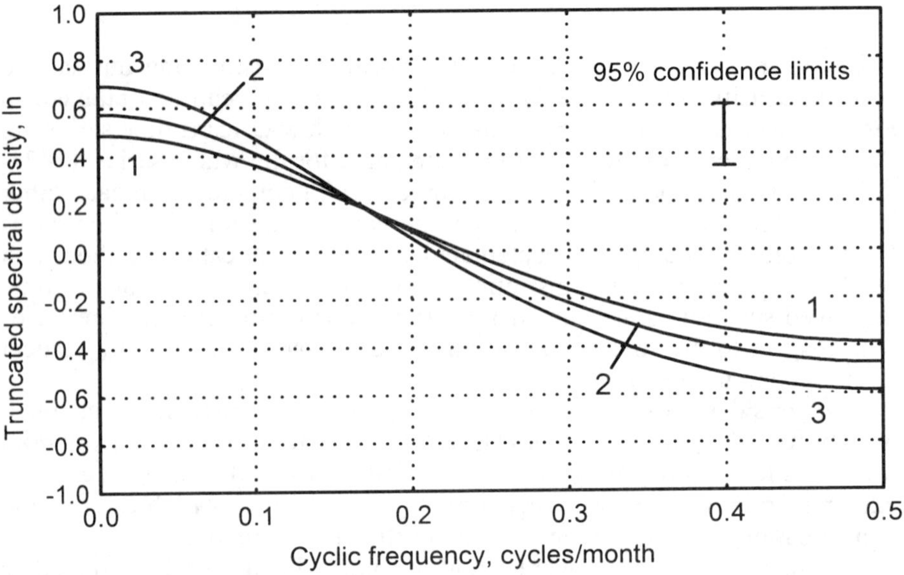

Fig. 3.4. A Fig. 3.3 but for frequency-truncated estimations

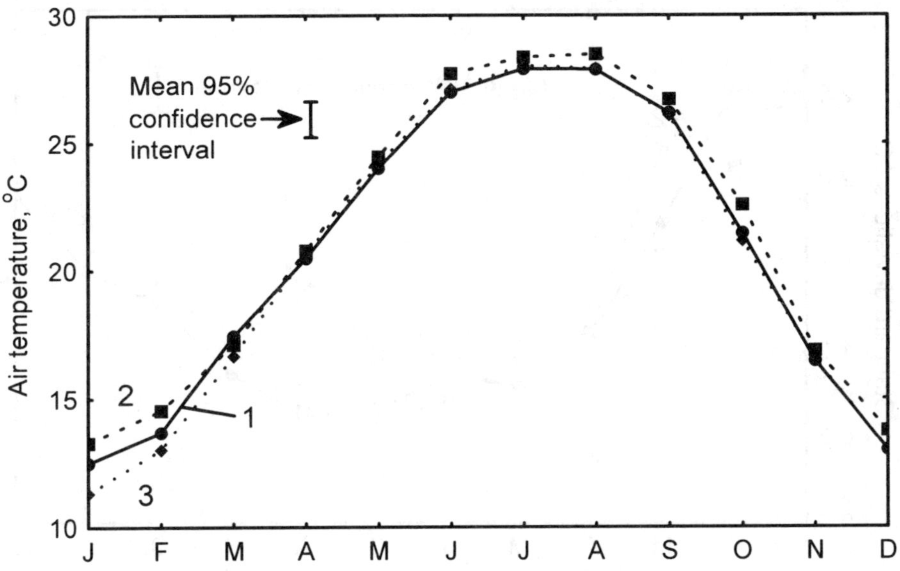

Fig. 3.5. Seasonal dynamics of monthly mean air temperatures in New Orleans. *1* 1874--1912; *2* 1913-1951; *3* 1952-1990

example, Figs. 3.3 and 3.4); there can be no question of frequency--truncated estimations.

Another important problem of the applicability of the correlation theory is the stability of mean values. In order to solve this problem, the Yaglom method of calculating the errors in mathematical expectation was used; it will be discussed in Sect. 3.3 of. Here, we will only note that application of this method to the series from Table 3.1, except for the sea level series, generally does not contradict the supposition of the stability of the mean value. In turn, this shows that we usually cannot detect statistically significant trends within the registrations of local climatic changes.

As a result, we can talk about the conditional applicability, within a certain range of frequencies and time scales, of the correlation theory to the description of local climate change. Evidently, this approach cannot be used for modeling extremely long climatic changes: we know that at time scales of thousands of years and more, local climatic changes are not stationary. We can treat climatic variations on these time scales as slowly changing stationary patterns.

Stochastic Models for Monthly Anomalies. Averaging climatic variables over a 12-month period is quite a natural procedure, and annual time series give important information on climatic changes. However, in order to examine variability within the range of periods from approximately 2

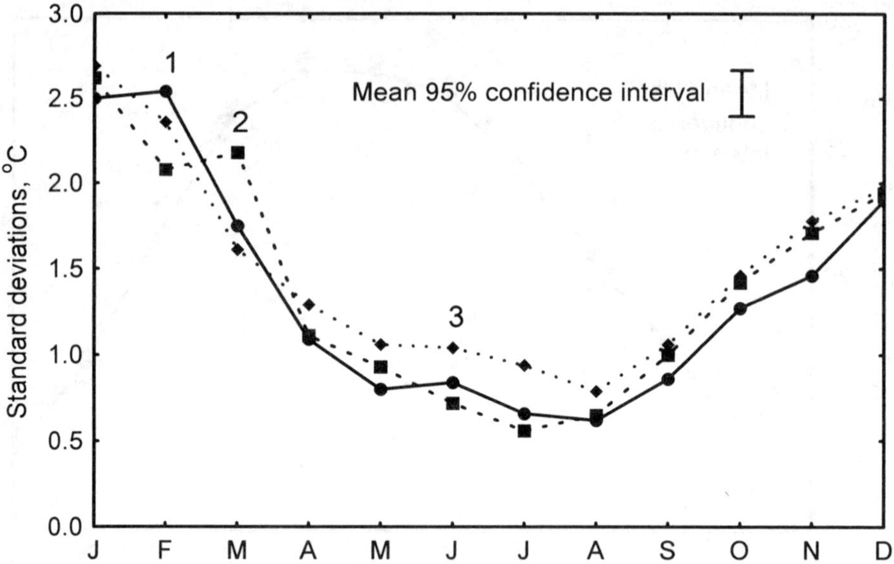

Fig. 3.6. Seasonal dynamics of monthly standard deviations. New Orleans near-surface air temperature record. *1* 1874-1912; *2* 1913-1951; *3* 1952-1990

months to several years, it is necessary to use monthly time series (which are more frequent than the annual ones). In this case, it is useful to transform the initial time series into the series of monthly anomalies with the help of formula (2.58). In this way we remove two types of nonstationarities: (a) nonstationarity caused by the seasonal dynamics of mean values m_i ; (b) nonstationarity as a result of 12-month variations in monthly standard deviations σ_i .

As an example, m_i and σ_i graphs for the New Orleans temperature record are presented in Figs. 3.5 and 3.6. It is evident that not only mean values but also standard deviations have a clear 12-month dynamics that has not considerably changed during the past 117 years. The same results were characteristic for practically all monthly time series from Table 3.1.

We can suppose that coefficients of correlation between the neighboring values of the resulting time series of monthly anomalies $\xi_l(t)$ from Eq. 2.58 also have seasonal dynamics. In other words, we expect that the following model could be useful for the description of monthly anomalies

$$\xi_l(t) = c_i * \xi_l(t - 1) + a(t) , \tag{3.1}$$

where t is time in months; c_i the seasonally varying coefficient, i is 1, 2, ... , 12, and $a(t)$ the sequence of non-correlated random values (white noise).

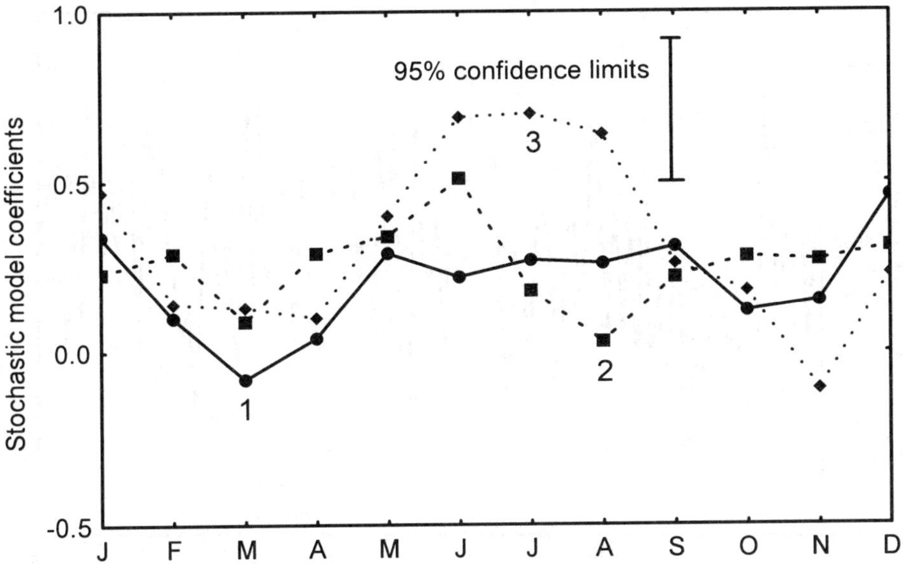

Fig. 3.7. Estimations of coefficients c_i for the seasonal model [Eq. (3.1)], New Orleans air temperature record. *1* 1874-1912; *2* 1913-1951; *3* 1952-1990

It is necessary to compare model (3.1) with the first-order Markov process [i.e., model (3.1) when c_i is constant] fitted to the same time series, $\xi_l(t)$. In Fig. 3.7 estimations of coefficients c_i are shown for different periods of time. Errors in correlation coefficient estimations are large; they are comparable to the seasonal and secular changes in c_i, and the whole pattern is not as clear as for parameters m_i and σ_i. Perhaps we can say that there are two maxima of c_i–in winter and in summer, and two minima–in spring and in autumn, although this is not evident for each 59--year estimation period. Recall that the New Orleans temperature record is one of few time series with such a large number of points (longer than 100 years). More typical are relatively short climatic time series. Therefore, in practice it is very difficult to obtain reliable parameter estimations for model (3.1). On the contrary, analysis of most of the monthly time series from Table 3.1 shows that the first-order Markov process model could be a (rough) first approximation.

Overview of Models Fitted to Various Climatic Variables. All in all, thousands of climatic time series were processed by the author of this book by hand, and many more were processed automatically. It was shown in previous chapters that many circumstances must be taken into consideration in to identify a stochastic model of a specific time series: stochastic analysis is a kind of art, and not something exclusively

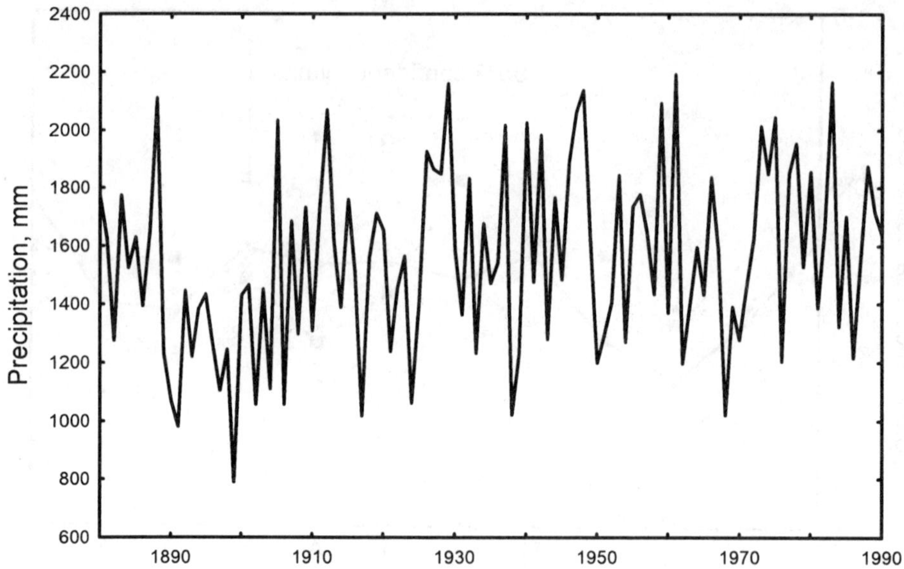

Fig. 3.8. Annual precipitation in New Orleans in mm. (Data from World Weather Disk 1994)

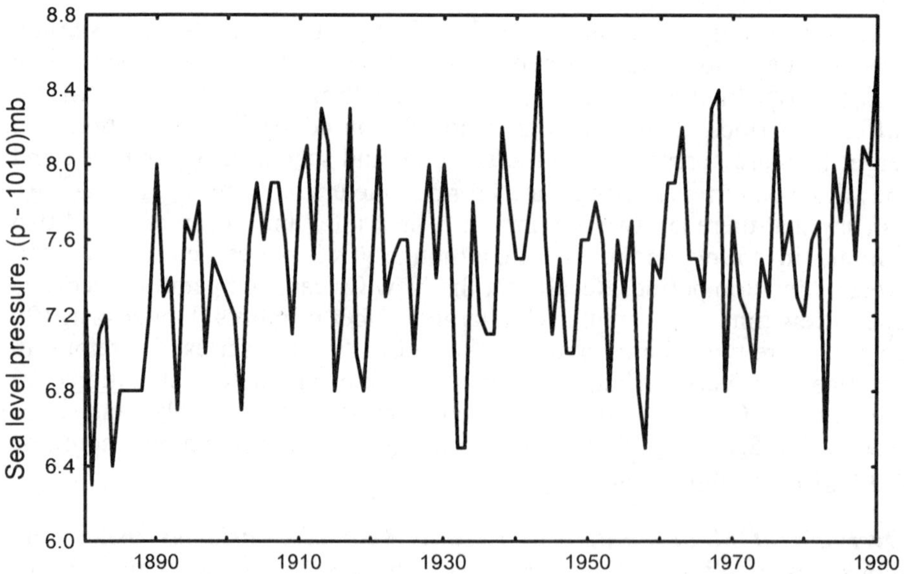

Fig. 3.9. Annual sea-level air pressure in New Orleans. (Data from World Weather Disk 1994)

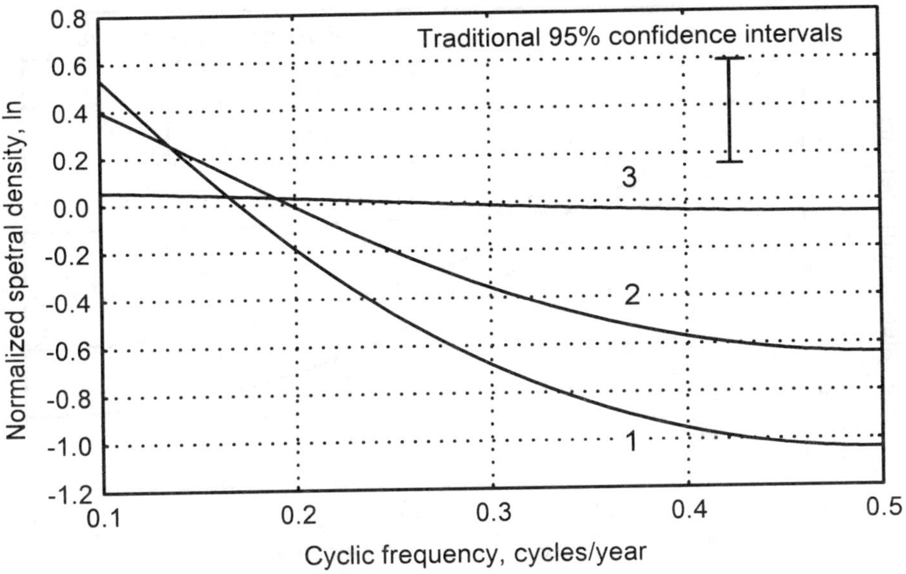

Fig. 3.10. Traditional MEM estimations of normalized spectral densities for climatic records in New Orleans, 1874 - 1990. *1* temperature, *2* air pressure, *3* precipitation

technical and standard (see Privalsky 1985). Therefore we included into Table 3.1 only those time series which have been analyzed by hand (i.e., with visual control of intermediate results), automatically processed series being only reference points.

An example of analysis of records other than temperature ones, is given in Figs. 3.8 - 3.11. In Figs. 3.8 and 3.9 the annual time series of New Orleans precipitation and sea level pressure are shown. None of these two time series manifests much resemblance with the New Orleans temperature record (see Fig. 1.12): unlike air temperature, precipitation and air pressure have not fallen off since the 1950s. In turn, there is not much in common between precipitation and air pressure: the coefficient of correlation between the two series equals only 0.1, that is less than the mean error of this estimation.

Stochastic models corresponding to time series in Figs. 3.8 and 3.9 are illustrated by Figs. 3.10 and 3.11. In both cases the Markov first-order process [i.e., AR(1) model] was chosen as the best. Here, traditional and frequency-truncated MEM estimations give almost the same spectra since the first-order autoregression coefficient is not very high. The precipitation record is satisfactorily described by the model with sample coefficient 0.03; hence, this model is close to the white noise, and its spectral density is close to the horizontal line. Both temperature and air pressure spectra differ statistically from the precipitation spectrum. On the

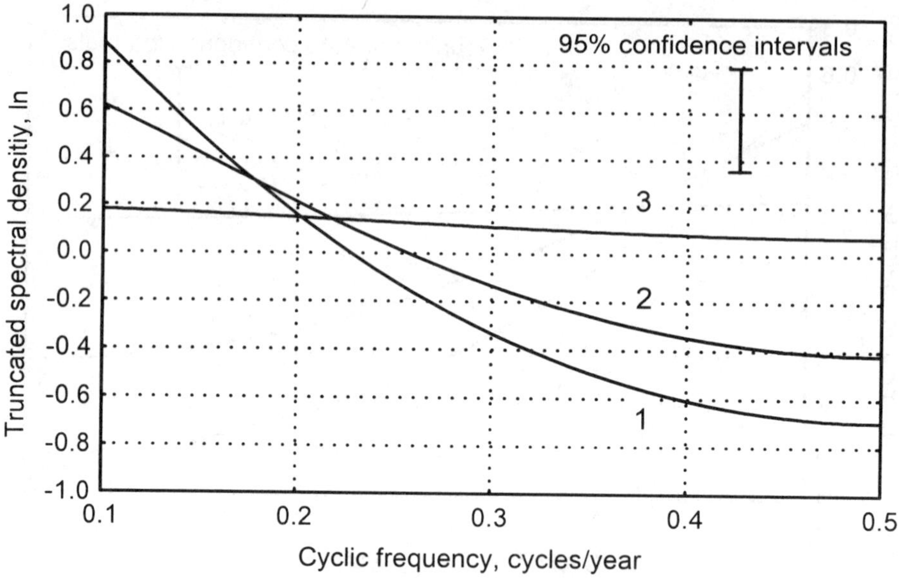

Fig. 3.11. Frequency-truncation MEM estimations of spectral densities. New Orleans climatic records, 1874 - 1990. *1* temperature, *2* air pressure, *3* precipitation

contrary, it is impossible to distinguish the precipitation and air pressure spectra from each other: much longer registrations are needed for this purpose. However, we can suppose that temperature sequences are more correlated in time than precipitation ones, and precipitation series are correlated weekly (if ever).

The above differences between stochastic models (and spectral densities) of the three main atmospheric parameters - annual temperature, air pressure, and precipitation - are the same for most meteorological stations: characteristic mean values of the AR(1) coefficient are 0.4 - 0.6, 0.3 - 0.5, and 0.0 - 0.3, respectively. However, usually it is difficult to establish some reasonable spatial pattern of AR(1) coefficients because of the limited length of registrations or because of the heterogeneous character of these patterns.

A summary of results for analyzed time series of the thick climatic subsystems is given in Table 3.1. The most important result is that the simplest and most fundamental models - the first-order Markov process [AR(1)] and white noise - are good first approximations for the description of almost all time series, with both annual and monthly resolution [in the latter case we mean time series of normalized anomalies, Eq. 2.58]. The only exception are series of sea level at tide gauge stations. Unlike registrations of other climatic variables, sea level records often manifest the presence of a monotonous trend (deterministic, nonrandom

Table 3.1. Main time series of the thick subsystem local variables and their stochastic models

Variable	Ann[a] / Mo[b]	Number of series	Source	Prevailing model type
Near-surface air temperature at meteorological stations	Ann, Mo	510	World ... (1994)	M[c]
Precipitation at meteorological stations	Ann, Mo	498	World ... (1994)	M, W[d]
Air pressure at sea level, air pressure at station level	Ann, Mo	485	World ... (1994)	M
Absolute atm. humidity, water content at aerological stations	Mo	82	Yarosh (1986)	M
Horizontal atmospheric water vapor transport	Mo	246	Yarosh (1986)	W
Sea surface temperatures averaged over 5° x 5° squares	Mo	195	Bottomley et al. (1990)	M
Air-sea fluxes at ship weather stations	Mo	60	Ariel et al. (1984)	W
Sea level heights at tide gauge stations	Ann	512	Permanent ... (1998)	T [e]
Sea level heights with removed deterministic trend	Ann	493	R.K.Kliege (1997, pers. comm.)	M
Evaporation, effective evaporation from land surface at water balance stations	Mo	34	Materialy ... (1965-1990)	M, W

[a]Time series of annual values. [b]Time series of monthly normalized anomalies calculated using formula (2.58); [c]The first-order Markov process [AR(1)]. [d]The white noise model [AR(0)]. [e]Model with monotonous deterministic trend

component). We will discuss this result in more details in Section 3.3; here, we mention only that the presence of a deterministic trend is obviously related to long-term vertical coastal movements of geological or geomorphological nature.

Another feature of climatic variations related to the general applicability of AR(1) and white noise models is the absence of periodicities. Our analysis does not prove the presence of periodical or quasi-periodical components within the time series from Table 3.1. Thus, our study does not corroborate hypotheses about the direct influence of periodical external processes on the climate system (see Sect. 1.2.2) or hypotheses about the internal feedback mechanisms creating such periodicities.

Finally, the distribution of model types over various types of climate variables in Table 3.1 is of major interest. All climatic variables can be divided into two groups: accumulation and transfer variables. The first group includes variables that are related to properties or substances, which can be accumulated within a climate subsystem: air temperature, air pressure, air humidity and atmospheric water content, sea temperature, sea level (after removing the deterministic trend). The second group represents variables that are related to water and heat fluxes within or between climate subsystems: evaporation, vertical turbulent heat exchange and other fluxes at the ocean-atmosphere interface, evaporation and effective evaporation from the land surface, horizontal water vapor transport in the atmosphere, and precipitation.

The major difference between stochastic models of the two types of climate records resides in the fact that time series of the first type, on average, are more correlated in time. Hence, the first-order Markov process models plays an important role in the description of variabilities of this type. On the contrary, transfer variables are often described by the white noise model or by the first-order Markov process with a small coefficient.

The transfer-accumulation contrast of climatic variability stochastic models is quite natural and can be easily understood. Indeed, local temporal variations of a single accumulation variable are relatively slow because of certain temporal inertia of the local accumulation process. On the contrary, a local transfer process is related to several or many local accumulation processes at different distant locations or within different media, so the spatial and intermedia randomness and heterogeneity is added to the randomness of a local process in this case. We will see later that spatial (global) averaging largely reinforces transfer-accumulation contrast of stochastic climate models.

El Niño Temperature Record. The area involved in ENSO events represents a special interest and the special case. As stated in Section 1.1.2, many researchers find periodical or quasi-periodical components in registrations of climatic variables related to ENSO: 3-7 years (Stone et al. 1998), 24 and 30 months (Blanke et al. 1997), 2-6, 8, 14 years (Kane and De Paula 1996), 13, 26 years (Kane 1997), 74, and even 320 years (Dovgalyuk and Klimenko 1996), etc. In turn, these findings are the cause of constructing many variants of models explaining assumed ENSO periodicities. In Section 3.1.2 we will analyze spatially averaged, longer ENSO time series; here we would like to discuss a processing of in situ registrations related to the El Niño phenomenon.

In Fig. 3.12 a time series of sea surface monthly temperature anomalies at 110° W, 0° N is shown. This record is usually considered as one of most

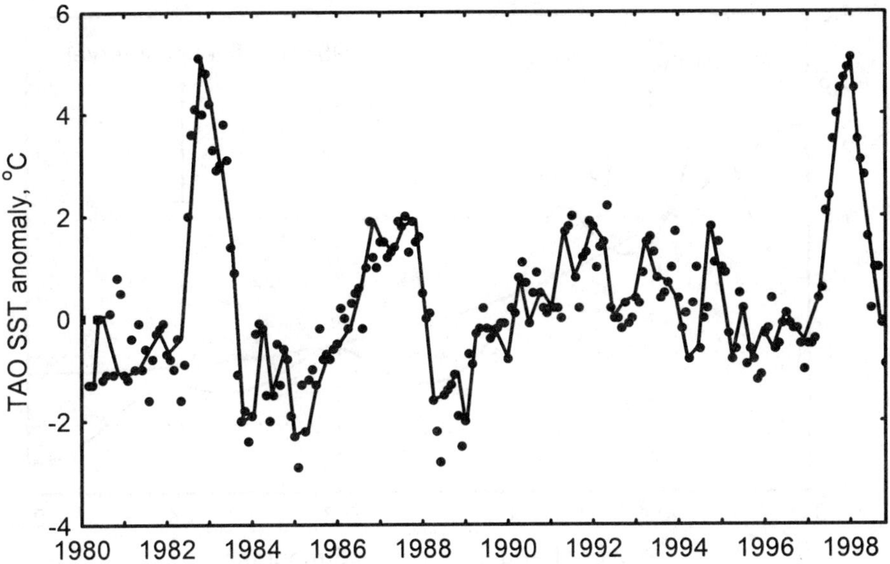

Fig. 3.12. TAO sea surface temperature anomaly time series, 0°N, 110°W. (Data from NOAA 1998). *Solid line* indicates a variant of the same time series with a 3-month resolution

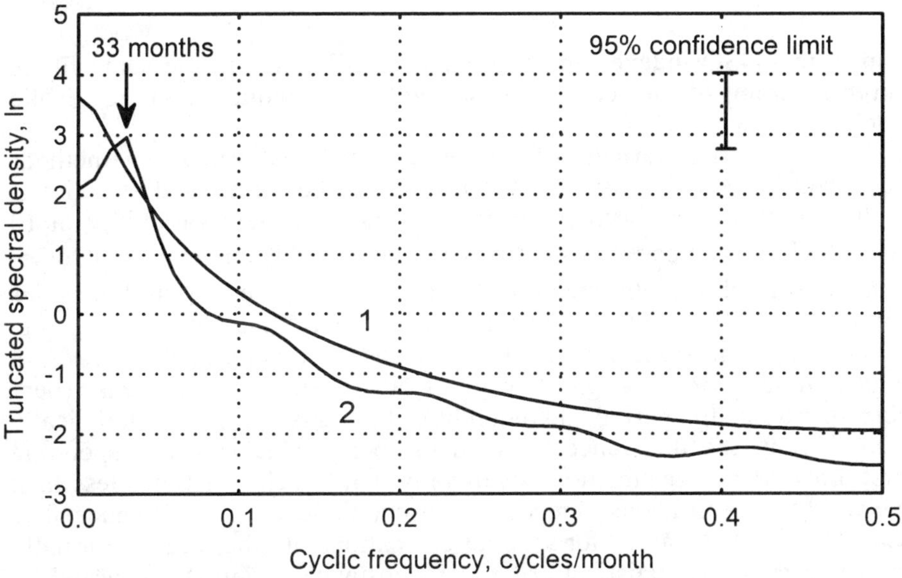

Fig. 3.13. Frequency-truncated estimations of El Niño sea surface temperature record. *1* 1st order model; *2* 11th order model. *Vertical bar* denotes 95% confidence interval for the second estimation

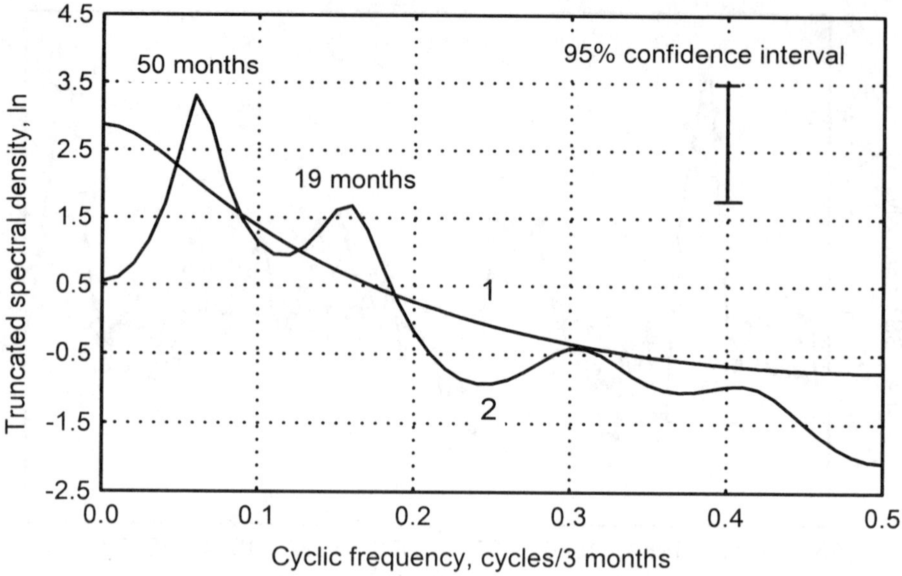

Fig. 3.14. Frequency-truncated estimations of El Niño sea surface temperature record with 3-month resolution. *1* 1st order model; *2* 8th order model. *Vertical bar* denotes 95% confidence interval for the second estimation

important ENSO indexes and a proof of oscillations. In the same figure another variant of the series is presented: with 3-month resolution (solid line).

The above two variants of the series were analyzed using methods described in Sect. 2.3. Briefly, the results can be formulated as follows: nothing unusual has been found in the time-frequency variability of El Niño SST as compared with series of the monthly anomalies of other climatic variables in other regions. In Fig. 3.13 truncated spectral densities for the monthly series are shown: AR(1) approximation (line 1), and AR(11) approximation (line 2). As in the case of other climatic indexes, a higher model order (e.g., 11) gives a certain spectral peak (here corresponding to period 33 months), but this peak is statistically insignificant: the difference between the peak value and corresponding first-order Markov estimation equals only 0.6, which is 2 times less than the width of a confidence interval of reasonable level of significance. The same result was obtained for spectral estimations of other orders, formally indicated by some statistical criteria; for instance, order 15 suggested by the CAT criterion.

Perhaps better results could be obtained using a lower resolution of the time series? In this case, frequencies of interest are not so close to the zero-frequency limit and, theoretically, could be better represented in the spectral pattern because a spectral peak can be obtained using lower model order. An appropriate pattern is presented in Fig. 3.14. We see that here 33-month peak is divided into 50-month and 19-month peaks. However, none of them is statistically higher, in comparison with the estimation error, than the Markov first-order spectrum at respective frequency.

These examples help to understand why so many periods are found in the El Niño SST registration: different authors use models with different orders (degrees of freedom) to process the same time series, and estimation errors are usually not taken into consideration. Note that results similar to ours were recently obtained using somewhat different methods by Karoly and Kestin (1998). Their conclusion, as well as ours, can be formulated as follows: "The dominant frequencies in the spectra do not appear to be constrained to specific frequency bands, and there is no evidence that the ENSO system has fixed modes of oscillation ... stochastic forcing may be an important component of ENSO dynamics".

3.1.1.2
Local Stochastic Dynamical Models

The most spectacular applications of the stochastic (stochastic dynamical) climate model theory were related to sea temperatures because the contrast in air-sea processes represents in essence a two-scale separation concept. Frankignoul and Hasselmann (1977) proposed a stochastic dynamical model of the upper oceanic layer where a simple model of heat content played the role of deterministic block forced by random atmospheric variables. Model results were compared with the temperature data at Ocean Weather Station I and demonstrated that the stochastic approach could successfully explain the main statistical properties of sea surface temperature anomalies in midlatitudes, as they reflect the response of the oceanic mixed layer to the quick changes in the air-sea fluxes (see also Frankignoul 1985, 1995).

Dobrovolski (1982) proposed a somewhat more sophisticated stochastic dynamical model of sea temperatures based on a modified Adem climatic model (see Adem 1991a). In a mixed-layer model the heat content equation was

$$h\rho_S c_S\left(\frac{\partial T_S'}{\partial t} + \overline{V}_Z * \nabla_Z T_S' + V_Z' * \nabla_Z \overline{T}_S - K_S * \nabla_Z^2 T_S'\right) =$$
$$\langle E_S\rangle' - \langle G_E\rangle' - \langle G_C\rangle', \tag{3.2}$$

where h is the thickness of the upper mixed oceanic layer; ρ_S the water density; c_S the specific heat at constant volume; $T_S' = \langle T^*\rangle - \overline{T}_S$, T^* being water temperature, angular brackets signifying temporal averaging over an intermediate time scale (approximately 1 month), and $\overline{T}_S$ being climatological mean temperature; V_Z the vector of horizontal sea currents; ∇_Z the two-dimensional gradient operator; K_S the arbitrary (exchange) Defant-type coefficient of large-scale horizontal turbulent thermal conductivity. The last three terms in Eq. (3.2) are anomalous heat losses due to radiation processes, evaporation from sea surface, and contact turbulent heat exchange with the near-surface atmospheric layer. In Eq. (3.2) equation for oceanic current speed

$$\overline{V}_Z + V_Z' = \frac{1}{h}\int_0^h \langle V_Z \rangle dz . \tag{3.3}$$

Formulas for heat losses due to radiation, evaporation, and contact vertical turbulent exchange with the atmosphere were

$$\langle E_S \rangle' = B_2'' T_a' + B_3 T_S' , \tag{3.4}$$

$$\langle G_E \rangle' = K_4 B |V_{aN}| (K_5 T_S' - U_N T_a'), \tag{3.5}$$

$$\langle G_C \rangle' = K_3 |V_{aN}| (T_S' - T_a'). \tag{3.6}$$

In Eqs.(3.4-3.6) T_a' is the anomalous near-surface air temperature; $|V_{aN}|$ the absolute value of climatological mean wind speed; U_N the climatological mean relative air humidity; and $B_2'', B_3, K_4, B, K_5, K_3$ the coefficients.

Thus the following processes are taken into consideration in the model: changes in the heat content of the upper oceanic layer because of varying temperature, horizontal heat transport by stationary oceanic currents and by anomalous drift currents, horizontal large-scale turbulent-like mixing; anomalous evaporation and vertical turbulent contact heat exchange at the ocean-atmosphere interface, and anomalous radiative balance at the sea surface. Atmospheric forcing and anomalous drift currents were constructed using the Adem atmospheric model and the Ekman formulas (see Dobrovolski 1992). In accordance with Hasselmann's stochastic climate models theory, pure atmospheric forcing (i.e., without ocean--atmosphere feedbacks) represented a monthly series of temporally uncorrelated random variables; statistical parameters of forcing were estimated using observational data.

Model (3.2) was applied for the description of SSTAs in characteristic sites: at the subarctic hydrological front (south of Newfoundland), at the

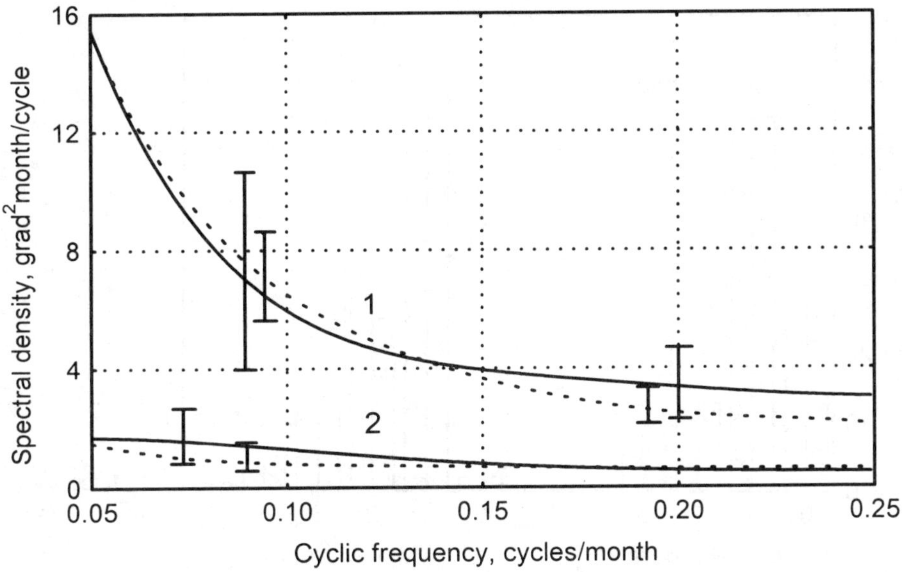

Fig. 3.15. Spectral densities of observed within 5° x 5° squares (*solid lines*) and modeled for the centers of respective squares (*dashed lines*) monthly sea surface temperature anomalies. Coordinates of squares' centers: *1* 42.5°N, 57.5°W; *2* 42.5°N, 17.5°W. *Vertical bars* denote 95% confidence intervals

same latitude but in the open ocean, in subtropical regions of the Northern Atlantic ocean, and others.

Spectacular results of the model integrations are shown in Fig. 3.15. It is clear that the model satisfactorily describes SSSTA variations both in the frontal zone (line 1 in Fig. 3.15) and in the open ocean at the same latitude (line 2). Moreover, the model describes and explains the dramatic difference between the two spectra: an enormous increase in spectral densities within the frontal zone. Experiments with varying structure of the model demonstrate that this difference is mainly caused by the third term in Eq. (3.2): thermal advection by anomalous currents. Later, this model was carried over to the regional and global case (appropriate results will be discussed in Section 3.1.2.

3.1.2 Regional, Spatially Averaged, and Two-Dimensional Patterns

3.1.2.1 2D Stochastic Patterns of Observational Data

A set of regional stochastic models was proposed by the author for the description of variability of monthly anomalies within the climate system.

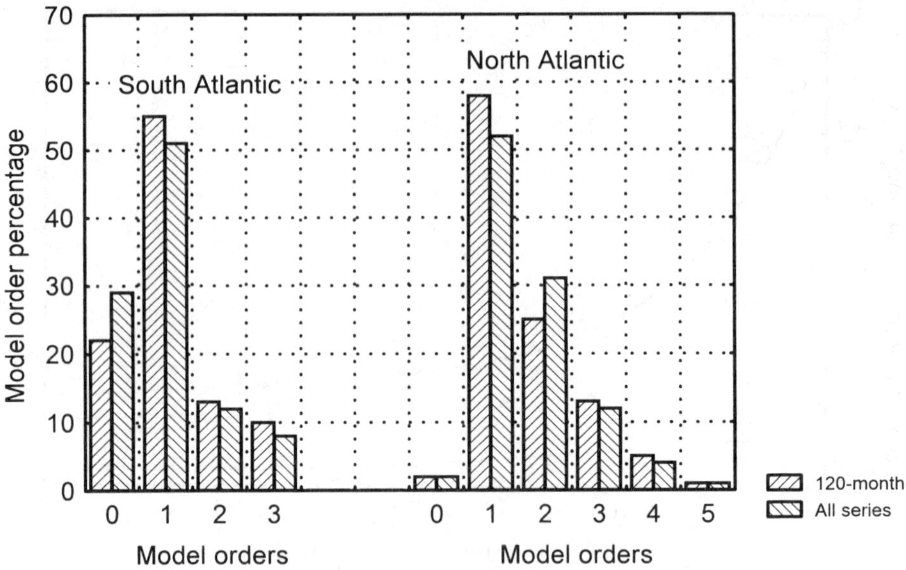

Fig. 3.16. Orders of stochastic (AR) models, formally detected by using the Akaike criterion. Monthly sea surface temperature anomalies for the North Atlantic (*left*) and South Atlantic (*right*)

This approach was applied to: the Atlantic Ocean, Indian Ocean, and extratropical parts of the Pacific Ocean. Also a more detailed model was proposed for the simulation of monthly anomalies of the climatic variables within the temperate zone of the Atlantic Ocean and adjacent parts of North America and Europe. Anticipating the description of these results, we can say that the stochastic approach in its simplest variant was a good first approximation to the modeling of the main regional features of observed variability almost everywhere save the eastern part of the tropical Pacific Ocean. In Section 3.1.2.2 we will discuss peculiarities of ENSO stochastic modeling.

In this section we will present results for the Atlantic Ocean as an example of the regional approach. All in all, 239 monthly time series of sea surface temperatures in 5° x 5° squares were analyzed: 154 for the North Atlantic and 85 for the South Atlantic (data from Bottomley et al. 1990 and from some other sources). Also 189 series of near surface air temperatures in 5° x 5° squares were processed (see Dobrovolski 1991, 1992).

Characteristic results for sea surface temperatures, with respect to model types, are presented in Fig. 3.16. Here, model orders obtained using formal application of the Akaike Criterion (see Sect. 2.4) are shown separately for the North and South Atlantic. Models of the first order

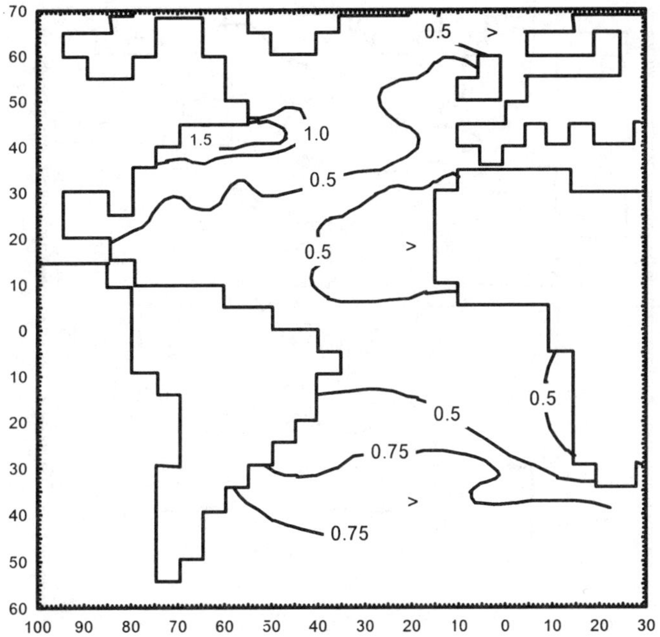

Fig. 3.17. Standard deviations of monthly sea surface temperature anomalies (σ_i) for January

prevail in both cases. Moreover, processing of the longest time series gives a larger percentage of the first order models. As stated in Chapter 2.4 and Section 3.1.1, even if Akaike's and other criteria formally indicate model orders larger than 1, the first-order model is a good approximation, given the width of confidence intervals for the model parameters. Note that this conclusion holds for all areas. We could not detect any reasonable spatial pattern of coefficients of the AR(1) model either. Spatial differences in this parameter do not exceed the errors of the estimations, the mean value of the coefficient being 0.55. Also note that seasonal variations of this parameter are not very large, and Eq. (3.1) decays into

$$\xi_1(t) = c_1\xi_1(t-1) + a(t)\,, \tag{3.7}$$

where ξ_1 is the monthly anomaly divided by the standard of the respective month σ_i - see Eq. (2.58), t is current time in months, c_1 is a constant coefficient, and $a(t)$ a sequence of uncorrelated random variables (white noise). Our analysis also demonstrates that probability distribution of

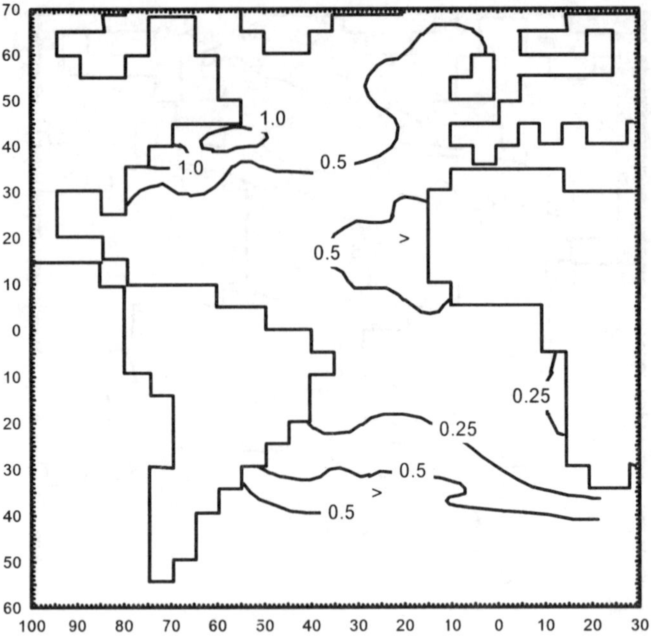

Fig. 3.18. Standard deviations of "residual white noise" (σ_a) multiplied by the mean standard deviations of monthly SSTA

ξ_1, in the first approximation, is close to normal (for more details see Dobrovolski 1991).

Thus, if we take c_1 approximately equal to 0.55 everywhere, we can obtain a very simple spatial pattern of the monthly SSTA variability where the only spatially varying parameter is the white noise standard deviation, σ_a. We can check this assumption in the following way: using the models of orders formally assessed with the help of statistical criteria from Chapter 2.3, we can calculate for each 5° x 5° square the value of the residual white noise standard deviation. If the above assumption holds, the spatial pattern of σ_a will resemble the spatial pattern of σ_i .

In Fig. 3.18 values of σ_a are shown; for the purpose of clearness these values are multiplied by (seasonal) mean standard deviations of monthly SSTAs. Thus, values in Fig. 3.18 specify a part of the total SSTA variance due to random forcing.

In turn, Fig. 3.17 demonstrates the spatial distribution of SSTA standard deviations for January. Recall that this pattern is rather typical for all seasons and remains qualitatively the same in summer, spring, and autumn: regions with high anomalies simply drift in north-south direction from cold to warm season.

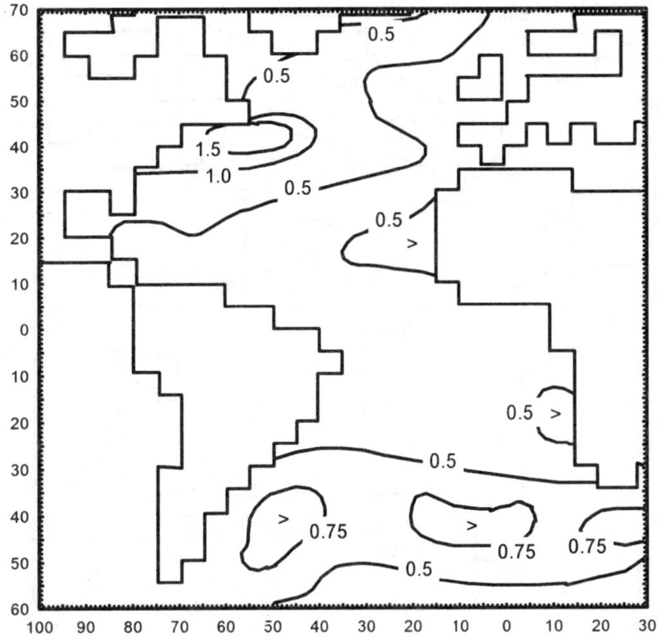

Fig. 3.19. Standard deviations of monthly sea surface temperature anomalies for January. Stochastic dynamical model run (120 months)

We can see that features of spatial distribution of residual white noise intensity and SSTA intensity are practically the same. The largest values are characteristic for the zones of subarctic and subantarctic hydrological fronts; a certain increase is observed in two regions of convergence near the African coast. The resemblance between patterns in Figs. 3.17 and 3.18 means the following: (a)The first-order model is really a good first approximation. (b) Zones of high random forcing and high SSTA answer to this forcing coincide with each other; thus the SSTA's origin seems to be, so to speak, local.

3.1.2.2
Stochastic Dynamical Regional Models

Two-dimensional stochastic dynamical model of ocean-atmosphere-land surface interaction was constructed by the author on the basis of the model of Adem (1991a). Atmospheric forcing v' was generated, using the Monte-Carlo method, in accordance with Hasselmann's recommendations: as a random, time-uncorrelated (at intermediate time scale, i.e., from month to month) fields. Following Adem's atmospheric model, v' was calculated in terms of the main independent atmospheric variable of the model,

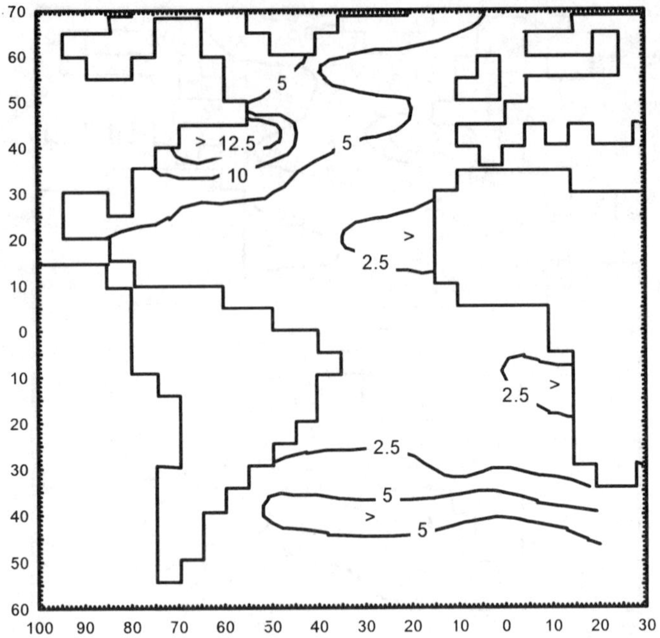

Fig. 3.20. Meridional temperature gradients of climatological sea surface temperatures in the Atlantic ocean. At °C/10° latitude. January

vertically averaged tropospheric temperature. Standard deviations and spatial correlations of forcing were assessed using observational data (see details in Dobrovolski 1992, 1994). We used a 5° x 5° grid and rectangular cylindrical cartographic projection.

The characteristic result of the model integrations - spatial distribution of January standard deviations of monthly SSTAs σ_i - is shown in Fig. 3.19. The model satisfactorily describes the main features of σ_i : SSTA intensification in the zones of subarctic and subantarctic hydrological fronts and in two convergence zones off African coasts, also characteristic σ_i values - more than 1 °C south of Newfoundland, more than 0.75 °C along 40°S, more than 0.5 °C in the two zones of convergence. Analogous patterns for summer, spring, and autumn also resemble the observed ones. Modeled SSTA spectral densities were also realistic and resembled those shown in Fig. 3.15.

While speculating about possible physical mechanisms responsible for the spatial structure in Figs. 3.17-3.19, analogy with the spatial pattern of horizontal climatological SST gradients immediately comes to mind. In Fig. 3.20 such a field is presented for winter (January) conditions. The resemblance between the pattern in Fig. 3.20 and those in Figs. 3.17 - 3.19

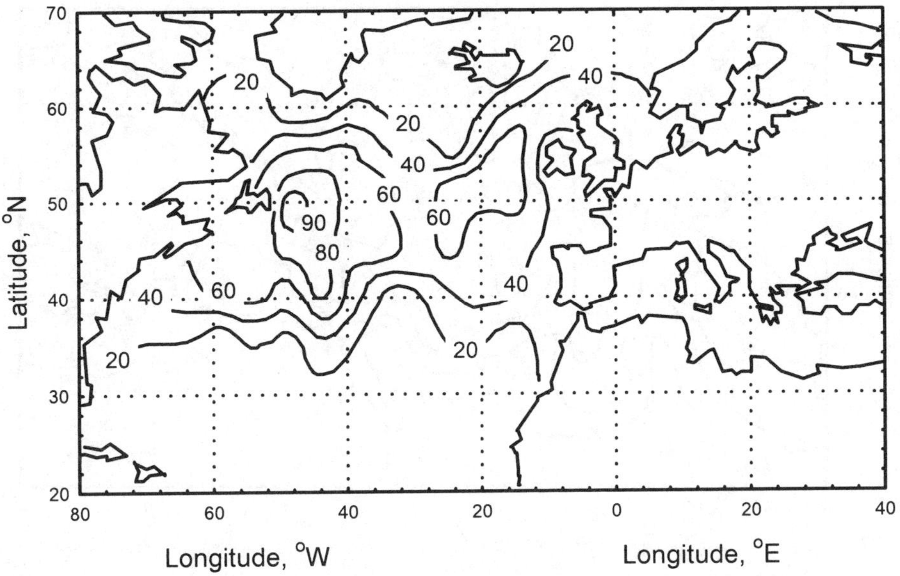

Fig. 3.21. Portion (in %) of SSTA standard deviation due to forcing by anomalous drift currents as compared to the total SSTA standard deviation. Results of stochastic experiments on Adem's interactive atmosphere-ocean-land surface model for winter conditions (January)

is striking and suggests that a shift in climatological SST gradients by anomalous winds might be a cause of the general SSTA pattern above.

In order to verify this and other hypotheses, sensitivity experiments were conducted. As an example, in Figs. 3.21-3.24 characteristic results of such experiments are given for a similar, although more detailed, model for a smaller region: the North Atlantic Ocean plus adjacent parts of North America, Europe, and Africa; a detailed description of this variant of the model is given in Dobrovolski and Rybak (1992), and Rybak (1992). In Fig. 3.21 the sensitivity of the model results to atmospheric forcing by anomalous drift currents is shown: the percentage of total SSTA standard deviations due to variations in the third term (in brackets) on the left side of Eq. (3.2). It is clear from Fig. 3.2 that the role of this mechanism is predominant in the zone of largest SSTAs - along the subarctic hydrological front where the values of horizontal SST gradients are high (cf. Fig. 3.20).

The second of the two most important mechanisms of the upper oceanic layer forcing by random atmospheric impulses is anomalous evaporation at the sea surface. Experiments aimed at the study of this mechanism (Fig. 3.22) show that the role of anomalous evaporation in forming sea temperature anomalies is not very significant in the zone of subarctic front but increases to the south. The role of anomalous vertical turbulent heat

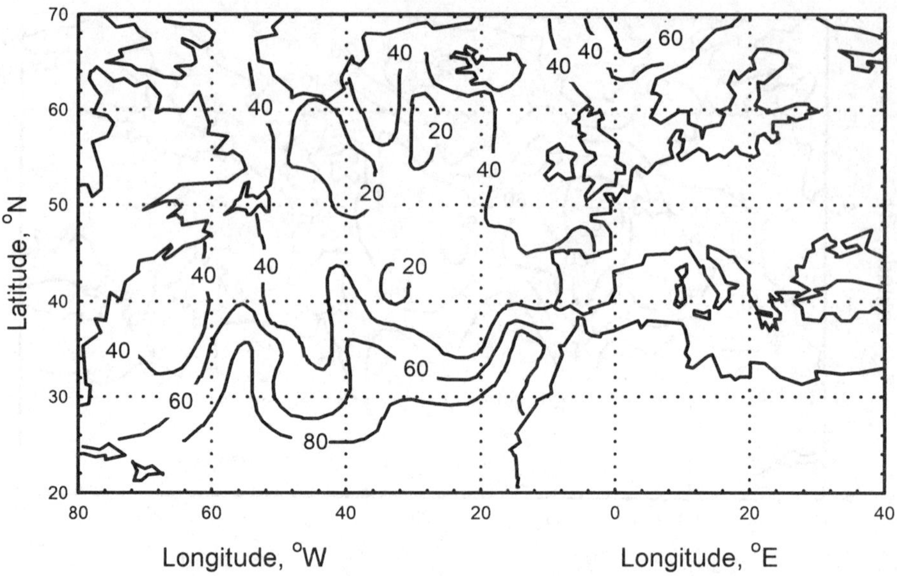

Fig. 3.22. A Fig. 3.21 but for anomalous heat fluxes due to evaporation at the sea surface

exchange at the ocean-atmosphere interface is analogous to the role of anomalous evaporation although not so important.

Generally speaking, there is a major difference between the roles of the two main types of ocean-atmosphere processes (anomalous drift currents on one hand, and evaporation and contact heat exchange on the other) in forming SSTA. In fact, anomalous drift currents can only increase SSTA variability, whereas anomalous evaporation and contact heat fluxes can either increase or reduce (dissipate) SSTAs. The specific effect of evaporation/contact heat fluxes on SSTA formation depends on the geographical position of the area under consideration and on the specific atmospheric and oceanic conditions. However, anomalous evaporation and contact heat exchange generally play the role of negative feedbacks rather than of SSTA forcing. This effect is most obvious in the zone of subarctic and subantarctic hydrological fronts. At the same time, forcing by anomalous drift currents still exists everywhere else (though less important), and resulting SSTAs are eventually dissipated through the anomalous fluxes across the oceanic surface. This role of sea-atmosphere fluxes was not evident before stochastic ocean-atmosphere experiments.

Another surprising result of the very first experiments with dynamic-stochastic climate models (started in the 1970s) was that large-scale turbulent - like horizontal thermal mixing in the ocean also played

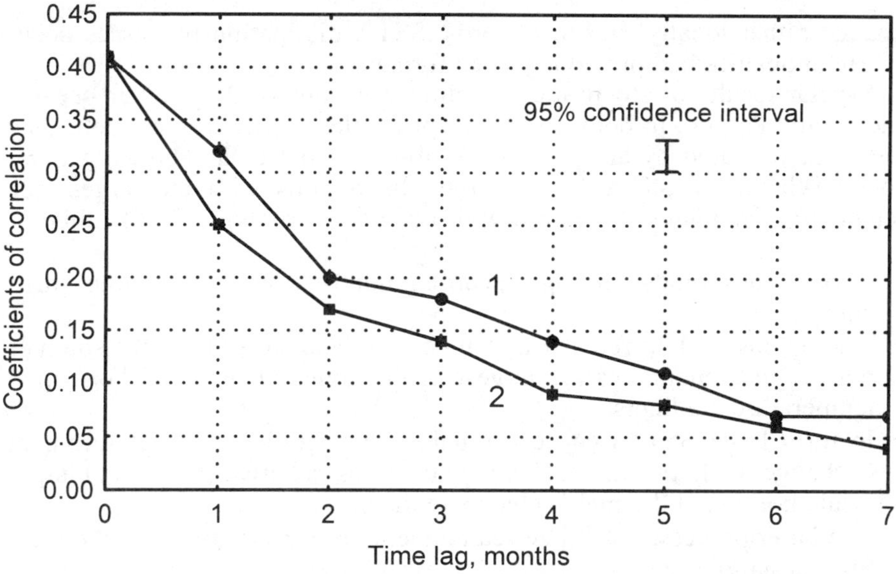

Fig. 3.23. Coefficients of correlation between monthly normalized anomalies of near surface air temperatures and sea surface temperatures. Results are averaged over 116 $5^{\circ} \times 5^{\circ}$ squares in the Atlantic Ocean. *1* Positive time lags (SSTA takes the lead over air temperature anomaly); *2* negative time lags

an important dissipating role with respect to large SSTAs. This horizontal thermal exchange, which represents a parametrization and averaging of subgrid oceanic processes, provided SSTA stationarity even in the experiments without ocean-atmosphere feedback fluxes. This mechanism works in the following way. Because of the limited spatial correlationship of atmospheric forcing, SSTAs of opposite sign are not far from each other: the characteristic correlation SSTA radius has an order of 1000 km. In turn, oceanic processes of horizontal heat redistribution smooth these contrasts and thus reduce both positive and negative sea surface temperature anomalies. Anomalous radiation processes also play the role of negative feedbacks with respect to SSTAs.

Finally, the role of quasi-stationary oceanic currents can be studied using the above dynamic-stochastic models. Experiments with and without the mechanism of SSTA "transport" by stationary currents demonstrated that the role of this mechanism in extra-tropical regions is not very important. Similar results were later obtained on stochastic models by Ostrovsky (1983), Ostrovsky and Piterbarg (1986), Piterbarg (1989), Rybak (1992), and others. (This is true with respect to SSTAs of spatial and temporal scales under consideration, i.e., more than several hundred km and more than several weeks). This means that SSTAs (at least, outside equatorial regions) do not float along main oceanic currents, but change rather locally. In other words, SSTA dissipation processes occur where anomalies had previously been formed.

Similar results of stochastic modeling were obtained for other oceans: Indian and Pacific. In both cases, except for the Pacific tropical belt, they were corroborated by analysis of the observational data. Thus, a general rough scheme of SSTA formation in the regions with the largest sea temperature anomalies (subarctic and subantactic fronts) looks like this:

- Anomalous winds during specific month cause anomalous drift currents.
- Anomalous drift currents bring temperature anomalies from the south or north; these anomalies are largest in the zones of maximal horizontal temperature gradients.
- In turn, large SSTAs cause anomalous evaporation, contact turbulent exchange with the atmosphere, anomalous radiative fluxes, and large--scale horizontal thermal exchange in the ocean.
- The latter processes stabilize sea temperature variations and make local SSTAs stationary.

As a result of the above scheme, relationships between near-surface air temperature anomalies and SSTAs seem to be different from those described in early works on the atmosphere-ocean interactions. Namely, SSTAs in most cases lead in the process of interaction. Figure 3.23 illustrates and corroborates this feature of climatic ocean/atmosphere variations.

The above results reflected the oceanic answer to random atmospheric forcing. Another problem of stochastic 2D climate models is the influence of oceanic anomalous processes (primarily, SSTAs) on the atmosphere. It was stated in Chapter 1 that it is not easy to detect the atmospheric answer even to tropical SSTAs; the reaction of the extratropical atmosphere to anomalous sea temperatures is a much more difficult problem. Note that typical standard deviation of monthly near surface air temperatures is several degrees Celsius (the value of this parameter is higher for the middle and upper atmosphere) whereas typical SSTA standard is only several tenths of °C, i.e., an order of magnitude smaller. This proves that atmospheric variability at time scales from 1 month to several years (and we cannot say much about features of longer scales of climatic changes using existing observational records) is determined mainly by internal atmospheric random processes.

Stochastic experiments with the coupled Adem ocean-atmosphere model generally corroborates this conclusion. The pure oceanic effect on the model monthly tropospheric temperature anomalies at intermediate vertical level (5 km) usually does not exceed a few tenths of a degree Celsius (Fig.2.4.16. in Dobrovolski 1992). However, a more specific atmospheric reaction, the sensitivity of the tropospheric water cycle to sea

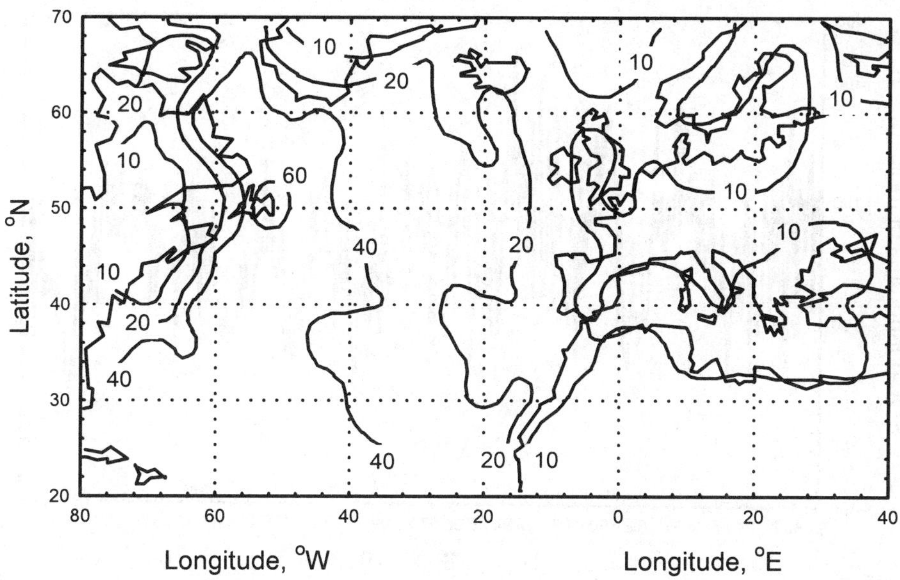

Fig. 3.24. Percentage of monthly precipitation anomaly due to anomalous evaporation. Results of stochastic experiments on a modified Adem ocean-atmosphere-land surface model for January

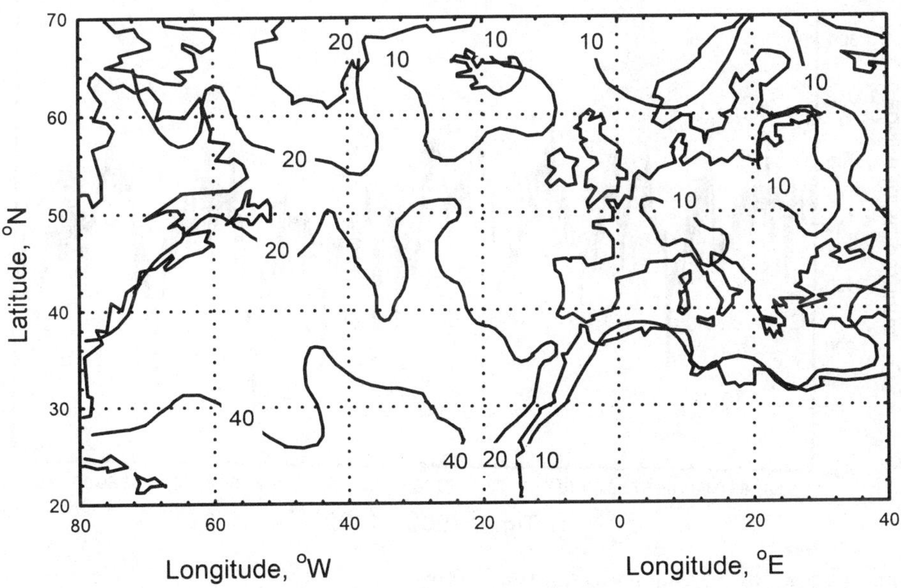

Fig. 3.25. A Fig. 3.24 but for July

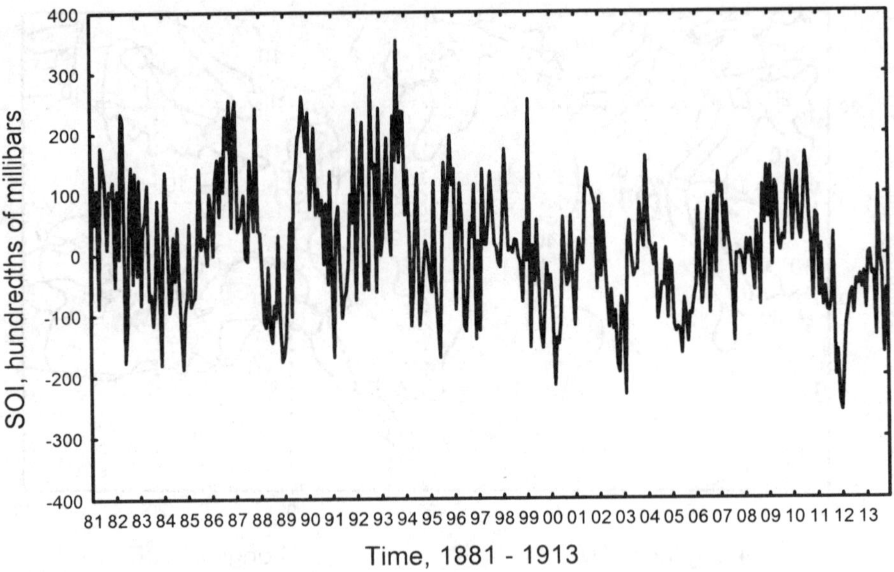

Fig. 3.26. COADS-based Southern Oscillation Index (SOI), 1881 - 1913

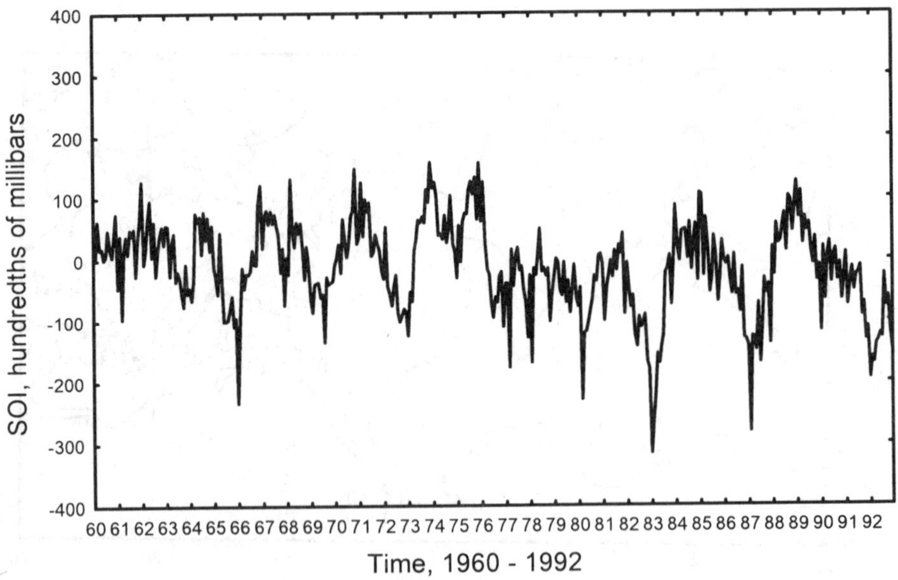

Fig. 3.27. A Fig. 3.26 but for period 1960 - 1992

surface evaporation anomalies, can be more pronounced. In Figs. 3.24 and 3.25 appropriate results are shown for winter (January) and summer (July) conditions. Figure 3.24 shows that in winter anomalous evaporation can play an important role in precipitation variability within the region of large SSTAs, the subarctic hydrological frontal zone. We also see sharp gradients in the sensitivity parameter along the North American coastline, and a reasonable pattern for the 10% curve, doubling the African and European coastline. Figure 3.25 demonstrates that the summer spatial pattern is more vague, and precipitation sensitivity to evaporation anomalies much less pronounced.

3.1.2.3
Stochastic Models of ENSO Events

It was shown in the previous section that dynamic-stochastic models in their simplest form (atmospheric white forcing of a simple model of the upper mixed oceanic layer) can describe observed features of climatic variabilities in extratropical regions and, to some extent, in the tropical regions of the Atlantic and Indian Oceans. However, stochastic modeling of ENSO seemingly requires a stochastic approach to more sophisticated ocean-atmosphere models.

Before discussing existing dynamic-stochastic models of ENSO events, it is necessary to demonstrate results of observational data analysis. In Section 3.1.1.1 it was shown that local sea temperatures within the ENSO area manifest no statistically significant oscillations. Here, we proceed to the analysis of regionally averaged parameter, the well-known COADS--based Southern Oscillation Index (SOI). The index is constructed in the following way. The Tahiti pole of the SOI is the area-averaged sea-level pressure anomaly for 20°N - 20°S, 180° - 70°W. The Darwin pole is the average sea-level pressure anomaly for the rest of the tropical band equatorward of 20° of latitude (20°N - 20°S, 70° - 0°W and 20°N - 20°S, 0° - 180°E). The SOI is calculated as the difference Tahiti minus Darwin.

In Figs. 3.26 and 3.27 two segments of the SOI series are presented: for the earliest longest period of uninterrupted observations (1981-1913), and for the latest period of observations of the same length (1960-1992). Data are from Todd (1996). It is clear that the SOI series is not homogeneous: the variability of the early segment is more important than in 1960 - 1992. Also, the temporal correlation of the first segment is lower than the correlation of the second segment (and this is corroborated by quantitative stochastic analysis). Seemingly, this nonhomogeneity of the series is related to the lack of observations from the 1880s to the 1910s.

However, there is one common feature in all segments of the SOI: the total absence of any statistically significant quasi-periodical or oscillatory components. This feature of the Southern Oscillation is an invariant of

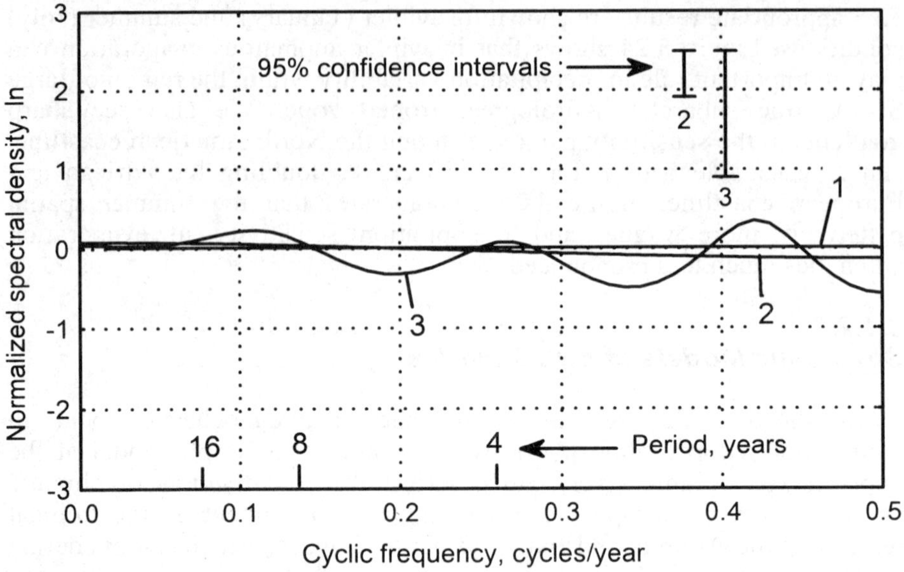

Fig. 3. 28. Estimations of SOI (July) spectral densities. *1* Zero order model (white noise); *2* 1st order; *3* 5th order; *4* 7th order

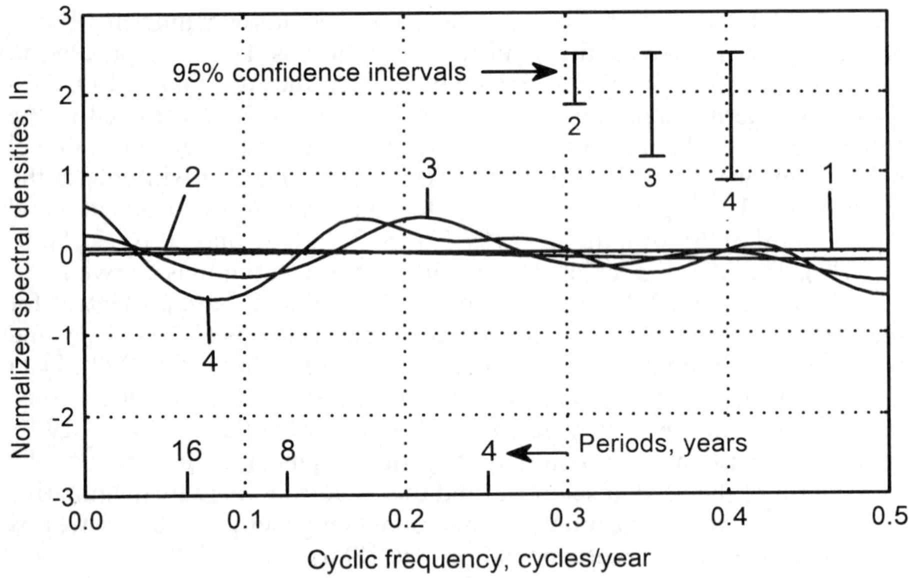

Fig. 3.29. A Fig. 3.28 but for time series of January SOI values

the way in which we preprocess the series: the results for an initial time series of monthly SOI values as well as for variants of series with 3-month, 4-month, and 12-month resolution and different variants of temporal averaging are all the same. As an example, in Figs. 3.28 and 3.29 spectral densities for two variants of 12-month resolution SOI series (July series and January series) are presented. Different criteria indicated different orders of stochastic models for spectral estimations (0, 1, 5, 7 for the July series and 0,1,5,7 for the January series), but spectral peaks in all cases were much less important than confidence intervals of respective estimations. Thus, a zero-order spectrum - horizontal line - is a good approximation for the SOI spectral density, and a time-uncorrelated sequence is a good approximation for the SOI interannual variability. Moreover, such perfect examples of a regional process which is *so* close to the white noise are relatively rare.

Thus, our calculations do not corroborate the conclusions of many works on the subject (see Table 1.1). The above results partly coincide with those obtained by Karoly and Kestin (1998, pp. 131-134) although we do not agree with their conclusions about the nonstationarity of stochastic SOI models. Actually, our results make the very theory of the ENSO questionable. In fact, if the ENSO pattern were as a set of finite number of chaotic nonlinear oscillators in the time-frequency domain, the energy of ENSO variability should be constrained to specific frequency bands.

The above contradiction makes it difficult to estimate results of existing models on random atmospheric forcing of ENSO events. Being dependent on existing ENSO models, these experiments inevitably give increased spectral densities within specific frequency intervals. For instance, Blanke et al. (1997) in their very interesting paper *Estimating the effect of stochastic wind stress on ENSO irregularity* write:

> Natural uncoupled atmospheric variability is thus a strong candidate for explaining the observed aperiodicity in ENSO time series ... . Addition of noise brings an irregular behaviour that is considerably richer in spatial patterns as well as in temporal structures. The broadening of the model ENSO spectral peak is roughly comparable to <that> observed.

We absolutely agree with the first two sentences, but express concern with respect to the third phrase: from our point of view, there is no ENSO spectral peak at all. We can make similar comments on other works on stochastic ENSO forcing, i.e., Stone et al. (1998). Thus, the problem of stochastic dynamical models of the equatorial ocean and the atmosphere is not yet completely solved, although in the near future we can expect major progress in this domain. Hasselmann's stochastic climate model theory seems an appropriate methodological tool in this case also. However, oceanic and atmospheric deterministic blocks are needed, that are more sophisticated, in comparison with extratropical models.

3.1.3 Globally Averaged Climate Variables

3.1.3.1 Global Water Mass Exchange. Global Mean Sea Level

Factors Contributing to Global Mean Sea Level Change. Before studying global mean temperature changes - an apparently major global climate variable - it is expedient to discuss the variability of the very redistribution of mass between climatic subsystems. Almost all the mass of thick climatic subsystems - oceans, ice sheets, also a considerable part of the atmospheric mass - is represented by their water mass. Unfortunately, there are no direct observations on the variability of the total oceanic and ice sheet water content at climatic scales, and there are only imperfect and short records of atmospheric water mass variability. However, indirect data (global mean sea level, precipitation, river runoff, solid glacial runoff) enable us to draw some conclusions about the features of global water exchange.

Global mean sea level (GMSL) represents the most frequently discussed variable related to global water exchange. Also, for a long time this variable was considered a manifestation of man-made global climate change and the greenhouse effect. However, recent discussion about Post Glacial Rebound (PGR) has caused much uncertainty, and many scientists think that "global sea level rise cannot be measured at all": see the discussion in AGU (1995), for instance. It is remarkable that the peak of publishing new variants of GMSL time series was in the 1980s, and later such publications were rare (if any). It is also significant that in the famous IPCC Second Assessment volume (Houghton et al. 1996) there is no observationally estimated GMSL curve at all.

We agree that there are many uncertainties in GMSL estimations and attribution. However, it seems to us that an accurate stochastic analysis of existing variants of GMSL curves and related time series can answer following important questions on the features of GMSL and global water exchange variability: Do the PGR effect and other geological processes considerably affect recent GMSL changes? Are there deterministic trends within the GMSL records? What are the laws of climatic sea level changes and global water mass redistribution?

Although over 30(!) types of processes potentially affecting sea level changes are known (Fairbridge and Jelgersma 1990), and the role of each process is subject to discussions (Dobrovolski 1992), in the present work we will follow a very simple classification of the mechanisms generating GMSL variations. We will consider three main types of processes:

1. Geological processes: mechanisms changing the shape of the oceanic basins and coasts (including the Post Glacial Rebound); mechanisms of forming juvenile, volcanic, mineralogic conversion waters, and other mechanisms of geological and geomorphological type. We suppose that these processes are relatively slow, and the characteristic time of the GMSL changes due to them is hundreds of years and more. Thus, geological processes are likely to create long-range trends within the time series of annual GMSL values. Indeed, it is difficult to imagine that the sign of globally averaged movement of oceanic coasts and bottom can be positive during several months or years, and then negative. More exactly, the GMSL reaction on this type of processes is likely to be linear: "PGR is linear over the tide gauge record, as are vertical crustal movements at plate boundary locations with long earthquake recurrence times" (AGU 1995). However, from our point of view, it is not possible to estimate "theoretically" the speed of such processes and the angle of such a linear trend. Therefore, our main task will be to check the hypothesis of the very presence of deterministic (linear) trend within GMSL series.
2. Eustatic causes of mean sea level changes: processes of redistribution of water mass between the ocean, the ice sheets, and the land surface mainly through mechanisms of evaporation, precipitation, and river and glacier (liquid and solid) runoff. Processes of this type are of a climatic nature, therefore in this case the whole climatic diapason of periods (frequencies) is relevant: from months to dozens of years, and more.
3. Steric changes. Initiated by the atmospheric thermal impact, these processes are supposed to be smoothed by the slow mechanisms of vertical thermal mixing in the ocean. Observational data on month-to-month and year-to-year variations of deep ocean temperatures (or "effective", vertically averaged oceanic temperature) are not available on the global scale. Coefficients of vertical turbulent mixing and vertical velocities in the ocean are poorly known: it was shown in Chapter 1 that estimations of these parameters can differ by an order of magnitude. The only available global mean sea temperature series is that of sea surface temperatures, so that we can only speculate about the character of steric GMSL variations. Existing GCM simulations of the greenhouse effect, for instance, suggest a monotonous (with trends) character of the steric sea level rise. Here our task will be: (a) to check the hypothesis of the deterministic monotonous (linear) trend within the GMSL series; (b) to compare GMSL models with those for existing series of global ocean and atmosphere temperatures.

Global Water Balance and GMSL. Following the above classification, we can write a simplified equation for month-to-month or year-to-year increments of the global mean sea level, ΔH, in the form:

$$\Delta H = \Delta H_{EU} + \Delta H_G + \Delta H_{ST}\ , \tag{3.8}$$

where ΔH_{EU} is the eustatic changes, ΔH_G the variations of the global mean sea level caused by geological and other external mechanisms, and ΔH_{ST} the steric changes. In turn, using the concept of constant global water mass (Kalinin 1968), we can propose a simple approximate equation for the eustatic component:

$$\Delta H_{EU} = \frac{-\Delta V_{SH} - \Delta V_{MG}}{S}\ , \tag{3.9}$$

where ΔV_{SH} is the increment of the ice sheets volume in water equivalent, ΔV_{MG} the increments of mountain glacier volume in water equivalent, S the area of the ocean surface that can be considered as a constant because relative changes in mean sea level at the time scales under consideration are small, i.e., centimeters and decimeters, and the resulting relative changes in the ocean surface area are negligible.

Finally, the equation for the global mean sea level increments, ΔH, can be rewritten in discrete form taking into consideration Eqs. (3.8) and (3.9):

$$\Delta H = -\Delta V_{SH} / S - \Delta V_{MG} / S + \Delta H_G + \Delta H_{ST}\ . \tag{3.10}$$

Consider each term in Eq. (3.10). Here, the first three terms represent the global water balance equation. We possess no direct measurements of changes in ice sheet volumes, although there are some indirect estimations. Thus, a series of total iceberg quantities were analyzed in Dobrovolski (1992) and it was shown that these series are close to realizations of uncorrelated random sequences. Also processing of paleoreconstructed global ice sheet volume often gives the same white noise-type model (see Chap. 4. Finally, analysis of mountain glacier volume records (Sect. 3.3) demonstrates that most of these records are satisfactorily described by the white noise model.

Therefore, we can hypothesize that terms $-\Delta V_{SH}/S$ and $-\Delta V_{MG}/S$ are close to the white noise process. It was stated above that geological or geomorphological GMSL changes are likely to be monotonous or linear. In turn, many uncertainties are related to the features of steric variability, ΔH_{ST}. We can expect that these changes can also be relatively slow, because processes of vertical thermal mixing and propagation are very slow, and their characteristic time might be of an order of more than 10^2-10^3 years. In any case, the steric component of the GMSL changes does

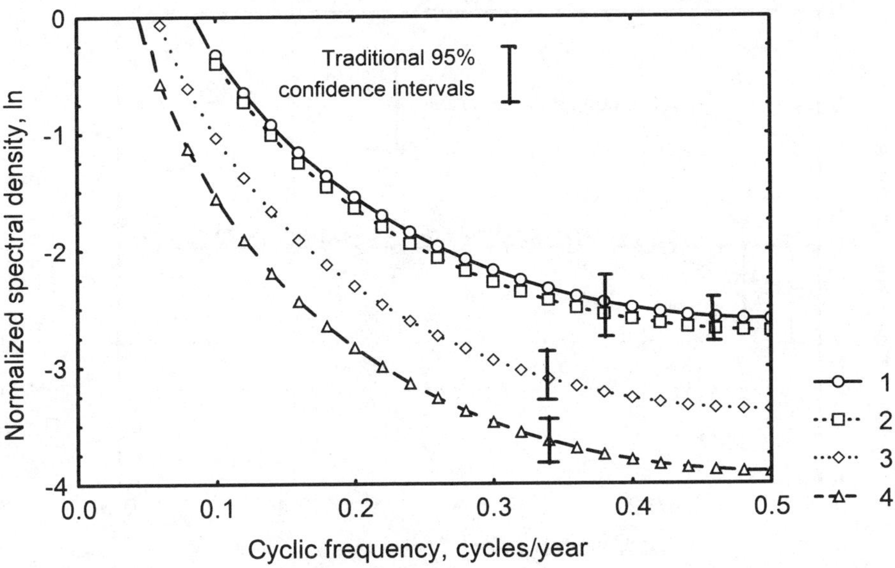

Fig. 3.30. Traditional spectra of GMSL series. (Data from: *1* Gornitz and Lebedeff 1987, *2* Fairbridge and Krebs 1962, *3* Barnett 1988, *4* Kliege et al. 1997). *Dotted line 3* coincides with spectral estimate of the discrete Wiener process realization

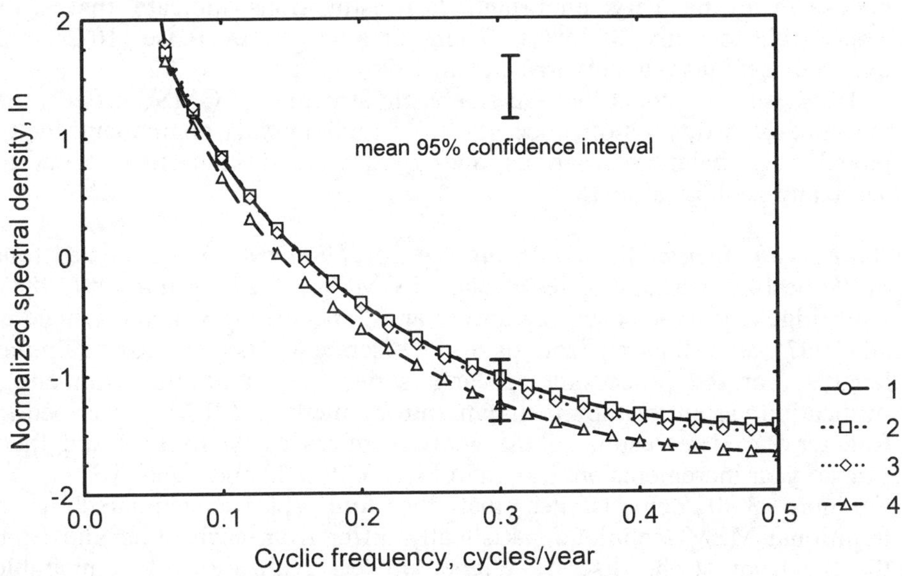

Fig. 3.31. A Fig. 3.30 but for frequency-truncation spectral estimations

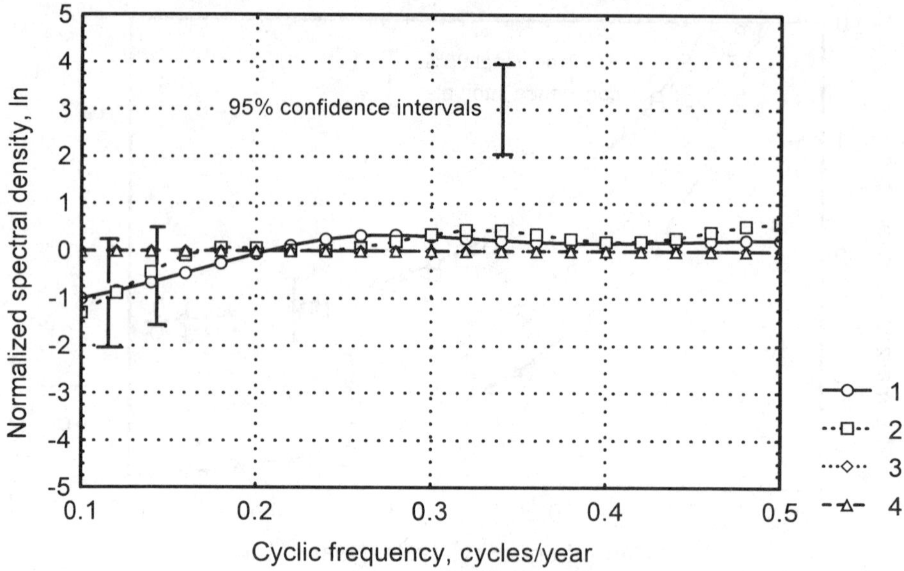

Fig. 3.32. A Fig. 3.30 but for year-to-year increments. *Dashed line* (*3*) coincides with the white noise (increments of the discrete Wiener process) spectrum

not seem to be very important: last estimations indicate that it is responsible for only 20-25% (2-7 cm) of total sea level rise (10-25 cm) over the past 100 years (Warrick et al. 1996).

Thus, we can expect that the stochastic structure of GMSL reflects the presence of two types of processes: (a) a random walk component due to global water balance variations, and (b) slow, trend-like changes due to steric and geological factors.

Analysis of Annual Global Mean Sea Level Records. We analyzed four of the best-known time series of annual GMSL values: by Barnett (1988), Fairbridge and Krebs (1962), Gornitz and Lebedeff (1987), and Kliege et al. (1997), as well as realizations of the discrete Wiener process of similar lengths. For the processing of each series we used the frequency--truncation variant of the maximum entropy method (MEM), described in Chapter 2.4. The results of the analysis of series themselves and their year-to year increments are shown in Figs. 3.30-3.32 and Table 3.2.

Figure 3.30 demonstrates that the four spectra estimated using traditional MEM technique statistically differ from each other and from the spectrum of the discrete Wiener process realization of comparable length (except for the Barnett series). However, the frequency-truncation method (Fig. 3.31) proves that all spectra practically coincide with each other and with the Wiener process spectrum. Analysis (Fig. 3.32) of

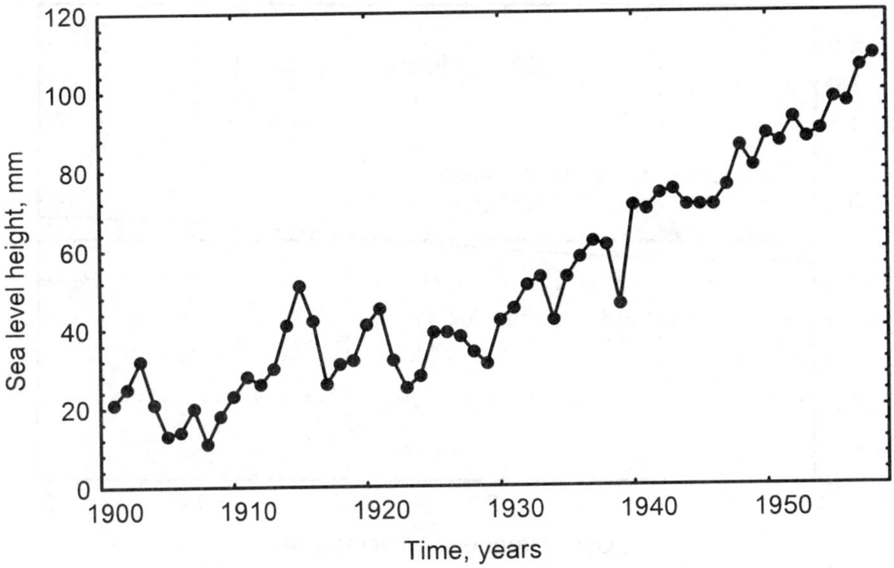

Fig. 3.33. Generalized Global Mean Sea Level time series

temporal GMSL increments also demonstrates the Wiener nature of the global mean sea level variations. Spectral densities of year-to-year sea level increments are close to the white noise spectrum, taking into account the width of confidence intervals.

Thus, analysis of both GMSL initial series and their temporal increments corroborates the presence of a Wiener-like component in GMSL variations. In order to check the hypothesis of a linear trend, mean values of GMSL temporal increments have to be studied. It is convenient to illustrate this issue using, as an example, a generalized GMSL series. We constructed such a series (Fig. 3.33), averaging the four records above. The resulting sequence qualitatively resembles each of the original records and represents a typical segment of global mean sea level "rise".

In short, results of the generalized series analysis were the same as those obtained for each of the initial four series: spectral density was practically the same as that of the discrete Wiener process, and the spectrum of the year-to-year increments resembled the white noise spectrum (Fig. 3.34).

The most interesting and surprising result of processing each of the five GMSL series under consideration is illustrated by Fig. 3.35. If GMSL series contain a significant monotonous (e.g., linear) trend, the mean value of the GMSL increment series must differ significantly from zero. In Fig. 3.35 increments of the generalized GMSL series are shown. As stated above, they are close to a sequence of uncorrelated random values;

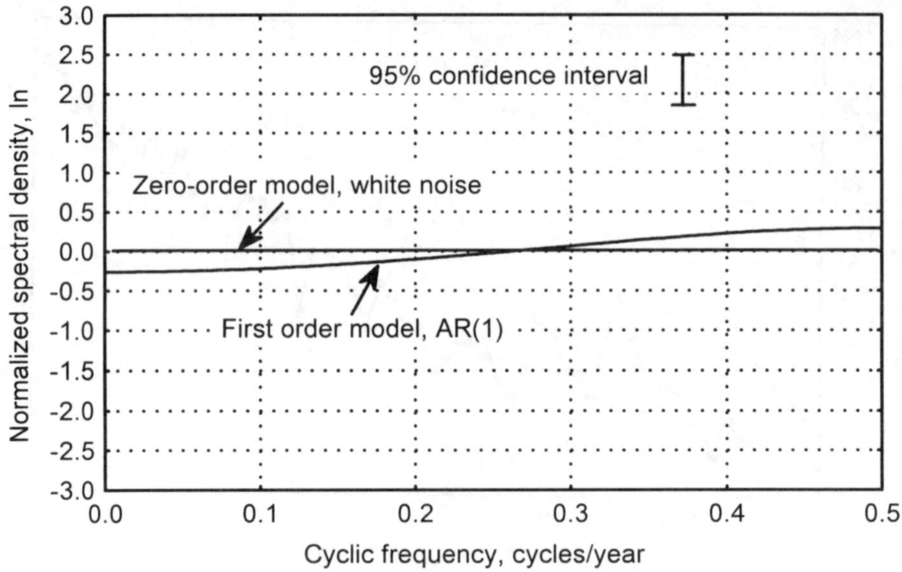

Fig. 3.34. Generalized GMSL series. Spectral densities of year-to-year increments: zero--order model (*horizontal line*), and first-order model (*inclining line*)

additionally, their distribution is close to normal: for instance, the coefficient of skewness is only -0.08, and the excess is 3.81. The mean value of the series is 1.5 mm/year; it is denoted by horizontal line No. 1 in Fig. 3.35. It is important to compare this value with the error of its estimation that can be expressed for uncorrelated normal random values by the followng formula:

$$\sigma_{\varepsilon} = \frac{\sigma_{\xi}}{\sqrt{N}}, \qquad (3.11)$$

where σ_{ε} is the standard deviation of estimation of the mean value, σ_{ξ} the standard deviation of the uncorrelated sequence, N the number of points within the segment under consideration. In our case, σ_{ε} equals about 1 mm/year, i.e., has the same order as the estimated mean value. Thus, the mean value of GMSL increments does not differ significantly from zero, and the presence of a monotonous (linear) trend is not corroborated by our analysis. Recall that results for each of the four original GMSL series (Table 3.2) were even more impressive. Mean values of GMSL increments were: 0.7 mm/year (Kliege et al. 1997), 0.6 mm/year (Fairbridge and Krebs 1962), 0.9 (Gornitz and Lebedeff 1987), and 1.1 mm/year (Barnett

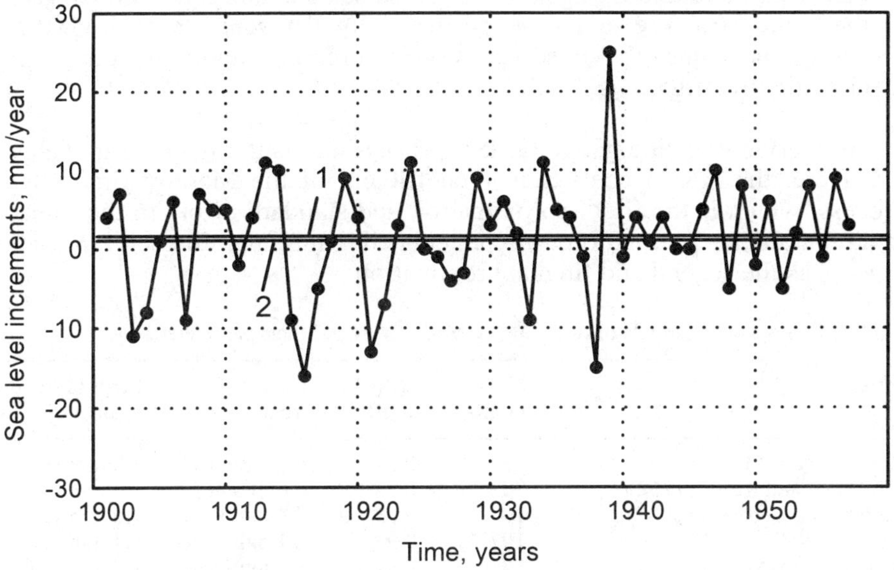

Fig. 3.35. Year-to-year increments of the generalized GMSL series. *1* Estimation of the mean value; *2* standard error of the mean value estimation

1988), whereas mean errors of these estimations were respectively 1.2, 1.3, 1.6, and 1.4 mm/year.

Mathematically speaking, this result is not so surprising: most of the short realizations of the discrete Wiener process (random walk) visually resemble processes with monotonous trends (see Fig. 2.16 and discussion in Chap. 2.2). Here, it is important to stress that accurate stochastic analysis, in this way, does not corroborate either the presence of geological trends, or the presence of steric trends within the global mean sea level changes. Important consequences of this result are the following:

1. Severely criticized during past years (because of supposed underestimation of the PGR and other geological effects), the tide gauge GMSL series seem to be not so bad.
2. Stochastic analysis shows that spatial averaging procedures, while obtaining GMSL series, effectively suppress local/regional crustal or geomorphological slow movements.
3. Procedures for removing the PGR and other slow global geological processes are not needed because these processes play no significant role with respect to estimated GMSL variations.
4. It will be shown below that stochastic structure and parameters of GMSL changes are similar to those estimated using satellite-based and paleoestimations. Moreover, it will be demonstrated in Section 3.1.3.2

that recent climatic changes of the global sea and air temperatures have the same structure and are governed by the same law as GMSL variations. Thus, the observed GMSL are an important source of climatic investigations.

An overview of tide gauge GMSL processing results is given in Table 3.2. Here, the AR(1) coefficient - coefficient of the Markov first-order process, which is the first approximation and standard errors in the linear trend parameter estimations are calculated using the discrete Wiener process as the second and final approximation.

Table 3.2. Stochastic models of tide gauge GMSL series and their parameters

Series	No. of points	AR(1) coeff.	Linear trend	Error in trend estimation
Kliege et al. (1997)	75	0.94	0.7 mm/year	1.2 mm/year
Fairbridge and Krebs (1962)	98	0.84	0.6 mm/year	1.3 mm/year
Gornitz and Lebedeff (1987)	103	0.83	0.9 mm/year	1.6 mm/year
Barnett (1988)	107	0.92	1.1 mm/year	1.4 mm/year
Generalized GMSL series	58	0.91	1.5 mm/year	1.0 mm/year

Satellite-Based TOPEX-POSEIDON GMSL Records. The above results were obtained using traditional data of coastal gauge measurements. It was important to check them using remotely sensed data on the sea level heights which take into account the whole surface of the ocean. Adequate data of this type became available only recently: one of the first publications of an accurately processed space-borne GMSL series with relatively small errors appeared in Hendricks et al. (1996) (Fig. 3.36).

In Fig. 3.37 the spectrum of this time series is shown together with the spectrum of realization of the Wiener process of the same length (74 points, Fig. 3.37.3). The appropriate stochastic model for the TOPEX-POSEIDON sea level series was as follows:

$$H(t) = 0.85\,H(t-1) + a(t)\,, \tag{3.12}$$

which is close to the expression for the Wiener process [Eq. (2.54)]. In Eq. (3.12) H denotes GMSL height taken in anomalies about mean value; t is the number of decade; a the white noise. Decade-to-decade increments of this series fit the white noise model even better than increments of the annual GMSL series. The observed distribution of increments is almost perfectly the normal one (Fig. 3.38). At the same time, the spectrum of increments series almost coincides with the horizontal line, which is the white noise spectrum (Fig. 3.39). Moreover, analysis of possible nonlinear properties of TOPEX increments using appropriate packs of programs (see, for instance, Sprott and Rowlands 1995) gives no differences

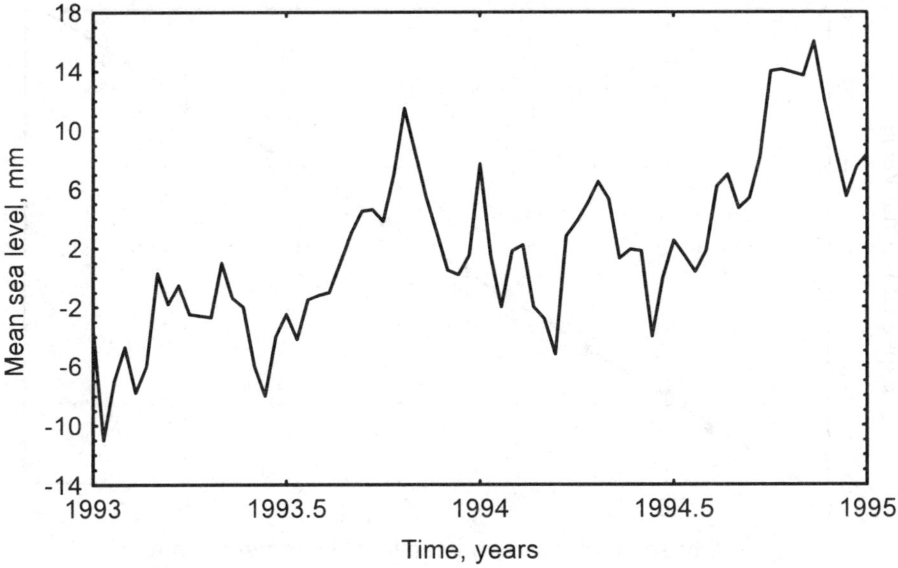

Fig. 3.36. Global Mean Sea Level. (Data from TOPEX-POSEIDON altimeter, Hendricks et al. 1996. Published with a kind permission of the American Geophysical Union)

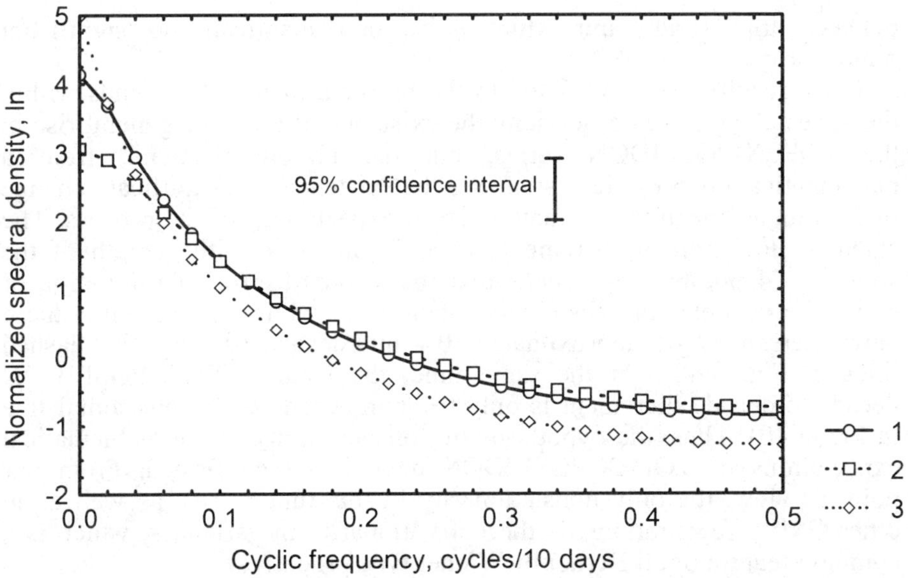

Fig. 3.37. Estimations of spectral densities: *1* TOPEX-POSEIDON GMSL series; *2* the same but with removed seasonal harmonics; *3* realization of the discrete Wiener process with the same number of points

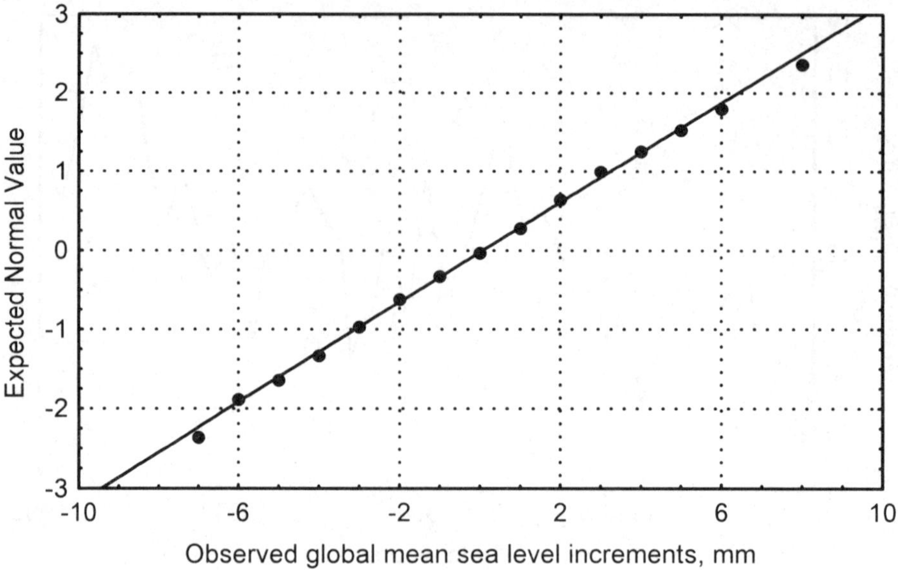

Fig. 3.38. Normal probability plot of the TOPEX-POSEIDON GMSL decade-to-decade increments

between this series and white noise and manifests no significant nonlinearity.

These results are also related to the problem of possible trends within the series. Again, we do not deny the existence of a visible general rise in the TOPEX-POSEIDON series, but the notion of trend, from a mathematical point of view, supposes the existence of a significant change in the time of mean value (mathematical expectation) of the process. The standard deviation of increments is 3.07 mm/decade, the length of the series is 74 points. This means that the standard error of estimation of mean increments of absolutely stationary white noise with these parameters *must* be approximately 0.4 mm/decade (denoted by dashed lines in Fig. 3.40). At the same time, the mean TOPEX-POSEIDON decade-to-decade increment is only 0.1 mm/decade (solid horizontal line in Fig. 3.40). Thus, the hypothesis of nonstationarity in the mathematical expectations of TOPEX-POSEIDON data is not confirmed. From our point of view, the only nonstationarity in this time series, as well as in other GMSL registrations, is the nonstationarity in variances, which is a common feature of all Bernoulli-Wiener type processes.

The problem of possible seasonal variations in the TOPEX-POSEIDON series is rather difficult. Indeed, the series is too short to remove the 12-month component statistically. At the same time, physical models at their present stage of development cannot adequately reproduce the impact of

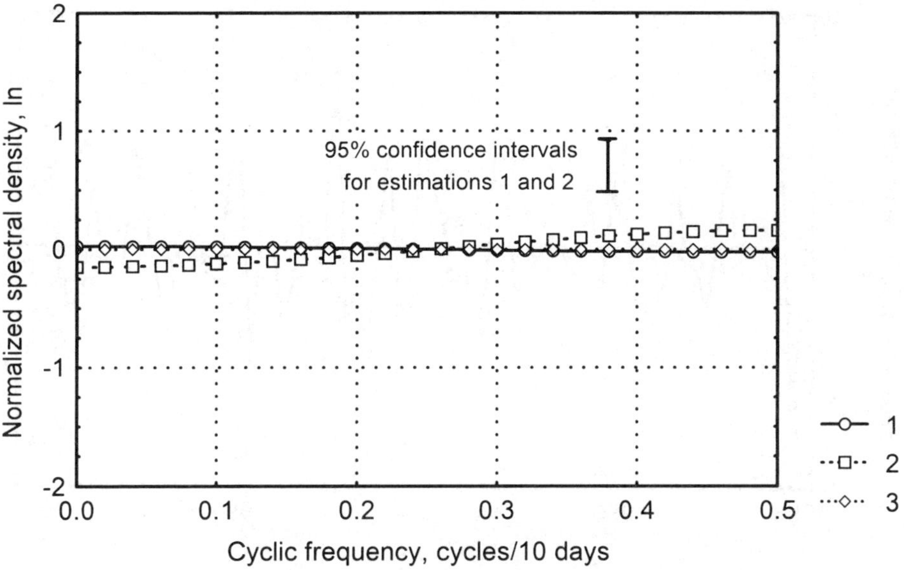

Fig. 3. 39. Spectral densities of TOPEX-POSEIDON GMSL decade-to-decade increments. *1* Initial series; *2* series with removed seasonal harmonics; *3* the white noise spectrum

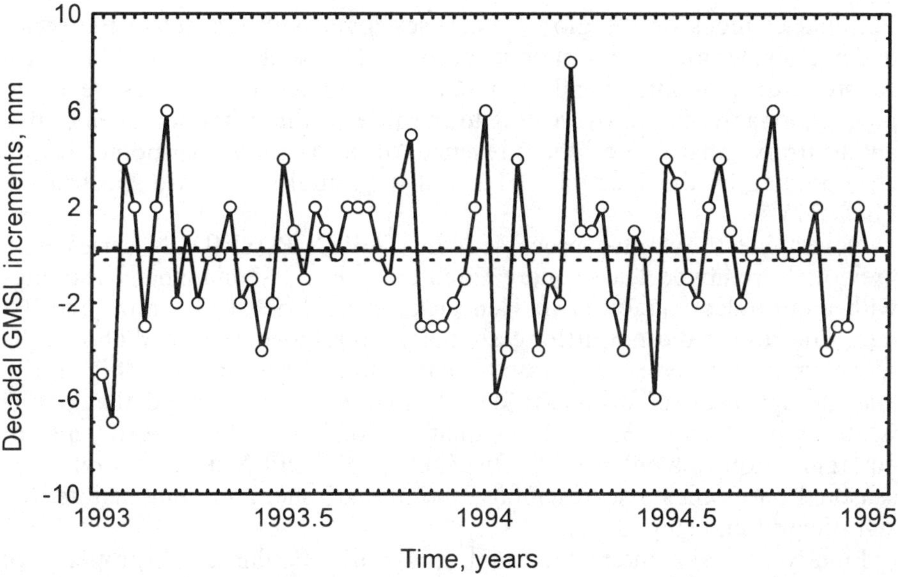

Fig. 3.40. Decade-to-decade increments of the TOPEX-POSEIDON Global Mean Sea Level time series (see text for details)

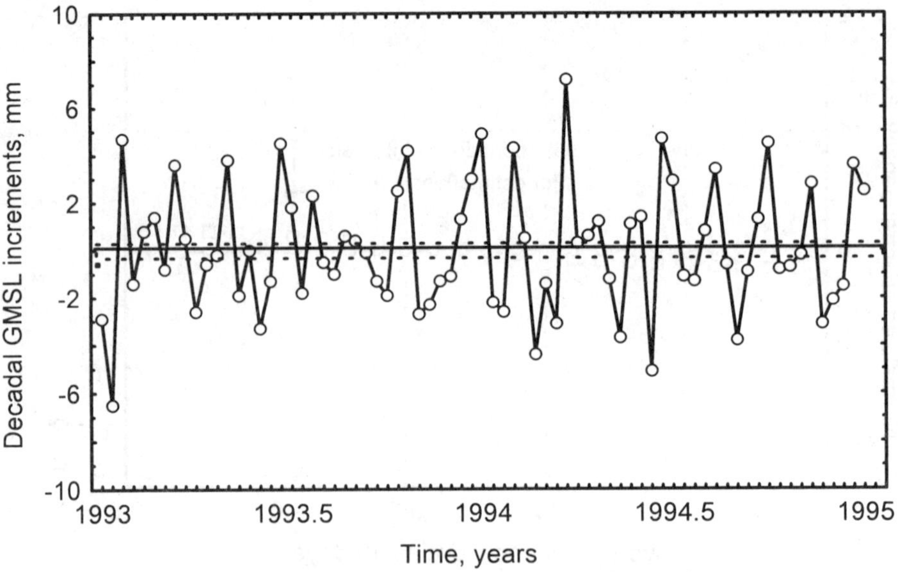

Fig. 3.41. A Fig. 3.40 but for TOPEX-POSEIDON GMSL series with removed seasonal harmonics

seasonal changes on the global mean sea level with respect to the series under consideration. The method used by Hendricks et al. (1996), least squares fitting of annual and semiannual harmonics to the series, seems to be questionable: it can overestimate the role of these harmonics and give an arbitrary phase. Yet the presence of a 6-month harmonic is not corroborated by the spectral analysis of the initial series (see spectrum in Fig. 3.37).

Nevertheless, we also processed the TOPEX-POSEIDON series with previously removed least square fitted annual and semiannual harmonics with amplitudes suggested in Hendricks et al. (1996): 3.6 mm, and 3.4 mm. The results did not differ essentially from those described above: the spectrum of this series was close to that of the Wiener process (Fig. 3.37), and the spectrum of decade-to-decade increments resembled that of the white noise (Fig. 3.39.2): estimation of the first coefficient of autoregression was only - 0.08. Distribution of TOPEX data with removed seasonal harmonics also resembled the normal one, and demonstrated no significant trend (Fig. 3.41).

Finally, a very interesting and, probably, fundamental property of GMSL changes is demonstrated by the TOPEX-POSEIDON data. If the Wiener process fits the decadal GMSL changes, we can calculate mean increments of sea level for all characteristic temporal scales using the expression:

$$M[abs\{H(t) - H(t+\tau)\}] = [\,t\,\sigma(\tau)]^{1/2}, \qquad (3.13)$$

where $M[\;\;]$ is the mathematical expectation operator (calculation of average value); t the time in decades; τ the characteristic temporal lag; and σ the standard deviation of decade-to-decade white noise sea level increments. If we take $\tau = 36.5$ (1 year), and the σ value from TOPEX-POSEIDON data, 3.0 mm/decade (2.8 for the series with removed annual and semiannual harmonics), the mean difference between sea levels will be approximately 1.8 cm (1.7 mm). This value is of the same order as the mean year-to year sea level increments calculated using the annual GMSL series. For instance, the mean year-to-year increment estimated for series from Barnett (1988) is 1.5 cm/year. This means that not only is the form of spectral density the same for the wide range of characteristic temporal scales, but also the value of the main parameter of the Wiener process persists. This property of parameters of the climatic system is called global climatic invariance and was introduced, with respect to atmospheric variables, in Ladoy et al. (1984).

Analysis of paleoreconstructions of sea level changes also corroborates the hypothesis on the temporal climatic invariance of GMSL changes. For instance, the range of mean sea level changes (±1.5 m) during the past 6,000 years assessed using 1,500 radiocarbon estimations of sea level (Newman et al. 1980) or taken from the work ofBard et al. (1991), is well described by extrapolating the Wiener process model: with help of Eq. (3.13) and the parameter σ obtained from the TOPEX-POSEIDON data (3 mm/decade), or from the annual GMSL series (1 - 1.5 cm/year).

3.1.3.2 Global Temperatures

Globally averaged temperature is usually regarded as a main climatic variable and a basis for climate theories. Historically, such time series were constructed on the basis of observations at land meteorological stations. Certain problems arise while constructing a sequence of spatially averaged temperatures. Thus, temperatures in certain regions can be biased, due to different methods of exposure (Nicholls et al. 1996). Portman (1993) and other authors find that the influence of urbanization on temperature estimations may be significant. Concern about a possible link between climate change and desertification is expressed by Jones (1994). Finally, the uneven distribution of meteorological stations over the globe and the lack of data during the early period of observations cause major difficulties.

Today, blended land-ocean global air temperature records are also constructed. Problems related to the creation of such time series are discussed, for example, in Nicholls et al. (1996). Unfortunately, we cannot

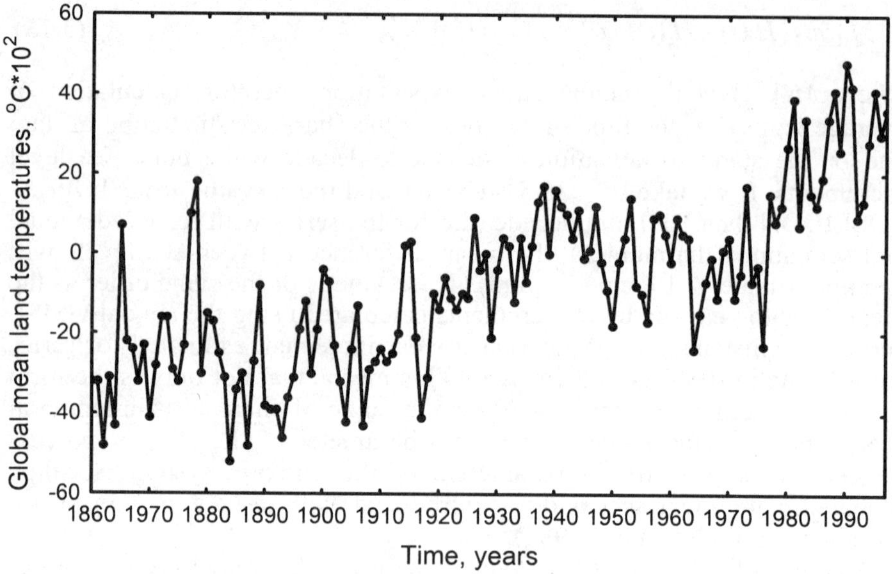

Fig. 3.42. Global near-surface air temperature (land only), 1861 - 1997. (Data from GISS 1998)

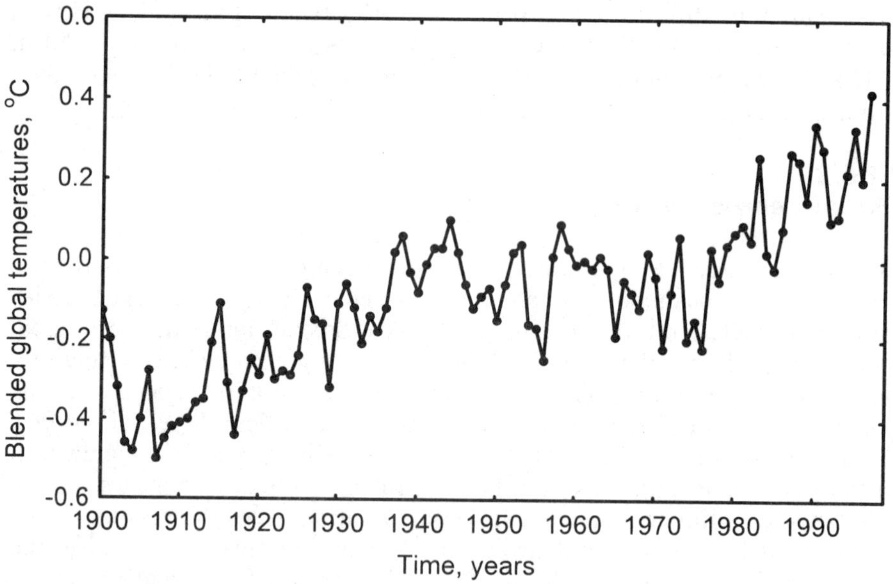

Fig. 3.43. Global Land + Sea Surface Temperature Index. 0.3 * Land + 0.7 * Sea, annual anomalies. (Data from NOAA 1998)

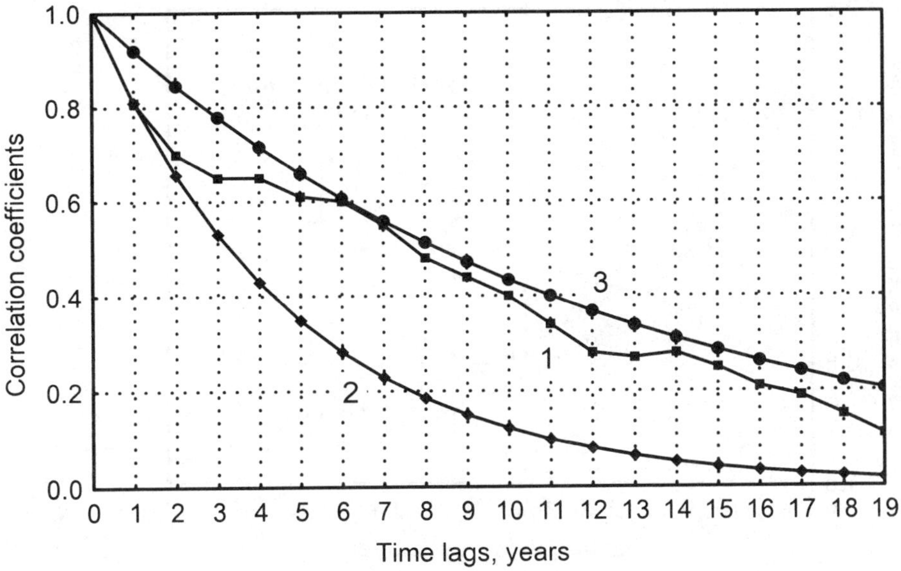

Fig. 3.44. Sample correlation coefficients, NOAA global temperature series (*1*); AR(1) model with coefficient 0.81 (*2*); AR(1) model with coefficient 0.92 (*3*)

say much about quantitative parameters of the errors in global temperature series; perhaps the only indication to be found in the literature is that such errors could equal at least hundredths of a degree Celsius.

In this section we will analyze global temperature records "as is", although some conclusions based on the structure of the series will be made with respect to possible errors within them. As an example of a global temperature record we will take two well-known series: the GISS global surface air temperature (land only) record (Fig. 3.42) and the NOAA global land + sea surface temperature index (Fig. 3.43). We also analyzed other variants of global temperature series (USDC 1994, Global Climate ... 1996, etc.), but the results of the analysis were similar to those described in this section.

Formally, the MEM analysis gives the first-order Markov process model [AR(1)], with a high value of the coefficient, as the best first approximation for the description of the above series. For instance, the equation for the NOAA series looks like this:

$$T_G(t) = c\,T_G(t-1) + a(t)\,, \tag{3.14}$$

where $T_G(t)$ is the centered global temperature at year t, $T_G(t-1)$ the temperature at year $(t-1)$, $a(t)$ the white noise, and c equals 0.81 (Yule-Walker scheme) or 0.84 (Burg scheme). However, analysis of sample

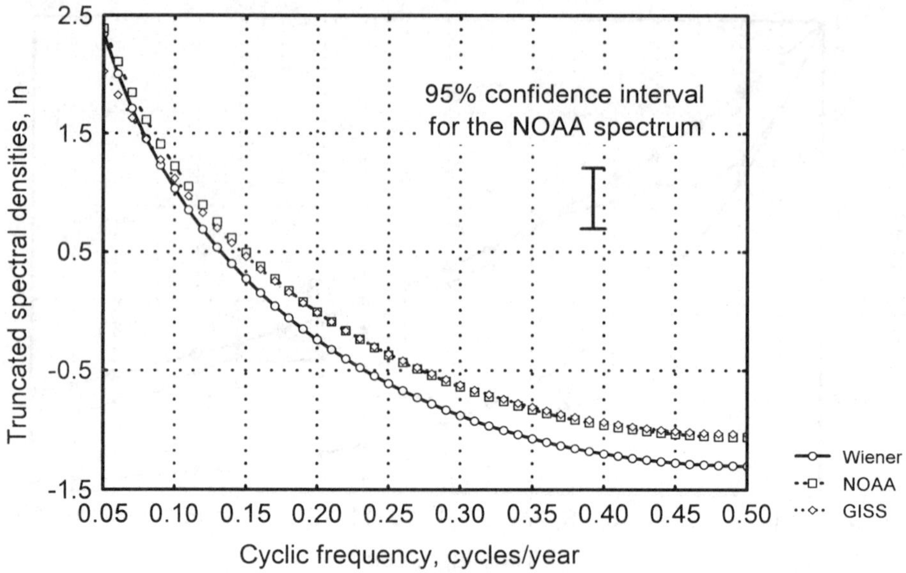

Fig. 3.45. Frequency-truncated spectral estimations for the global mean air temperatures and the 98-point realization of the discrete Wiener process

correlation coefficients (Fig. 3.44) shows that a higher first-order Markov coefficient better describes global temperature covariances. In this figure covariances of the AR(1) model with coefficient 0.92 (curve 3) are much closer to the observed covariances than those obtained using the AR(1) model with sample coefficient 0.81.

At the same time, we know that a segment of the discrete Wiener process with 98 points (the length of the NOAA series), analyzed with the use of the MEM techniques, gives the first-order Markov [AR(1)] model with an average value of coefficient 0.92 (and not closer to 1.00, because of limited series length). For this reason it is necessary to compare the global temperature series with realizations of the discrete Wiener process (random walk). Because of the great importance of the global temperature issue, we will try to accurately use different methods of such a comparison. Lack of space makes us demonstrate results for only one, the NOAA series, results for other records being similar.

First of all, it is convenient to compare appropriate estimates of spectral densities. In Fig. 3.45 spectra of the two series are presented together with the spectrum of a segment of the discrete Wiener process of similar length. It is clear that all three estimates of spectral density coincide with each other, given the width of the confidence interval. (In Fig. 3.45 we do not present spectral densities for frequencies of less than 0.05 cycle/year, i.e., for periods of more than 20 years, because nothing can be said about such

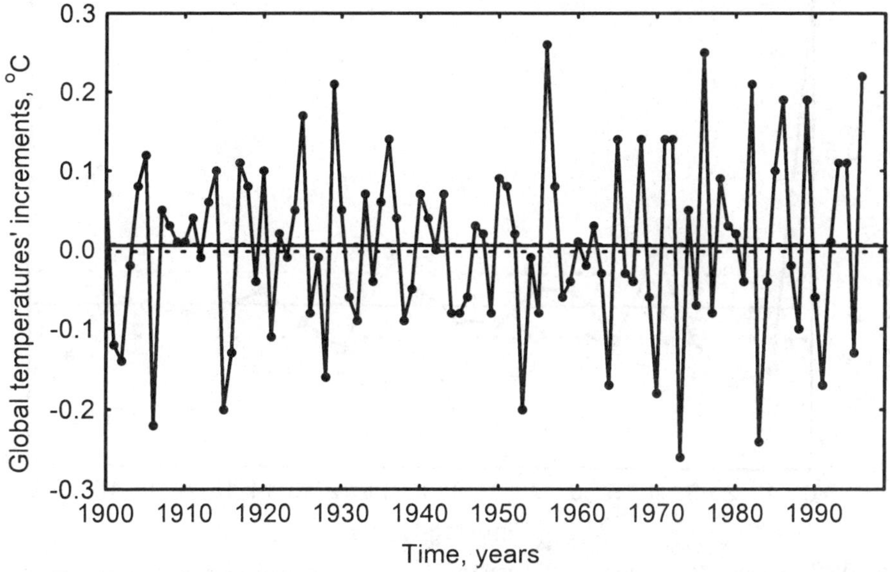

Fig. 3.46. Year-to-year increments of the NOAA global temperature series. *Solid horizontal line* denotes the mean value of increments m_i . *Dashed lines* denote standard errors of m_i estimation

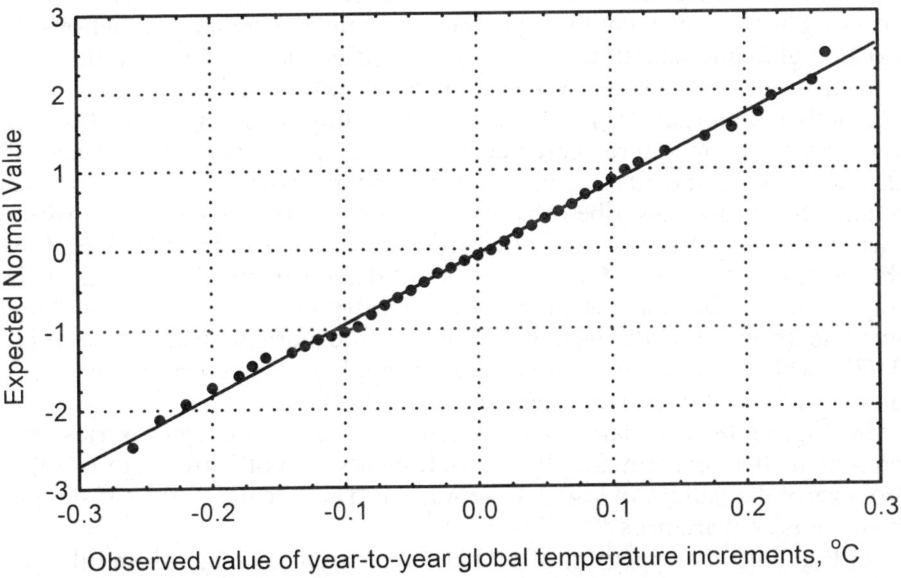

Fig. 3.47. Normal distribution plot of the global NOAA air temperature series

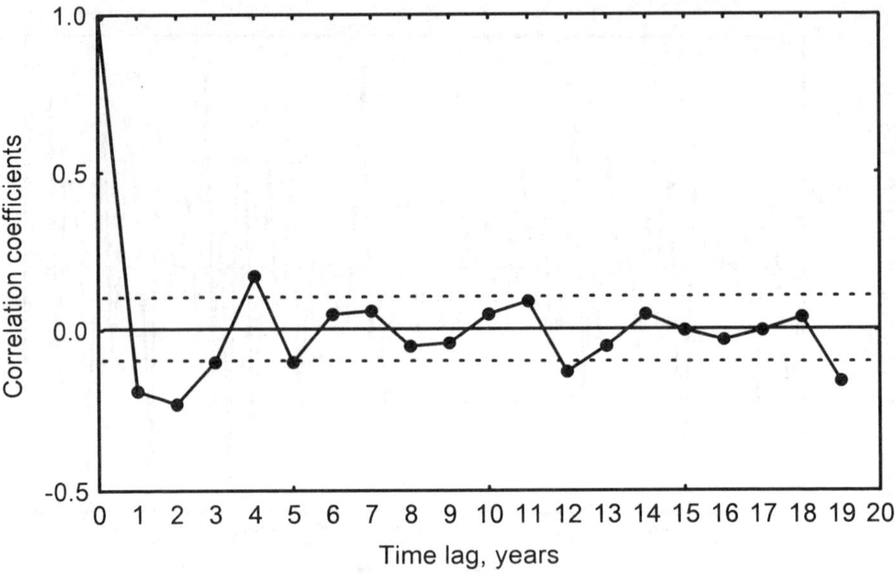

Fig. 3.48. Sample correlation coefficients of the NOAA global temperature increments. *Dashed lines* denote standard error of estimations

frequencies, taking into consideration the limited series length: 98-137 years). Thus, we can say that spectra and sample correlation functions of existing global temperature records do not differ statistically from those of a realization of the discrete Wiener process (random walk).

Another important stage of the global heating study is the analysis of year-to-year temperature increments (Fig. 3.46) (the very appearance of this series gives no idea of significant temperature rise at all). If global temperatures are described by the random walk model, year-to-year increments must be close to realizations of white noise. In Figs. 3.46-3.49 the results of the appropriate analysis are demonstrated. First of all, it is important to note that the probability distribution of global temperature increments is perfectly well described by the normal distribution (Fig. 3.47), which proves the possibility of applying the correlation (two-moment) theory of random processes.

In Fig. 3.48 correlation coefficients of the increments series are presented. It is evident that these coefficients do not differ significantly from zero: the pattern in Fig. 3.47 represents a classical example of sample white noise covariances.

In Fig. 3.49 the spectrum of global temperature increments and that of white noise are compared. It is evident that the difference between the two spectra does not exceed the width of a confidence interval.

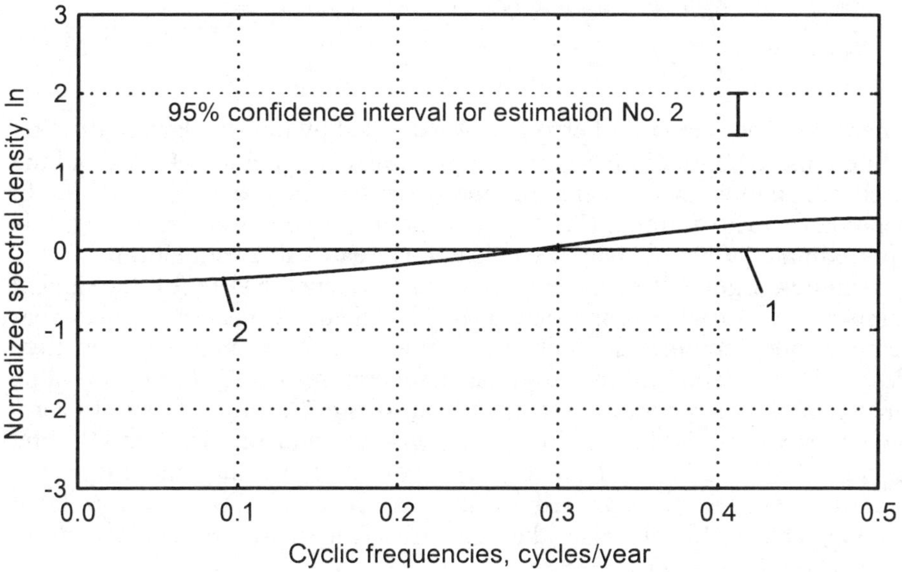

Fig. 3.49. Spectral densities: *1* white noise; *2* global NOAA temperature increments

The following considerations can also explain the existing small difference between the two spectra (higher values of observed spectrum at high frequencies and lower values at low frequencies). We can expect certain errors to be present within the global temperature series. In the first and simplest approximation, these errors can be regarded as uncorrelated random values. In this case, the model for observed global temperatures is

$$T'_G(t) = T_G(t) + a_2(t), \tag{3.15}$$

where $T'_G(t)$ is the observed global temperature at year t, $a_2(t)$ the second white noise, i.e., errors in observations, and $T_G(t)$ the discrete Wiener process

$$T_G(t) = T_G(t-1) + a(t). \tag{3.16}$$

It is easy to show that the spectrum of year-to-year increments of the process $T'_G(t)$ will qualitatively resemble spectrum No. 2 in Fig. 3.48. Also neighboring members of T'_G series will be negatively correlated with coefficient

$$\overline{R}(\tau) = -\frac{l^2}{1+2l^2}, \tag{3.17}$$

where l is the standard of errors $a_2(t)$ divided by the standard of the first white noise $a(t)$. In our case, σ_a is approximately equal to 0.11 °C, and the first correlation coefficient of increments series equals - 0.19. In accordance with formula (3.17), the standard of supposed errors, $a_2(t)$, is approximately 0.06 °C, which seems to be a reasonable estimation.

Another argument with respect to the Wiener nature of the global temperature variations is related to the mean value of year-to-year temperature increments. Indeed: if there is a monotonous positive deterministic trend within the global temperature record, the mean value of year-to-year temperature increments must significantly differ from zero. This mean value is denoted by the horizontal solid line in Fig. 3.46 and equals 0.006 °C/year. However, we proved above that the increment record is very well described by the sequence of normal uncorrelated random values. In this case, the standard error of the increments' mean value estimation equals

$$\varepsilon(m_i) = \frac{\sigma_i}{\sqrt{N}}, \tag{3.18}$$

where σ_i is the standard deviation of year-to-year global temperature increments that equals 0.11 °C/year in our case, and N the number of points within time series (98). Thus, ε (m_i) equals about 0.011°C/year, which is larger than the above observational estimation of 0.006 °C/year.

Hence, our analysis does not corroborate the presence of a monotonous (or linear) trend within the global temperature series. Note that the same result was obtained for all other variants of global temperature records. In other words, the *typical* consecutive sum of centered normal uncorrelated values (*typical* random walk or the discrete Wiener process with the above values of parameters) *must* have a visible tendency of even larger rise (or lowering).

We can use one more method for estimating the significance of the deterministic trend within the global temperature record. In fact, in most cases, researchers consider annual global temperatures, while estimating the significance of global heating, as uncorrelated values. However, year-to-year changes in global temperatures are strongly correlated. Therefore, it is expedient to apply Yaglom's formula for the error in the mean value of a time series:

$$\sigma(m_T) = \sigma(T_G)\sqrt{\frac{2T_1}{N}}, \tag{3.19}$$

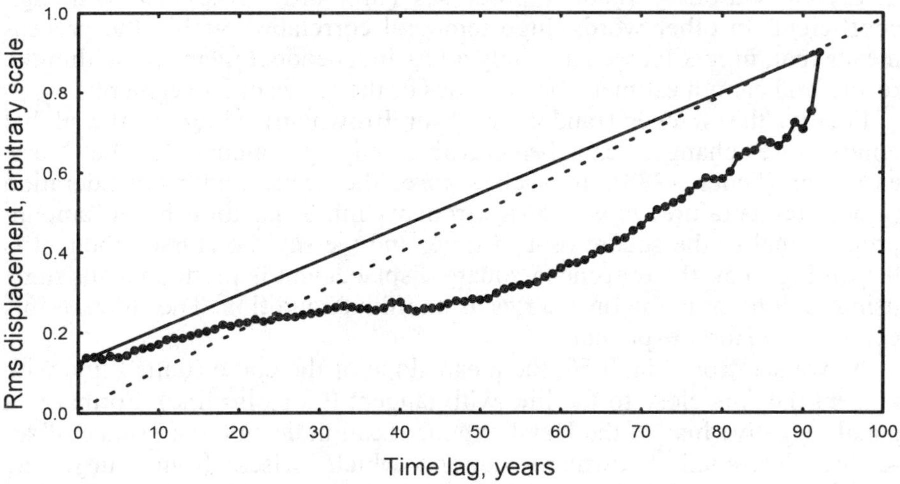

Fig. 3.50. The Hurst exponent fitted to the NOAA global temperature record

where $\sigma(m_T)$ is the standard deviation of the mean value estimation, $\sigma(T_G)$ the standard deviation of the initial series, N the number of points within the series, and T_1 the "correlation time" (integral time scale) of process T_G. Equations for T_1 in continuous and discrete cases are, respectively (Yaglom 1981):

$$T_1 = \int_0^\infty \left|\overline{R}_1(\tau)\right| d\tau \,, \tag{3.20}$$

$$T_1 = \frac{1}{2} + \sum_{\tau=1}^{\infty} \left|\overline{R}_1(\tau)\right| , \tag{3.21}$$

where $\overline{R}_1(\tau)$ are correlation coefficients, and τ the time lags (in our case, years).

If there is a monotonous trend within the global temperature record, different segments of this sequence must have significantly different mean values. For instance, the mean value of the second half of the NOAA record (+0.026 °C) must differ statistically from the mean value estimated using the whole series (-0.091 °C, i.e., the difference is 0.117). However, formula 3.19 gives the mean error of the latter value as 0.176. Thus, the difference in time-averaged values of global temperatures can be satisfactorily explained even from the point of view of the correlation theory of stationary random processes (first order Markov) with high coefficient. In other words, high temporal correlation within the process means that, in reality, we have only a few independent members within the record and cannot estimate mean values of the series or its segments.

Finally, the Wiener (random walk or Brownian) nature of the global temperature changes can be corroborated by calculating the Hurst exponent (Feder 1988). In such a case, the value under consideration (global temperature) moves away from its initial position by an amount proportional to the square root of time, and we say the Hurst exponent is 0.5. In Fig. 3.50 the root-mean-square displacement is plotted versus time, using each point in the time series as an initial condition. The slope of this curve is the Hurst exponent.

As we see from Fig. 3.50, the mean slope of the curve (dotted line with tangent 0.43) is close to the line with tangent 0.5 (solid line). Formally, a small negative bias of the Hurst exponent can indicate the presence of so--called fractional Brownian motion, which arises from integrating correlated (colored) noise. In such a case, an exponent less than 0.5 would indicate antipersistance (past trends tend to reverse in the future). However, the difference (0.07) is small enough and can be explained in our case by the presence of random observational errors within the time series [see formula (3.17) and related discussion].

As a conclusion to this section, we can formulate the following statements/hypotheses on the manifestations and nature of the global temperature changes:

1. Global temperature variations, during the instrumental period, i.e., during about 100 years, manifest no stable, stationary value (level, or regime), which could attract global temperatures. This means that negative temperature feedbacks on the global and temporal scale under consideration are negligible. Also it will be shown in Chapter 4 that the absence of stationarity in global temperature fluctuations is characteristic for time scales up to several thousands of years. Heat capacities of main, thick climatic subsystems - ice sheets and the ocean - seem to be practically unlimited, in comparison with year-to-year or decade-to-decade portions of anomalous heat received or lost by these subsystems.
2. However, global temperatures also do not manifest the presence of any deterministic (monotonous) trends. Thus, the hypothesis of monotonous, man-made changes in global climate (e.g., greenhouse effect) is not corroborated by our analysis.
3. The above calculations and analysis show that natural factors of global temperature variability act like random, time-uncorrelated impulses. Possible global effects of such mechanisms as ENSO or volcanic eruptions, together with other possible mechanisms of climate change, do not differ statistically from white noise. Our approach does not

corroborate the presence of nonlinear or linear oscillations within the global temperature records, because such oscillations would inevitably highlight spectral density at specific frequency bands, which is not demonstrated by accurate time series processing. In brief, global temperatures are likely to be a random outcome of a sum of a practically unlimited number of natural climatic mechanisms rather than the result of a few deterministic processes.

4. The best stochastic model, explaining the above and other features of global temperatures, is the discrete Wiener process (random walk). Thus, the best physical model for the description of global temperatures' behavior, seems to resemble the Brownian motion model. These aspects of the climate theory will be discussed in the next three sections.

3.1.3.3
"Minus Two" Law of Climatic Variability

It follows from the equation for spectral density of the Markov first-order process [Eq. (2.52)] that if this model approaches the discrete Wiener process (i.e., if coefficient $c \to 1$), spectral density $f(\lambda)$ is quasi- - proportional to λ^{-2} for frequencies $\lambda << 0.5$ cycles/discreteness interval. This means that in bilogarithmic scales spectral density is close to the straight line with tangent -2 . Hasselmann (1976) described this "minus two" law of climatic variability and demonstrated that it was adequate if climatic feedbacks were not very important. Later, Hasselmann and some other scientists, analyzing observed spectra, expressed doubts about the "minus two" degree law. For instance, Ladoy et al. (1984) proposed a -5/3 variant.

However, our analysis of globally averaged state (accumulation) parameters of thick climatic subsystems - sea and air temperatures, sea level (also paleoreconstructed ice volume and some other related parameters) shows that the -2 law is valid for the description of global climatic variability within a wide range of temporal scales. From our point of view, the spectrum angle in bilogarithmic scales can often be underestimated because of: (a) the limited length of climatic records; (b) the presence of random errors within these records. Both causes tend to underestimate spectral density at low frequencies. Also (c) errors in estimations of stochastic parameters were high if the processes under consideration were close to nonstationary ones (which is the case of global accumulation climatic parameters), so it was difficult to compare spectral estimations of different records with each other.

The low-frequency truncation procedure (Chap. 2.4) helps to decrease the errors of type (c) and shows that spectra of different climatic variables very well fit each other (e.g., see Fig. 4.3). Accurate estimations of both recent and paleoclimatic records (for time scales up to several dozens of

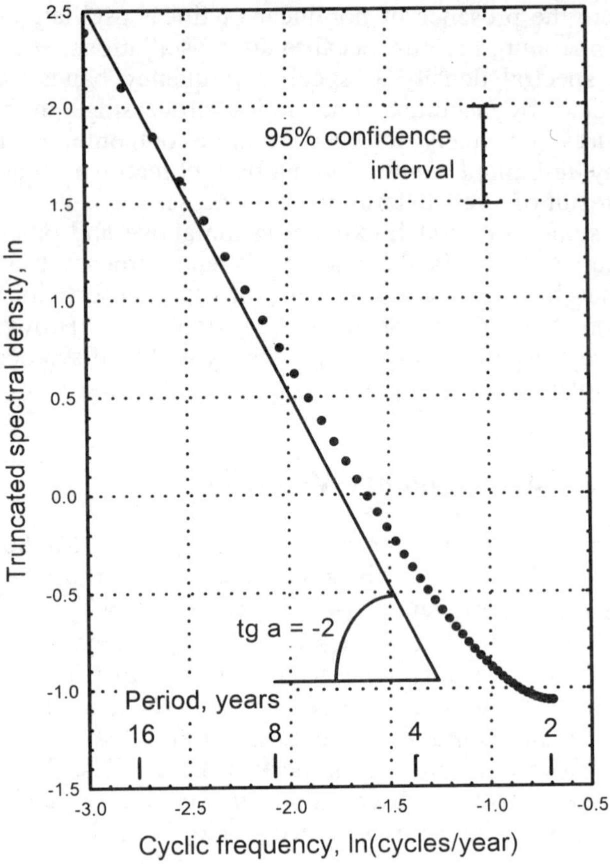

Fig. 3.51. Spectral density of the NOAA blended air-sea global temperatures record in bilogarithmic scales. *Straight line* denotes the -2 spectrum

ka) demonstrate a universal applicability of the law. Examples of spectra in bilogarithmic scales are presented in Figs. 3.51, 3.52, and 4.3. In Fig. 3.51 the spectral density of the NOAA blended air-sea global temperatures asymptotically approaches the -2 line. In Fig. 3.52 the bilogarithmic spectrum of the generalized global mean sea level record is even closer to the straight line with such a tangent. The same feature of climate variability is demonstrated by the Vostok deuterium and methane paleorecords (related to the palaeotemperature record) in Fig. 4.3. It is important to stress that all other series of globally averaged accumulation climatic parameters analyzed in our investigation, other variants of annual temperature and sea level series, sequences of monthly global temperature anomalies, manifested the same law of climatic variability.

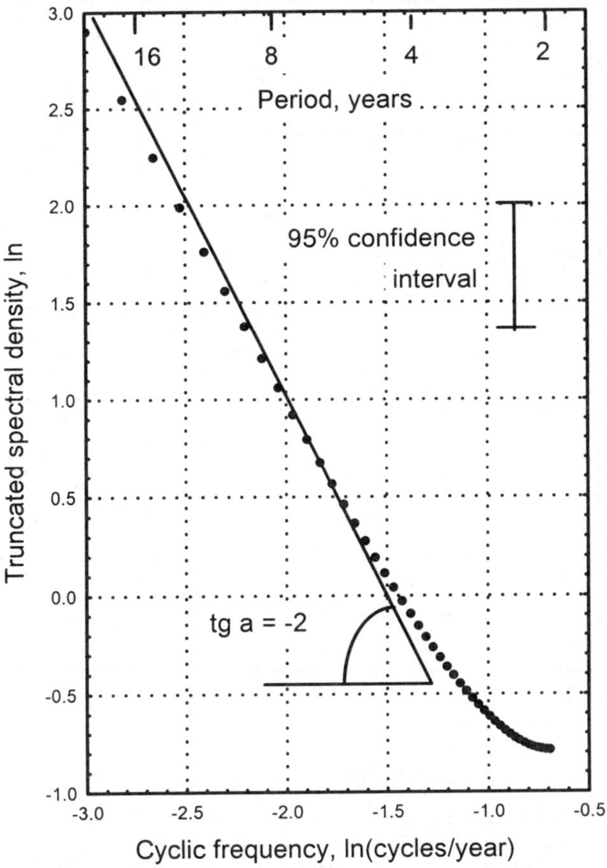

Fig. 3.52. Spectral density of the generalized mean global sea level series in bilogarithmic scales. *Straight line* denotes -2 spectrum

3.1.3.4
Stochastic Dynamical Models of Global Temperatures

The first stochastic dynamical model describing general observed features of global temperature spectrum was proposed by Lemke (1977). This model was an application of Hasselmann's stochastic climate model theory to the zonally averaged Budyko-Sellers energy balance model. Later, these results were criticized: "the models were very simple and the results have been much debated ..." (Frankignoul 1995). However, from our point of view, given existing uncertainties in oceanic vertical velocities and coefficients of oceanic vertical turbulent mixing (of one order of magnitude, see Sect. 1.4.1), the simplicity of a dynamic-stochastic model in this case is an advantage, and not a shortcoming.

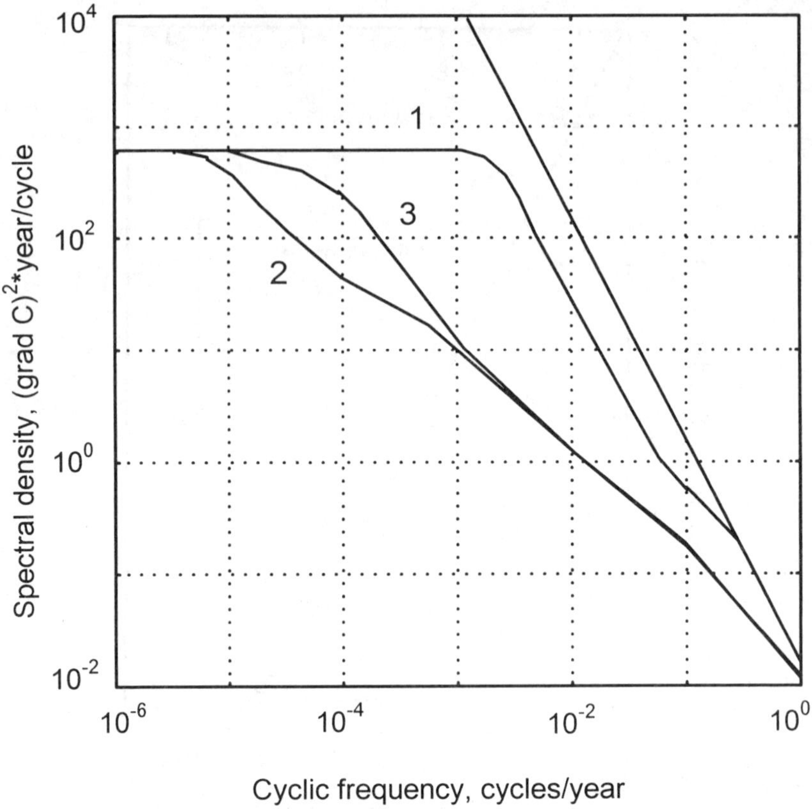

Fig. 3.53. Spectrum of the surface temperature (55°N). Zonally averaged global stochastic dynamical model by Lemke (1977). *1* One-layer oceanic model; *2* two-layer model with upwelling; *3* two-layer model without upwelling. (After Lemke 1977, p.364, Fig. 5)

This problem can be illustrated in the following way. Coupled large atmosphere (GCM)-oceanic models, in principle, can also be regarded as a sort of stochastic dynamical model: they are stochastized internally by synoptic and other high-frequency variabilities generated by the model itself. Theoretically, we could expect them to be an ideal variant of stochastic dynamical models, because they include the description of a largest possible number of physical processes. However, experiments on long-term global temperature simulations, using coupled models, do not satisfactorily reproduce the observed temperature spectrum.

We analyzed spectra of globally and annually averaged surface air temperatures as simulated by the six longest-running CMIP (Coupled Model Intercomparison Project) models (McCravy 1997), and the results were as follows. The mean tangent of the spectrum line (in bilogarithmic scales) between frequencies of 0.25 and 0.05 cycles/year, averaged over six spectra, was -0.87, i.e., approximately two times less than observed (Fig. 2.51). Values for specific models were: -0.83 (ECHAM1+LSG); -1.01 (ECHAM3+LSG); -1.31 (ECHAM4+OPYC); -0.89 (GFDL); -0.71 (NCAR GSM); and -0.47 (UKMO HADGM2). For lower frequencies, between 0.05 and 0.01 cycles/year, results were approximately the same.

Thus, we see that coupled models essentially underestimate (by two times and more) the rate of log-spectral growth with decreasing frequency and give an absolutely different law and pattern of global climatic variability, in comparison with the observed one. This means that coupled models dramatically (by orders of magnitude) underestimate the observed natural long-term variability of climate. It is clear therefore that the coupled models' ability to assess the role of natural processes in climatic changes is at least doubtful. In Chapter 1.4 we already listed possible causes of coupled models' shortcomings with respect to the description of long-term climate changes.

Returning to early, simple stochastic dynamical models, we can say that their skill in describing observed global temperature variability is much better than that of contemporary coupled models. In Fig. 3.53 model spectra of the surface temperature from Lemke (1977) are presented. It is interesting that the simpler the variant of the Lemke model, the better are the results: his one-layer model spectrum is much closer to the -2 line than spectra generated by two-layer models with and without upwelling.

3.1.3.5
Local-Global Polarization Phenomenon

In the above sections it was shown that, in general, our results corroborate Hasselmann's -2 degree law of climatic variability. At the same time, now we can say much more about the applicability of this law: it is true mostly for accumulation (state), globally averaged variables (parameters) of thick climatic subsystems: global sea temperatures, air temperature, sea level (also, within a certain range of frequencies, for the ice sheet volume - see Chap. 4). On the contrary, local accumulation (air and sea temperature, atmospheric air pressure, atmospheric water content) and transfer (heat and mass fluxes) parameters of thick subsystems demonstrate quite different stochastic patterns.

Thus, an important natural phenomenon takes place with respect to the stochastic properties of climate: a local-global polarization of stochastic models of the thick climatic subsystems. A brief description of the local-global polarization phenomenon is as follows (see Figs. 1.8 and 3.54).

On local scales (at meteorological and ocean stations, also within relatively small regions), temporal climatic changes in parameters of the atmosphere and the ocean are described, in the first approximation, by the fundamental first-order Markov process model with different values of its

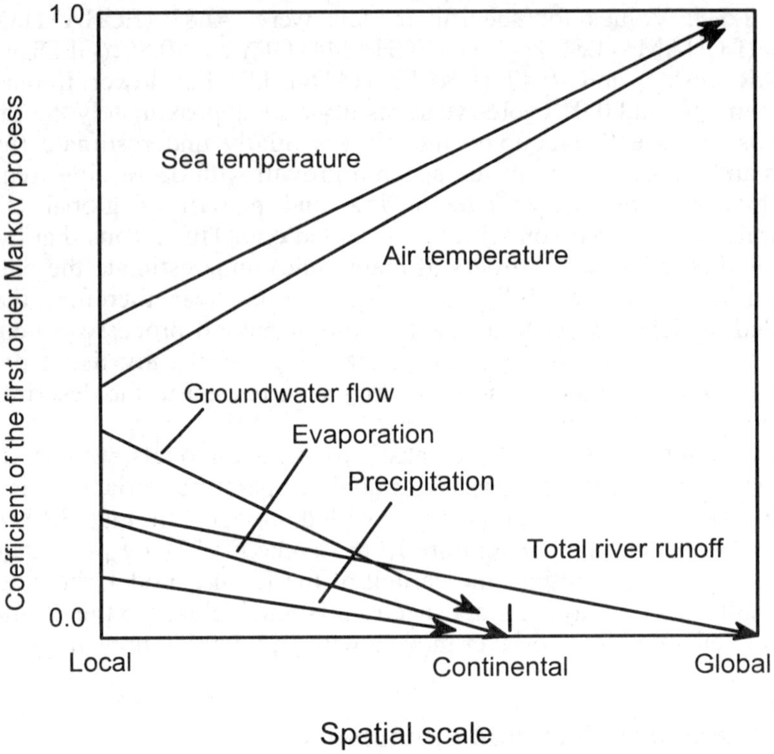

Fig. 3.54. Scheme of the local-global polarization phenomenon. *Arrows* denote changes in the first-order Markov process coefficient (annual time series) under changing spatial scales of climate process averaging

coefficient, or by more sophisticated models; but if we increase the spatial scale of the processes, i.e., if we take the time series of parameters (variables), averaged over large regions or globally, we see that the models of the thick subsystem variables are divided into two groups. Models of accumulation (state) variables - air and sea temperature, sea level, and some others - tend to the Wiener process, and the model coefficient becomes a unit coefficient. At the same time, models of transfer variables (precipitation, runoff, solid glacier runoff, evaporation, vertical fluxes of turbulent and latent heat), on a global scale, seem to become the white noise model, and the coefficient becomes zero.

This phenomenon is illustrated by Fig. 3.54. On a local scale, coefficients of the first-order Markov process [AR(1)], describing individual annual time series, vary within a large range from 0 to values close to 1. The mean values of the first-order Markov process coefficient are approximately 0.1 for annual precipitation series (Ratkovich 1976) and our estimations; 0.1 -0.2, 0.2, 0.33 for, respectively, evapotranspiration, total river runoff and groundwater flow series (Ibid); 0.3-0.5 for air and sea surface temperature (our estimations). Estimations of the coefficient on a continental scale are zero for time series of averaged precipitation and evaporation - our estimations using data by L.P.Kuznetsova (1991, pers. comm.) for North America and Europe; also zero for river runoff at global scale (Dobrovolski 1992) using data from (Mirovoi...1974). Values of the AR(1) coefficient are close to 1 for globally averaged sea and air temperatures, also for the global sea level (Sects. 3.1.3.1 and 3.1.3.2).

Thus, a net separation of accumulation variable models, on the one hand, and of transfer variable models, on the other, occurs while increasing spatial scales. In turn, intermediate-type models, the AR(1) models with intermediate value of the coefficient, disappear in the process of spatial averaging. In other words, we can say that global averaging acts like a filter that enhances only two, polarized types of stochastic behavior and suppresses other types of models.

Seemingly, the local-global polarization phenomenon is due to the spatial stochastic structure of atmospheric variables, which, in turn, is affected by the restricted radius of synoptic processes in the atmosphere (see Fig. 1.8). Indeed, ocean-atmosphere and land-atmosphere feedbacks depend on the coefficient of variation of the processes under consideration, which is the function of relation TA/C where TA is the mean value of anomalous heat or water flux between climatic subsystems or their parts, and C is heat or water capacity of the appropriate climatic subsystem (or its part). When we move from the local spatial scale to the global scale (i.e., if we increase the spatial scale of process averaging), we see that capacities of the parts of climatic subsystems increase almost linearly with increasing A, area under consideration. However, mean values of anomalous heat and water fluxes between subsystems increase with much less speed. The latter is approximately proportional to $A^{1/2}$, because of the restricted radius of spatial correlation of atmospheric processes. Finally, on a global scale, the contrast between capacities of climatic subsystems and anomalies of fluxes between them is enormous (in comparison with local processes). Thus, very small (resembling the white noise) globally averaged fluxes between the ocean and the atmosphere create very small relative changes in the total heat and water content of these climatic subsystems. Global-scale feedbacks do not appear under these conditions. Therefore, the temporal variability of global oceanic and atmospheric heat and water content is sufficiently described by the Wiener process model and the -2 degree law of climatic variability without feedbacks. The local-global polarization is a key mechanism of relations between local and global climatic changes (below we will provide some quantitative estimations related to this mechanism).

Before starting a description of thin layers, we would like to note that several interesting oceanic and atmospheric stochastic and dynamic-stochastic models are reviewed in an important paper (Frankignoul 1995).

3.2 Variabilities of Thin Climatic Subsystems

3.2.1 Analyzed Data

3.2.1.1 37 GHz Polarization Difference and Related Data

As stated above, global-scale data on thin earth layers - vegetation cover, floodable areas, snow cover, sea ice - have become available only recently. We analyzed different types of data: at local, regional, and global scales. Because of the relative shortness of available records, sequences of thin layer parameters were taken in the form of monthly time series.

The first type of analyzed monthly series, 37 GHz Polarization Difference temperatures (PD) (to obtain these data see Choudhury and DiGirolamo 1994) is divided into three groups. The first includes PD for characteristic floodable areas: the Manaus area (with a surface of about 10,000 km^2), the Ihla Bananai area of the Araguaia River, the Buenos Aires area within the La Plata basin, and the Paraguai Parana river area. The surface of the last three areas is about 490 km^2. It is supposed that 37 GHz PD series in these cases are related to the changes in flooding conditions: surfaces of flooded areas, heights of rivers, etc.

The second group of analyzed remote sensing data represents 37 GHz PD for nonfloodable areas: nonfloodable area to the SE of Ihla Bananai in South America; the Sahara desert in Egypt, (25.5 - 27.5° N, 25.5 - 27.5° E), the Sahel zone mostly in Niger (14 - 16° N, 5 - 15° E), and tropical savanna in northern Botswana (19 - 20° S, 25 - 26° E). In Choudhury (1989) and other papers it was suggested that time series of this type reflect changes in the density of the vegetation cover.

The last group of satellite-based data investigated are PD temperatures averaged for land surface within latitudinal belts of the global scale: 0 - 40° N and 55° S - 75° N (Fig. 5a,b).

Finally, we analyzed atmospheric carbon dioxide concentrations at Mauna Loa. The hypothesis was suggested in Choudhury and Di Girolamo (1994) that variations of this parameter, with the anthropogenic monotonous trend previously removed, fit the changes in biomass and water equivalent of the global vegetation cover which, in turn, are satisfactorily described by the global changes in land 37 GHz polarization difference temperatures. (See details in Dobrovolski and Choudhury 1996).

Each series above was analyzed in nine variants: (1) in its initial form, (2) with seasonal dynamics of monthly mean values removed, (3) with both seasonal dynamics of monthly means and monthly anomalies standards removed, and (4 - 9) divided into two segments of equivalent lengths. For Sahel 37 GHz PD monthly anomaly data, the series of month-to-month increments was also analyzed. In turn, for each variant of the analyzed series 40 variants of stochastic models were proposed. Thus, all in all, 117 variants of the time series were investigated, more than 4,000 variants of stochastic models were tested, and about 1,000,000 parameters of stochastic and statistical models were visually considered and compared.

3.2.1.2
Snow and Sea Ice Remotely Sensed Data

The following monthly satellite-based data sets have been used: land snow area summed for the Northern Hemisphere (IPCC 1993), and global sea ice area (P.Gloersen 1995, pers. comm.). A detailed description of appropriate methods for obtaining and processing remote sensing data can be found in Foster and Chang (1993; for snow cover), and in Gloersen et al. (1992), Parkinson and Gloersen (1993) (for sea ice). The lengths of the above series were, respectively, 228 months (Jan. 1973-Dec. 1991) and 96 months (Jan. 1979-Dec. 1986).

It is difficult to assess absolute values of errors in the above globally averaged data sets because of their unique character, i.e., because of the absence of appropriate time series based on conventional measurements. Nevertheless, we can use some analogs of the above parameters to estimate their accuracy. Thus, conventionally measured mean seasonal dynamics of the Northern Hemisphere snow area (although containing some errors, too) can be such an analog with respect to the accuracy of the estimations of long-term variations in the remotely sensed variant of this parameter. Data from Foster and Chang (1993) show that the mean difference in appropriate monthly estimations of the area of the Northern Hemisphere snow cover is about 12% of its maximum monthly value.

Data on sea ice area are, perhaps, the most correct, in comparison with remotely sensed data on parameters of other earth surface covers, because of the large contrast in the microwave emissivities of sea ice and liquid water. For example, the mean difference between local ice concentrations obtained from the SSMI (special sensor microwave imager) and those obtained from aircraft range from -3 to +5% (Cavalieri et al. 1991). In turn, ice-concentration differences between SSMI and SMMR (scanning multichannel microwave radiometer) had a bias in the means of 0.2%, and a standard deviation of 2.5 - 5% (Gloersen et al. 1992). For the purposes of the present work we used a 96-month (1979-1986) time series of the total global areas of squares with more than 50% surface coverage by sea ice from (P.Gloersen 1995, pers. comm.). This time series was obtained using observations from the Nimbus-7 SMMR and sea ice algorithm coefficients from SMMR-SSMI overlap data. Although there are no comparisons between global SMMR ice area estimations and conventional estimations, the consistency of local remotely sensed data and the comprehensive results of stochastic analysis of global sea ice areas are encouraging.

3.2.1.3 Related Satellite Based and Conventional Data on Global Air and Sea Temperatures

In order to compare variabilities in thin climatic subsystems with those of thick subsystems, the results of additional analysis of monthly air and sea temperatures series are described in this section. Unlike land surface parameters, global data sets on sea and air temperatures can be obtained by both remote sensing and conventional methods. It was shown in Nyoku and Brown (1993), Reynolds et al. (1989), and other works, that remotely sensed monthly sea surface temperatures (SST) still contain considerable errors, and a blended SST product, obtained by merging satellite and in situ measurements, is the best possible approximation. Evidently, in situ SST data sets are characterized by high errors also - see discussion on this subject in Dobrovolski (1992). Besides, if it is possible to estimate errors in local, satellite-based and conventional SST measurements, treatment of globally averaged estimations is extremely difficult because of their unique character.

Nevertheless, stochastic analysis of globally averaged SST series obtained using different methods from Reynolds et al. (1989: length of series was 84 months, January 1982 - December 1988) and Parker (1995): development of data is documented by Parker et al. (1994: 171 months, January 1981 - March 1995) shows that the difference between the mentioned variants of SST global series is not great with respect to the type of stochastic model and the form of the spectra. Anticipating discussion of the results, we would like to stress that both satellite-based and conventional global SST series as well as blended series, give the same discrete Wiener process model. This means that different variants of stochastic models of global SST are somewhat invariant and do not depend dramatically on specific errors of measurements. Moreover, the type of stochastic (Bernoulli-Wiener) models of global monthly SST anomalies is the same as the type of models identified for the series of global annual temperatures (Dobrovolski 1992).

The same can be said of the data on another extremely important parameter of the state of climatic system, the tropospheric temperature. In the present work we have analyzed microwave monthly global mean tropospheric temperatures from (Fifth ... 1994). The length of the series was 180 months (Jan. 1979 - Dec. 1993). As will be shown, the normalized spectrum of this time series practically coincides with the spectrum of oscillations of global, conventionally measured sea surface temperatures.

3.2.2 Comparison of Results for Remotely Sensed and Conventional Data

3.2.2.1 Comparison of Results on Local Scales

As stochastic analysis, in general, presupposes investigation of stationary or stationarizable time series, in the following parts of this work we shall deal mostly with normalized monthly anomalies (the third variant of the above-mentioned series processing). In some cases, analysis of initial time series, although incorrect from certain points of view can give some helpful information, and analysis of segments of the initial series is important. The problem of monotonous trends in time series will also be discussed.

In order to check the applicability of remotely sensed data to the analysis of land surface conditions, we compared statistical parameters calculated both for conventional (if any) and satellite-based data. Thus, we calculated stochastic and statistical parameters for both the Rio Negro river heights in Manaus and 37 GHz polarization differences (monthly anomalies normalized by appropriate monthly standards) for an area including Manaus. All in all, 12 independent parameters and 3 dependent parameters (excess, excess about normal, and season with relatively low month-to-month autocovariances) are indicated. Taking reasonable confidence intervals into consideration, the results for both time series practically coincided. In order to verify whether these results are stable, the same calculations were implemented by us for the halves of each series. Taking into account the bias of estimations due to the length of the time series (in the case of mode, fourth moment and excess) these results resembled both each other and those for the whole series. Also, seasonal changes in standards of anomalies and spectra of the two series were similar.

Generally speaking, a detailed comparison of stochastic and statistical properties of satellite-based (37 GHz polarization differences) and conventional (river gauge heights related to the surface of flooded areas) observations in the natural zone investigated (floodable area of tropical rain forests) shows that the main basic features of flood events can be obtained from remote-sensing analysis. This result can be relevant for regions with poor or absent network of conventional river rain gauges as well as for large regions.

3.2.2.2 Globally Averaged 37 GHz Polarization Differences Data. Concentration of Carbon Dioxide in the Atmosphere

It was shown in Choudhury (1989), Giddings and Choudhury (1989), Choudhury and Di Girolamo (1994), etc. that for nonfloodable areas, and for

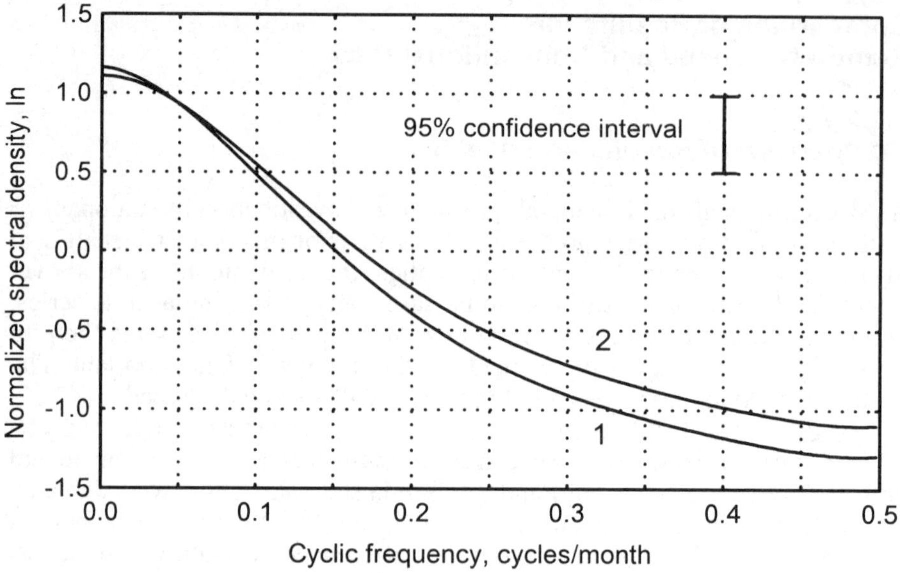

Fig. 3.55. Normalized spectral densities of monthly anomalies: *1* Mauna Loa CO_2 concentrations; *2* 0°-40°N 37 GHz polarization differences

large regions 37 GHz polarization difference data are related to the water equivalent of vegetation cover. It is stated in the papers cited that on the global scale this parameter can, in turn, reflect the concentration of carbon dioxide in the atmosphere. In the present work, we try to verify this hypothesis. For this purpose the same calculations as in the previous section were implemented for three series: Mauna Loa carbon dioxide concentrations with previously removed anthropogenic trend, and 37 GHz data averaged for land within the latitudinal belt O° - 40° N and 55° S - 75° N. As in the case of remote sensing data for Rio Negro river heights and the Manaus area, parameters for these three series practically coincided. For example, the median for 37 GHz data (0° - 40° N) and for carbon dioxide is - 0.1 and 0.0, respectively; mode: 1.6 and 1.6; fourth moment: 2.0 and 1.9; half the year with relatively high month-to-month autocovariances: "cold" (October through March) and "cold"; difference of appropriate autocovariances for cold and warm period: 0.2 and 0.1; order of stochastic models: 1 and 1; spectral peaks for series without removing seasonal dynamics: only on frequency 1 cycle/12 months in both cases. Also the first trial coefficient for both processes was practically the same: 0.50 and 0.55 (for 55°S - 75°N PD series estimation of this coefficient is 0.55).

In Fig. 3.55 spectra of the first order for 0-40°N GHz polarization difference and the Mauna Loa series are compared. It is evident that the two spectra are similar. The resemblance between the 55°S-75°N GHz PD spectrum and the Mauna Loa spectrum is even better: they coincide absolutely.

As conclusion to this part, the following hypothesis can be suggested. Even if current values of 37 GHz data and values of conventionally observed hydrological and related parameters do not always coincide, their general (stochastic) patterns can be similar. Above we considered, as a case study, two extreme variants of spatial scales: the local one (Manaus) and the global one (global vegetation water equivalent obviously related to soil moisture, carbon dioxide concentration and other parameters) and showed that these suggestions can be corroborated using parametric analysis of time series. Below we will apply this hypothesis to the 37 GHz data for regions of different types and areas, and try to investigate general features of the relevant processes.

3.2.3
Results of Stochastic Analysis of Local and Regional Hydrological Changes

3.2.3.1
Results of 37 GHz PD Data Analysis for Floodable Areas

Some results related to the floodable areas have already been discussed above. Here, we continue with discussing results for floodable areas with more pronounced interannual variability.

One of the most important features of time series for many floodable areas is the pronounced seasonal dynamics in the main parameters of the processes. For instance, in Fig. 3.56 seasonal dynamics of month-to-month normalized autocovariances is presented (sample estimations). Obviously, seasonal changes in both parameters for La Plata and Bananai regions are large, even if we take into consideration the limited lengths of the series and appropriate large errors in sample estimates. As a result of these and other calculations, a hypothesis can be suggested that first-order autoregressive models with seasonal changes in the coefficient can fit, in the first approximation, many 37 GHz PD series. On the other hand, it is clear that 37 GHz polarization data for floodable areas where floods do not occur every year (unlike floods within the Manaus area) give more statistically significant information. Yet, information for regions with irregular floods is much more important. Perhaps these results can be relevant for elaborating a strategy for future monitoring within the EOS and similar projects.

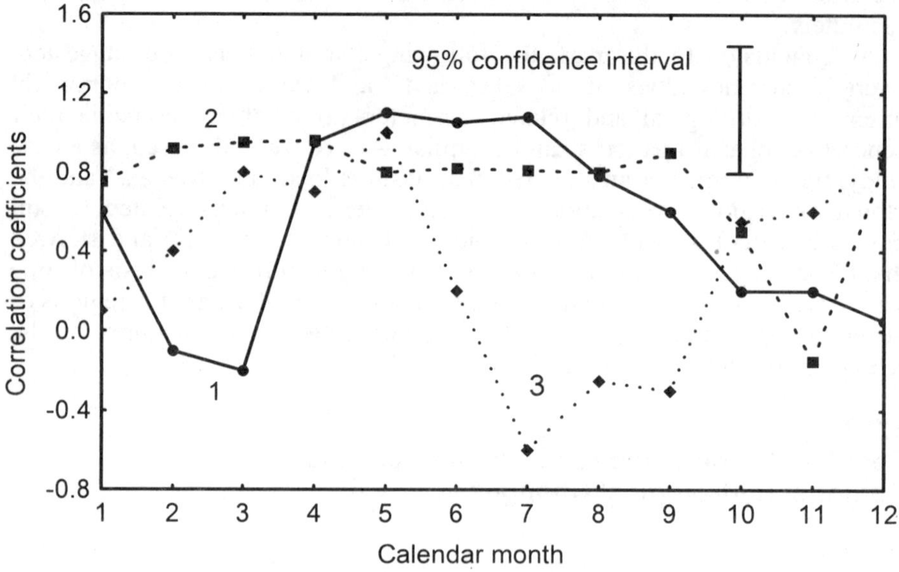

Fig. 3.56. Seasonal changes in month-to-month correlation coefficients of 37 GHz polarization anomalies for floodable areas: *1* La Plata region near Buenos Aires; *2* Parana region, Paraguai; *3* Bananai region (Araguaia river). *Vertical bar* denotes mean error of estimations

3.2.3.2 Results for 37 GHz PD Data on Vegetation Cover in Different Natural Zones

Comparison of Results for Sahara, Tropical Savanna, and Sahel Regions. Analysis of data of this kind seems to be one of the most interesting. Indeed, information on floods can be obtained using conventional measurements (although they are not available in many regions), whereas information on water equivalent of vegetation cover, related to soil moisture, for large regions can be obtained only by using satellite-based methods.

As stated above, we considered 37 GHz polarization difference data for three relevant regions with different natural conditions: the area within the Sahara desert in Egypt, the tropical savanna region within Botswana, and the Sahel region in Niger. We started our analysis by estimating spectra of the analyzed series without removing seasonal dynamics of mathematical expectations and monthly standards. It is interesting that the most humid region (tropical savanna) demonstrates the sharpest 12-month spectral peak (with no other significant peaks). At the same time, there are no statistically significant peaks in the Sahara spectrum, and peaks on 1 cycle/12 months and zero frequency for the Sahel series cannot be significantly distinguished from each other.

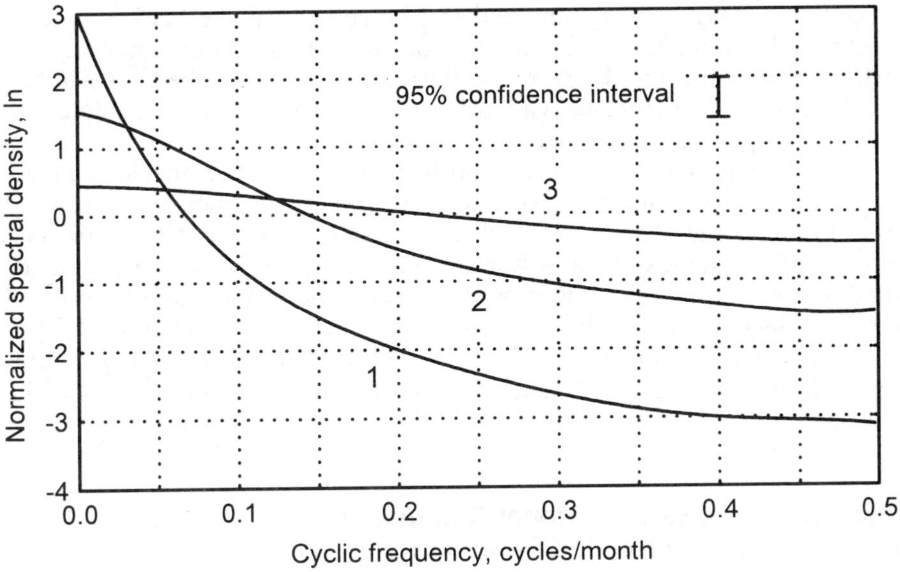

Fig. 3.57. Spectral densities of 37 GHz polarization difference monthly anomalies: *1* Sahel, *2* tropical savanna, *3* Sahara.

Seasonal dynamics of monthly standards and autocovariances is of great interest. This dynamics is much less pronounced in comparison with floodable areas. If we consider confidence limits of a reasonable level of significance, seasonal dynamics of standards and month-to-month covariances is not important. Thus, in the first approximation, we can use a model of a stationary random process for the description of monthly anomalies of the three processes investigated, and obtain estimations of spectral densities and other appropriate parameters.

Detailed analysis of the three series shows that their monthly anomalies are well described by the Markov first-order processes. Appropriate spectral densities for the three regions are shown in Fig. 3.57. It is relevant that more humid (tropical savanna) and more dry (Sahara), thus, more "stable" regions possess flatter spectra than that of the Sahel. In turn, the Sahel spectrum demonstrates an extremely high peak at low frequencies. Perhaps this is the property of “intermediate” regions.

The Sahel series, from our point of view, is of great interest, and its analysis gives surprising results. It can be described, in the first approximation (if we use a stationary model), as the first-order Markov process with coefficient 0.91. This value is close to 1, which is why it is important to verify whether a model with a coefficient higher than 0.91 or another type of model better fits this series. For this purpose it is useful to consider estimations of correlation coefficients. The Markov process with coefficient 0.93 better fits the form of sample autocovariances (especially when the time lags are large: more than 13 months) than the Markov process with sample estimation of the first coefficient and the Markov process with iterative estimation of the first coefficient (0.91). Thus, the first Markov coefficient tends to increase (values 0.89; 0.91; 0.93) while proceeding to a better description of the process.

Another important fact in the identification of the Sahel stochastic model is solving its characteristic equation of autoregression. For all variants of model orders, one of the roots is always close to or practically on the unit circle on the complex plane. It was shown in Chapter 2 that in this case the time series is close to a nonstationary, but stationarizable one: the series of the first increments of the process investigated can be described by a stationary model. Thus, the model for Sahel humidification changes seems to be close to the discrete Wiener process, at least, at the time scales under consideration.

3.2.4 Results of Analysis of Global Changes in Hydrological and Related Parameters

In Fig. 3.58 truncated normalized spectra of state parameters for the atmosphere and ocean are shown (recall that appropriate time series are taken in the form of monthly anomalies, and some of them are satellite-based, in order to compare results with those for thin earth layers). For most of the frequency range spectra practically coincide. The divergence among spectra becomes statistically significant for frequencies less than approximately 0.03 - 0.04, i.e., only for 6-8% of the total frequency range where estimations of the individual spectra become unreliable because of the limited length of time series.

Thus, Fig. 3.58 shows that spectra of different records of global atmospheric and oceanic temperatures obtained using different methods coincide with each other and with the spectrum of the discrete Wiener process. In order to check the applicability of this model to the description of investigated processes, one must also process series of appropriate month-to-month increments. In accordance with the Wiener model, they are expected to be noncorrelated in time, i.e. to demonstrate zero spectrum. Indeed, calculations show that time series of increments relating to thick climate subsystems are described by the white noise process or are close to it.

Truncated normalized spectra of parameters of the thin subsystems, snow area in the Northern Hemisphere (which includes almost all seasonally variable snow cover on the earth), mean global vegetation cover density (37 GHz PD), and global sea ice area are also shown in Fig. 3.58. As in the case of thick subsystems, the spectra for the thin subsystems in Fig. 3.58 coincide with each other, taking into consideration the confidence limits of estimates, apart from zero and neighboring frequencies. Monthly anomalies of the

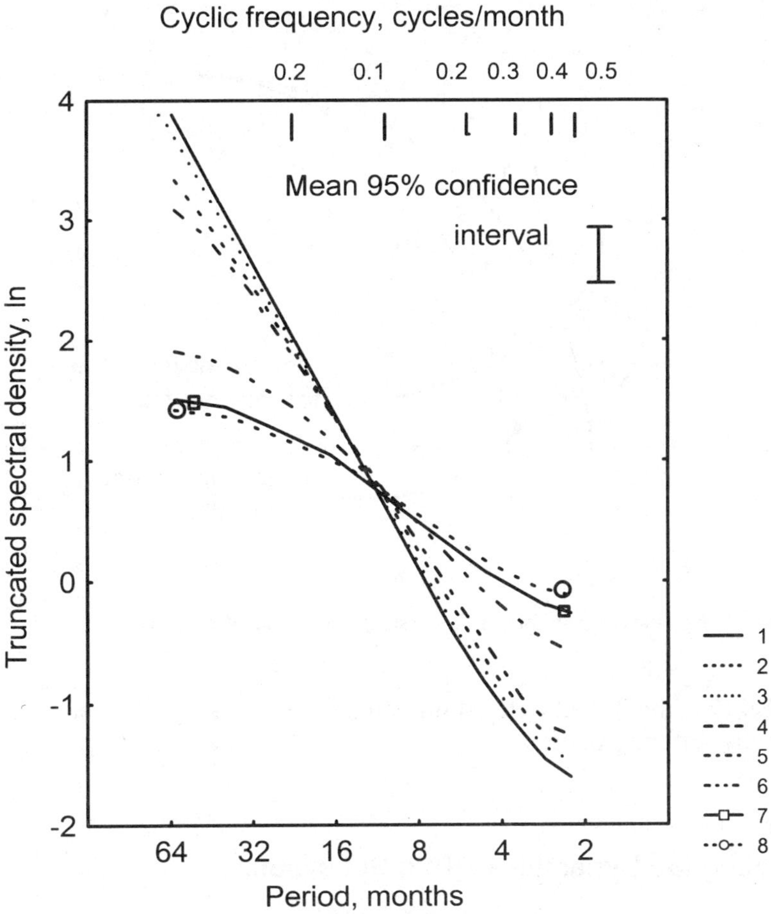

Fig. 3.58. Spectra of variables of thick subsystems (1-5), and variables of thin subsystems (6-8) in bilogarithmic coordinates. *1* Realization of the discrete Wiener process; *2* global microwave air temperature; *3* satellite sea surface temperature anomalies for the Northern Hemisphere (0-60°N); *4* global in situ sea surface temperature anomalies; *5* blended (satellite and in situ) mean sea surface temperature anomalies for the Northern Hemisphere (0-60°N); *6* global sea ice area; *7* global 37 GHz polarization difference (density of vegetation cover); *8* Northern Hemispheric snow area. Data from: Reynolds et al. (1989) (*3*, *5*); IPCC (1993) (*8*); Choudhury and Di Girolamo (1994) (*7*);USDC (1994) (*2*); P.Gloersen (1995, pers. comm.) (*6*); Parker (1995) (*4*). *Straight line* denotes the -2 degree spectrum

investigated parameters of global thin subsystems , unlike those for thick subsystem parameters, are satisfactorily described by the Markov first order process with coefficient 0.5 - 0.6.

The spectrum corresponding to the -2 degree law is also presented in Fig. 3.58. It is evident that the first group of spectra is close to the straight line with tangent - 2 and thus corroborates the minus two law of climatic

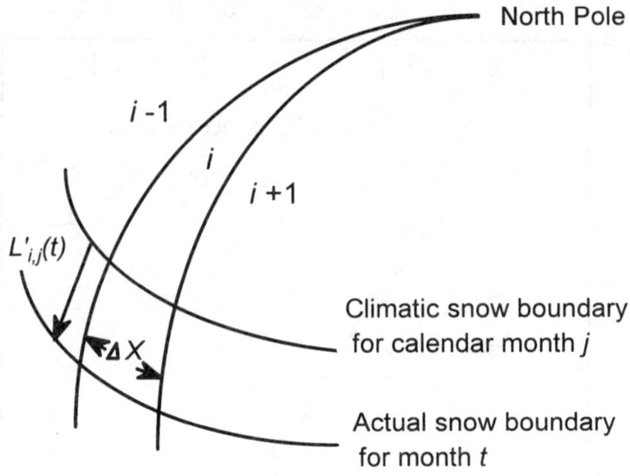

Fig. 3.59. Schematic diagram for the simple stochastic model of snow cover

variability without feedback. At the same time, thin subsystem spectra give a distinctly differing pattern.

3.2.5 Modeling the Dynamics of Thin Subsystems

In order to illustrate and understand the main features of local and global changes in the major state parameters of thin climatic subsystems, the following simple stochastic model can be suggested.

The main thin subsystems are confined to one (snow cover) or two (sea ice) latitudinal belts. In principle, global vegetation cover also can be presented as two latitudinal belts: within midlatitudes of the Northern Hemisphere and within equatorial and tropical latitudes. We will consider as a first example snow cover which has only one unstable, southern boundary.

Consider a meridional strip i (Fig. 3.59) with east-west width ΔX approximately equal to the characteristic size of atmospheric synoptic eddies, i.e., about 1,000 km. Investigations summarized in Dobrovolski (1992) show that the correlation radii of monthly anomalies of atmospheric temperature, precipitation, air humidity, and horizontal atmospheric vapor transport are of similar magnitude. Assume that within this strip and near the climatological snow boundary, conditions at the land surface are homogeneous.

Initially, ignore seasonal variations and suppose all months to be a specific calendar month j. Thus, we are looking for a stochastic equation for the anomaly of the distance between the North Pole and the southern edge of the snow cover, $L'_{i,j}(t)$, where t is time. As stated above, intermediate (about 1 month) and longer time scales are investigated, so 1 month is taken as the temporal interval.

On intermediate time scales there are considerable negative feedbacks in the interactions between the atmosphere and the underlying surface (Adem 1991a). If the boundary of snow, sea ice, or vegetation cover in a specific month differs substantially from its climatology, there will be a tendency to return towards normal through the mechanisms of radiation processes, evaporation and evapotranspiration, vertical turbulent heat exchange, large scale ocean turbulence (for sea ice), snow and ice melting or formation, and vegetation degradation or expansion (Piterbarg 1985; Dobrovolski and Rybak 1992). For the snow cover as well as for the other main thin climatic subsystems, this feedback in its simplest linear form can be described by the following equation:

$$L'_{i,j}(t) = k\,L'_{i,j}(t-1), \tag{3.22}$$

where values of approximately constant coefficient k lie between 0 and 1.

Now superimpose an atmospheric forcing onto Eq.(3.22). Following the concept of two-scale weather-climate separation by Hasselmann (1976), we will consider an atmospheric forcing term $a_{i,j}(t)$ at intermediate temporal scale as a temporally uncorrelated random variable. The mathematical expectation of $a_{i,j}(t)$ (white noise) for specific i and j will be zero.

A simple model of hemispheric snow cover proposed by R.D.Koster (1995, pers. comm.) and empirical models of the same type for other thin subsystems can give an idea of how to construct specific equations for $a_{i,j}(t)$. For example, in Koster's model, the temperature-related forcing of the snow cover is either the area (or extent) of anomalous air temperatures where there has been no snow, or the snow area under the air with positive temperatures. In turn, the meridional extent of air temperature anomaly can be written in terms of the mean climatological air temperatures and the magnitude and sign of the actual anomalies at each latitude. Yet, as stated above, we try not to construct specific exact equations for each thin subsystem but to present a common simple approach to the type of equations governing their variabilities.

Thus, the equation for the snow extension anomaly for month t will be:

$$L'_{i,j}(t) = k\,L'_{i,j}(t-1) + a_{i,j}(t). \tag{3.23}$$

As ΔX is chosen to be approximately equal to the horizontal correlation radius of atmospheric processes on intermediate temporal scales, the $a(t)$ for different longitudes will be uncorrelated and will result in a global atmospheric forcing $A(t)$ which is supposed to be white noise also. The result of the integration along the lateral boundary of the cover can be represented as follows:

$$S'_j(t) = k S'_j(t-1) + A_j(t)\,, \qquad (3.24)$$

where S'_j is the monthly anomaly of hemispheric snow cover area; the global variance of atmospheric forcing $A_j(t)$ is equal to the sum of local variances $a_{i,j}(t)$.

Table 3.3. Seasonally changing coefficients of autoregresssion k of global monthly snow anomalies. Calculations were made using data from IPCC (1993)

Season	Coefficient k	Coefficient minus mean annual value
Winter	0.55	-0.06
Spring	0.74	0.13
Summer	0.74	0.13
Fall	0.45	-0.16

It is evident that, in reality, (3.24) is an equation of a stochastic process with seasonally varying parameters. Nevertheless, analysis of observational data suggests that seasonal changes in coefficients k are not dramatic, at least they have an order of the error bars on k itself. As an example, estimations of coefficient k for global snow area anomalies and for different seasons are presented in Table 3.3. It is clear that the differences between the estimations of k for different seasons and its mean annual value are comparable with the mean error of these estimations (about 0.14). Seasonal dynamics in standard deviations of $S_j(t)$ and $A_j(t)$, can be removed using the procedures described in Chap. 2. Finally, we obtain an equation for seasonally normalized monthly anomalies of the areas of thin subsystems:

$$s'(t) = k s'(t-1) + a'(t)\,. \qquad (3.25)$$

The form of Eq. (3.25) is thus identified by us as appropriate for the description of changes in the global area of climatic thin subsystems and related parameters. Using the above considerations, we can understand the difference between models for the main parameters of thin and thick subsystems.

3.2.6
Local-Global Polarization Phenomenon and Thin Climatic Subsystems

Two conditions are necessary for the existence, on the global scale, of stochastic processes with negative feedback related to the interactions of the atmosphere and underlying surface at intermediate and longer time scales. Firstly, a negative feedback of the above type must exist on the local and regional scale; secondly, these processes must not be "lost" in the process of global averaging; in other words, the relative range of global variations governed by the negative-feedback mechanisms must be significant.

It is convenient to measure the relative variability of the global processes by their coefficient of variation, i.e., the standard deviation of the parameter divided by its mean value. After the global summarizing of Eq. (3.23) which describes regional snow area anomalies, the standard deviation of the anomaly will increase by $(B/\Delta X)^{1/2}$, and the mean value of the snow area will increase by $B/\Delta X$, where B is the characteristic length of the variable part of the snow cover boundary. Thus, the coefficient of variation will decrease by:

$$CVD = (B/\Delta X)^{1/2}. \tag{3.26}$$

Estimations of mean values of parameter *CVD* for the snow cover area as well as for the sea ice area and the vegetation aerial density (calculated using the same formula) are presented in Table 3.4. The coefficient of variation for these climatic subsystems will decrease only by three to five times, and the process (3.25) will still be "visible".

Table 3.4. Parameters of global decrease of coefficient of variability *CVD* and CVD^2 for different climatic subsystems. Estimations using Eq. (3.26) for thin and Eq. (3.27) for thick subsystems

Subsystem	*CVD*	CVD^2
A. Thin subsystems:		$B/\Delta X$
Snow cover area	3.5	12.2
Sea ice area	4.3	18.3
Vegetation cover spatial density	5.1	26.0
B. Thick subsystems:		A/S_X
Upper oceanic layer heat content	21.3	454
Tropospheric heat content	25.5	649

The local and regional variations of the state parameters of thick climatic subsystems like sea surface temperature anomalies or air temperature anomalies, at intermediate and larger temporal scales, are in the first approximation described by the same type of model with negative feedback (Adem 1970, 1991a; Dobrovolski and Yarosh 1980; Dobrovolski 1992, 1994). However, global integration of the ocean surface temperature and air temperature (more precisely, the heat content of the troposphere and oceanic upper layer) is carried out over the whole surface of the ocean. In this case, the approximate formula for the parameter of decreasing of variation coefficient will be:

$$CVD = (A / S_x)^{1/2}, \tag{3.27}$$

where A is the total area of the troposphere or the total oceanic area, and S_x is the area of a circle with diameter equal to the characteristic integral spatial correlation scale ΔX (approximately 10^3 km). Estimations of CVD for thick subsystems are also presented in Table 3.4, and are equal to 21 and 26.

Thus, the globally averaged stochastic variations of primary variable of the thick subsystems such as mean oceanic and atmospheric temperature, of the negative-feedback type [Eq. (3.25)], are very weak and perhaps almost nonexistent at the investigated time scales. In this case, other mechanisms (radiative processes, volcanic eruptions, processes involving deep oceanic layers or global ENSO-type events, different nonlinear processes, anthropogenic effects) can play important roles. Yet, for the temporal scale under consideration, all these mechanisms, together with the processes of interactions within the climatic system, appear to create only small relative changes in the globally averaged thick subsystems whose behavior is therefore described, in the first approximation, by the no-feedback Wiener--type processes (Hasselmann's -2 degree law of climatic variability without feedback).

In other words, the behavior of the parameters describing the thin and thick climatic subsystems can be explained as follows. In the case of thin climatic subsystems, only a few groups of cyclones and anticyclones near the lateral frontier of the cover are important. That is why temporal changes of globally averaged parameters (e.g., area of snow cover) have the same features as the changes at regional level within the size of an individual eddy. At the same time, the mean states of the ocean and the atmosphere at the intermediate and longer temporal scales are determined by such a large number of atmospheric and oceanic eddies that the variation of the mean state is related to the specific features of none of these groups of eddies.

Consequently, the characteristic timescale or integral time scale (Yaglom 1987) of the changes in land snow cover area and in the areas of other thin subsystems is equal to approximately 1.5 months and is obviously related to the intermediate time scale (or a frequency at which the spectrum of

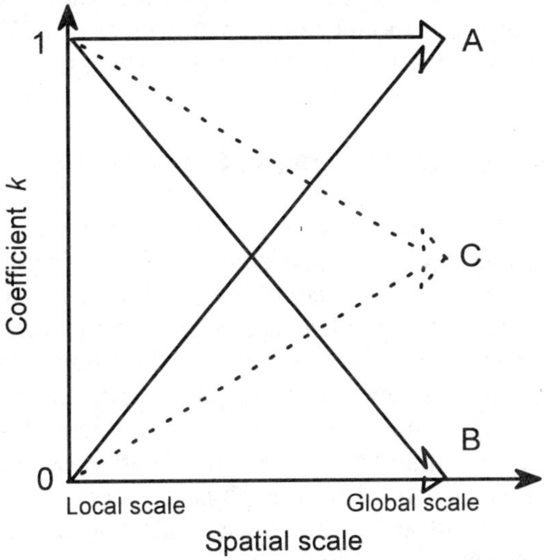

Fig. 3.60. Illustration of the local-global contrasts in temporal changes in the climatic subsystems. *A* Bulk (temperature, water equivalent) parameters of thick and thin subsystems; *B* transfer parameters of thick and thin subsystems; *C* area or density of thin subsystems; *k* coefficient of the Markov first-order process

synoptic variability becomes flat), whereas the characteristic time of changes in mean global temperature seems to be at least of the order of dozens of years (we do not exclude that in certain cases local events of, say, ENSO type, volcanic eruptions, etc. can affect global temperature at other time scales).

Strictly speaking, in the real climatic system atmospheric forcing on a local scale, $a(t)$ differs slightly from white noise because of inertia within the combined atmosphere-surface system. Nevertheless, it can be shown that the different character of spatial integration over thick and thin subsystems would lead to the above described difference in local-global patterns in this case also.

Finally, Fig. 3.60 hypothetically illustrates the local-global contrasts for thin and thick subsystems. At local and regional scales, intermediate-scale (monthly) temporal variations of different parameters of climatic subsystems are described in the first approximation by the first-order Markov process with different values of coefficient k in the range (0, 1). As we increase the scale of spatial averaging of the process, k tends to 1 for bulk (accumulation) processes, e.g., temperature of ocean and atmosphere, perhaps water equivalent of land snow cover (this last hypothesis is suggested by the results of stochastic experiments on a 3D model of the global hydrological cycle, Dobrovolski et al. 1995), to 0 for transfer processes (increments of accumulated water or heat), and to some intermediate value for the area of thin covers (land snow, sea ice, vegetation cover). Respectively, the model tends to the model of Wiener process, to white noise, or remains the first-order Markov process.

3.2.7 Discussion on the Global Climatic Subsystems

The above analysis demonstrates, from our point of view, the existence of global climatic subsystems whose behavior differs from that of both the atmosphere and the ocean: snow cover, vegetation cover, and sea ice. These three earth covers are not as thick as the ocean and the atmosphere, the areas of their spatial extent are relatively small, and their boundaries are extremely unstable. Unlike the ocean and the atmosphere, their most significant global parameter is not temperature, heat, or water content, but area of spatial extent. Although at small temporal and spatial scales these climatic subsystems are governed by different physical processes, their behavior at climatic time scales and at the global spatial scale has essential common features. The temporal variability of their global areas is much more stationary than that of the variability of global air and ocean temperature, and the relative role of the high frequencies of their oscillations is much more important than that of global temperature variations.

Only analysis of remotely sensed data on the state of thin climatic subsystems can bring answers to important questions: what are the best-fitting stochastic models for the description of their variabilities? Does their behavior differ significantly from the behavior of the global temperature, and what is the stochastic model for the latter? How does the whole climatic system work? We hypothesize that the answers to these questions can be as follows.

1. The fundamental model of the stationary first-order Markov process with different values of the coefficient (globally averaged mean value is 0.4-0.6) governs local monthly anomalies of the thin climatic subsystems area variations. The mean integral time scale (correlation time) of local variations in snow cover, sea ice, and vegetation cover is about 1.5 months, and is evidently related to the characteristic time scale (about 1 month) of the internal atmospheric forcing. Surprisingly enough, the same stochastic model of the first-order Markov process, with the same mean value of the coefficient, is valid for the description of temporal variability in the monthly anomalies of globally averaged areas of thin subsystems. Thus, a local-global invariance (or the lack of local-global contrast) of the models and their coefficients takes place in the case of the thin subsystems.
2. On the contrary, a sharp contrast between, on the one hand, local and, on the other, global variabilities of primary (accumulation) parameters of all

thick climatic subsystems, takes place. On the local scale time series of monthly anomalies of air and sea surface temperatures are governed, in the first approximation, by the same stationary first-order Markov process model with different values of coefficient (for local air temperatures this coefficient is relatively small). But time series of globally averaged sea surface temperature and global air temperature are essentially nonstationary, and are very close to realizing the discrete Wiener process model. By the way, no statistically significant deterministic monotonous trends were found by us in the analyzed global and local time series; there were only nonstationarities in the mean variances of the processes, of the type of random walk.

3. The mechanism which governs the above processes can be described as follows. In accordance with Hasselmann's two-scale weather-climate separation hypothesis, synoptic processes in the atmosphere create at climatic frequencies (1 cycle/month and less, i.e., far from main synoptic frequencies, 1 cycle/several days) a variability which has much more flat spectral than the form of spectra near synoptic frequencies. Correspondingly, this random "weather" forcing on the climatic scale possesses only small temporal correlation. Specific mechanisms at the sea and land surface, which transform this initial input forcing, involve different physical processes: evaporation and evapotranspiration, vertical turbulent heat exchange, snow melting, etc. But the result of the transformation of the atmospheric forcing, at intermediate (1 cycle/month) and lower frequencies, in the first approximation, is possibly similar: input and output "signals" form a sort of negative feedback. Thus, anomalous heating of the sea surface during a specific month (for instance, by anomalous winds and currents from south) causes anomalous evaporation, vertical turbulent cooling of the surface, and anomalous radiative vertical fluxes. In turn, these mechanisms tend to smooth and, eventually, to eliminate the initial anomaly. Just the same happens at the land surface with the thin climatic subsystems. Abnormal snow or vegetation cover eventually submit to climatic conditions which are not favorable for maintaining snow or vegetation and tend to return the situation to the norm by the mechanisms of, respectively, snow melting and droughts.

Mathematically speaking, these processes of, say, linear negative feedback, are described by the first-order Markov process (first-order autoregression process) with coefficient between 0 and 1. Because of the relatively small area of the most important in these cases, lateral parts of thin subsystems (covers), global averaging of snow or vegetation variations involves only a few groups of independent atmospheric eddies that cannot totally eliminate the anomalous effects of each other. Finally, the negative feedback mechanism and appropriate stochastic model is still "seen" and works on the global scale.

On the contrary, global averaging of the sea surface and air temperature is performed over the whole area of the ocean or the globe. During a specific month, hundreds of independent anomalous groups of atmospheric (and oceanic) eddies form global mean air and sea surface temperature, and eventually almost eliminate the anomalous effect of each other. At the same time, the heat capacity of the ocean and the atmosphere grows almost linearly with the area of the part of thick subsystems under consideration. Under these conditions, during the procedure of global averaging, a global negative feedback mechanism of the above type governs only small relative anomalies. Finally, on the global scale, it is practically destroyed by the great number of random local nonlinearities. Residual global temperature anomalies, without sufficient feedbacks, follow the pattern of the random walk with independent temporal increments.

At first sight, realization of this nonstationary process resembles a process with a monotonous deterministic trend. However, the difference between these two types of nonstationarity is crucial, and is easily detected using the analysis of temporal increments. The mean value of the temporal first increments (differences) of the process with monotonous deterministic trend is statistically significant, i.e., the absolute value of the mean increment is more than the error of its estimation. At the same time, the mean value of the first increments of the discrete Wiener process does not differ statistically from zero.

3.3 Changes in Local and Discrete Climatic Objects

3.3.1 Rivers and River Runoff

Historically, land surface hydrology was the first earth science where stochastic methods and models were massively used. A review of early applications of stochastic models in hydrology can be found in the fundamental monograph by Kritsky and Menkel (1950). Seemingly, these authors were the first who used the first order Markov model for the description of river runoff variations (this model is often called the simple Markov chain in hydrology). Later, a large variety of models were proposed; however, the first-order Markov process is still considered as a good first approximation to the modeling of annual river discharges. Taking into consideration that stochastic modeling in hydrology was described in detail in many review papers and books (e.g., Ratkovich 1976, 1993; Muzylev et al. 1982; Ratkovich and Bolgov 1997) and also was considered in the previous monograph by the author (Dobrovolski 1992), here we will give only a brief description of the main results relating to this subject.

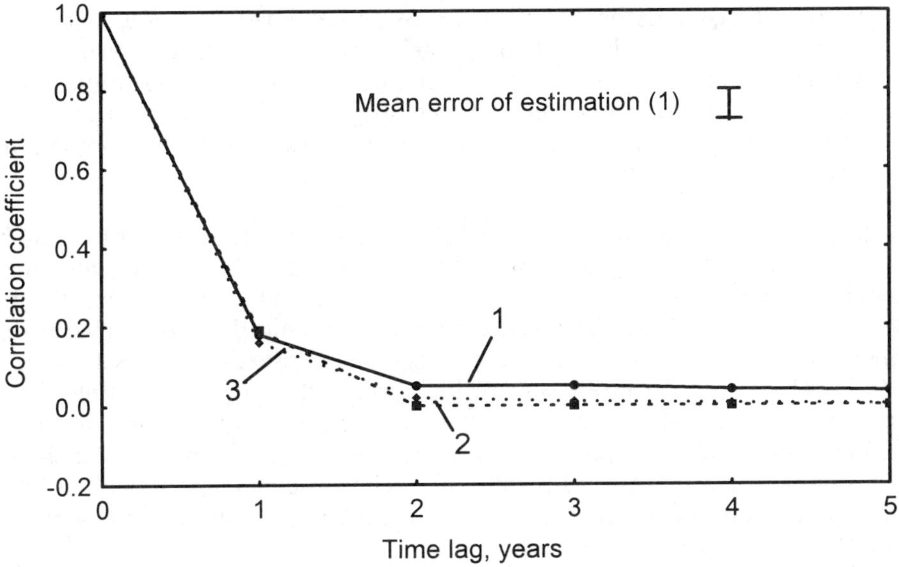

Fig. 3.61. Correlation coefficients of annual river runoff series. *1* Mean estimation for 344 observed runoff series (Ratkovich 1976); *2* and *3* theoretical estimations using two variants of a stochastical dynamic runoff model, (Dobrovolski 1992)

In Ratkovich (1976, 1993) over 400 time series of annual river discharges all over the world were studied. Also about 120 time series of large rivers' annual and monthly discharges - data from NASA (1995) - were processed by the author of the present monograph. Results of these studies were similar and can be summarized as follows.

1. If we take into consideration only those fragments of series where man--made changes of the runoff (dams, irrigation, etc.) are supposed to be not very important, variations in annual river discharges can be regarded as stationary. Changes in the mean value (mathematical expectation) of runoff do not exceed possible errors of these estimations. The same conclusion can be made with respect to the stationarity of the correlation function and spectral density.
2. The correlation function of an annual runoff series falls off so rapidly (Fig. 3.61) that only correlations with a time lag of 1 year can be regarded as statistically significant. Globally averaged correlation coefficient with unit time lag $\overline{R}_1(1)$ equals approximately 0.2. Rivers of dry regions manifest larger $\overline{R}_1(1)$, seemingly because of a larger role of groundwater component in the total runoff.
3. The first-order Markov process model (the simple Markov chain) usually can be considered as a good first approximation to the description of

annual runoff series. The probability density function of river discharges is strongly asymmetrical, so that somewhat better results can be obtained if we use not initial, but appropriately transformed runoff series. For instance, transformation of initial runoff into runoff probabilities (or normalized probabilities) removes the above-mentioned asymmetry, and processing of such time series is more consistent with the correlation theory of random processes.

In Fig. 3.61 correlation coefficients of annual runoff, averaged over 344 rivers of the world, are presented. In order to describe theoretically the globally averaged stochastic pattern of individual river runoff, in Dobrovolski (1992) we proposed a simple stochastic dynamical model of annual river discharges, based on a conceptual watershed model (Salas et al. 1981). The following processes are included in the simplest possible way into the model: precipitation, evapotranspiration, infiltration, surface, groundwater, and total flow, and water accumulation within the watershed. To obtain globally averaged results of stochastic dynamical modeling, we also used global, time-averaged data on the components of land surface water balance from Lvovich (1986).

The resulting theoretical globally averaged autocovariances of the total river runoff are shown in Fig. 3.61 (curves 2 and 3 correspond to two variants of the stochastic dynamical model). It is evident that theoretical values of $\overline{R}_1(1)$ perfectly well fit the observed data. It is shown in Dobrovolski (1992) that the groundwater component of the runoff, on average, plays the key role in forming the specific correlation and spectral structure of runoff changes. Also a simple stochastic dynamical model of monthly river discharges is proposed in Dobrovolski (1992), which does not need much a priori information about hydrological parameters; when compared with observational spectra, this model proved effective.

The problem of river runoff, averaged over large regions and globally, is very important while considering the role of this component of world water balance in stochastic structure of climate change. Seemingly, data on spatially averaged runoff from the fundamental monograph (World Water Balance ... 1974) still remain the most accurate. We processed about 30 time series of total runoff, averaged over large regions, continents, and the globe from this work, and the results were as follows:

1. Spatial averaging generally decreases asymmetry in annual runoff values. Therefore, the applicability of the correlation theory becomes more evident for spatially averaged series.
2. This procedure also decreases temporal nonstationarities (already insignificant for individual series).
3. The first-order Markov process coefficient [close to $\overline{R}_1(1)$], being essentially nonzero for individual runoff series, diminishes for large regions and for the entire globe. Thus, a relatively small region - the

watershed of a European inland river flow - demonstrates the high value of $\overline{R}_1(1)$ (about 0.5), whereas a zero correlation coefficient is typical for a runoff series averaged over continents.

4. Finally, on a global scale, temporal correlation of annual runoff values is absolutely insignificant, and the white noise model is appropriate for the description of global runoff.

3.3.2 Mountain Glaciers

Mountain glaciers, as well as other parts of the cryosphere, atmosphere, and hydrosphere, are affected by numerous regular (deterministic) and irregular (chaotic) processes and factors, so that it is suitable to use statistical methods in the studies of long-term temporal and spatial variabilities of glaciers.

Two approaches to the investigations of glacier variations, taking into account their irregular nature, must be discussed.

1. *Statistical studies* of spatial-temporal changes. In these works an attempt is made to use deterministic models to describe internal processes within each glacier, spatial patterns unifying several glaciers within a mountain area, or several mountain areas within the whole hemisphere, and then to calculate the statistics of the parameters of these models, to compare these statistics with each other. Investigations of this kind have given important results for relationships between the dynamics of different glaciers, conclusions about the applicability of different variants of glacier models and methods of computing balances, and the nature of climate-glacier system. Most of the existing works dealing with mountain glacier statistics use similar approaches. Their results are described in Sects. 1.1 - 1.6. of the review paper (Reynaud and Dobrovolski 1998).
2. *Stochastic studies.* The aim of these investigations is to establish relationships between current and past states of glacier parameters using explicit models for random components of processes and basic concepts of the theory of random processes. A comparison of these models with stochastic models of various processes within the climatic system can help to understand fundamental features of the climatic system and its atmosphere-glacier subsystem. Thus, the study of glacier dynamics is implemented in a somewhat opposite direction, in comparison with the first approach: from random features to deterministic features.

Time series of annual cumulative specific net balances of 26 mountain glaciers were processed by us (Table 3.5). Data were taken from the

Table 3.5. Results of stochastic analysis of changes in mountain glaciers net specific balances

1 Glacier	2 Country	3 Mean increments	4 Rms incr.	5 $k'(1)$	6 Order incr.	7 Model
Abramov	Tadjikistan	- 0.5	0.6	0.2	0	W/T-
Alfotbren	Norway	0.3	1.3	-0.1	0	Wiener
Aus. Broggerbreen	Svalbard	-0.4	0.3	0.0	0	W/T-
Careser	Italy	-0.6	0.7	0.4	0-1	W/T-
Djankuat	Russia	-0.1	0.6	-0.3	0	Wiener
Engabreen	Norway	0.7	1.2	-0.2	0	W/T+
Grasubreen	Norway	-0.1	0.7	-0.2	0	Wiener
Gries	Switzerland	-0.4	0.9	0.2	0	W/T-
Gulkana	USA, Alaska	-0.2	0.4	0.0	0	W/T-
Hintereisferner	Austria	-0.4	0.5	0.2	0	W/T-
Kara-Batkak	Kirgizstan	-0.4	0.5	0.3	0-1	W/T-
Kesselwanferner	Austria	0.0	0.4	0.3	0	Wiener
Kozel'skii	Russia	-0.2	1.1	-0.1	0	Wiener
Mid.Lovenbreen	Svalbard	-0.3	0.3	-0.1	0	W/T-
Malyi Aktru	Russia	0.0	0.5	0.1	0	Wiener
Nigarsbreen	Norway	0.4	1.1	-0.1	0	W/T+
Place	Canada	-0.7	0.7	0.1	0	W/T-
Sarennes	France	-0.6	0.9	0.2	0	W/T-
South Cascade	USA	-0.5	0.9	0.0	0	W/T-
Silvretta	Switzerland	0.1	0.8	0.1	0	Wiener
Sonnblickees	Austria	-0.1	0.7	0.2	0	Wiener
Storbreen	Norway	-0.2	0.6	0.1	0	W/T-
Storglaciaren	Sweden	-0.3	0.5	0.2	0	W/T-
Saint Sorlin	France	-0.5	0.9	0.1	0	W/T-
Ts.Tuyuksuiskii	Kazakhstan	-0.4	0.5	0.4	1	W/T-
Wolverine	USA, Alaska	-0.1	1.2	0.1	0	Wiener

Glacier Mass Balance Bulletin (Glacier ... 1993). The shortest series (length less than 20 years) were not taken into consideration. Thus, balance variabilities of most glaciers with a long period of appropriate observations (described by Hoelzle and Trindler 1998) were studied.

The frequency-truncation variant of the maximum entropy method (MEM) of spectral and stochastic analysis was used in the present work. In Table 3.5 results of these calculations are shown. Here, mean values of year-to-year increments of glacier specific net balances are presented in the third column; standard deviations of increments in 4th column; coefficients of the AR(1) model for increments in 5th column; order of the model for increments series in 6th column; Model - type of model for the description of initial time series of net balances - in 7th column. ("Wiener" means the discrete Wiener process; "W/T" - the discrete Wiener process or model with trend).

For almost all the time series of year-to-year increments of net balances, except Tsentral'nyi Tuyuksuiskii and, perhaps, Kareser and Kara-Batkak, the zero-order model, i.e., white noise, was a good approximation (however, the above net exception can be explained by the sample variability of estimations). The following geographical feature can also be deducted from Table 3.5: glaciers with the largest natural variability of year-to-year changes in balances are those with pronounced Wiener or positive type of glacier variability (Alfotbren, Engabreen, and Nigarsbreen in Norway, Kozel'skii in Kamchatka, Wolverine in Alaska). For all these glaciers, standard deviation of year-to-year increments is more than 1 m/ya. They are situated near major atmospheric and oceanic frontal zones with high interannual variabilities of atmospheric processes.

Yet the following problem was essential: do series of increments contain statistically significant, nonzero mean value? In other words, is there a linear trend within initial time series of net balances? In order to answer this question we used very simple considerations. If the absolute value of mean increments is less than the error of estimation of mean white noise with the appropriate standard deviation (uncorrelated process with zero mean), a monotonous (linear) trend in the series of balances is not significant statistically, and changes in balances are described by the discrete Wiener process model. On the contrary, if the absolute mean increment is much more than the mean error of estimation of white noise mean value, the trend (positive or negative) can be significant, and centered glacier balances are described by the following model:

$$B(t) = B(t-1) + a\,(t) + c\,t\,, \tag{3.27}$$

where c is the constant (positive or negative), and t the time in years (the middle of the record being considered as zero year). Also we used somewhat more sophisticated method (Yaglom method described above) for the analysis of the linear trend errors in Tsentral'nyi Tuyuksuiskiy glacier.

All in all, nine series are well described by the discrete Wiener process model, because the trend parameter is less than mean error of its estimation. Also four or five other series, on average, fit the same model because of the sample variability of estimations. Thus, only 11 records (42% of the time series) demonstrate a significant negative trend, 1 or 2 records contain a significant positive trend, and 13-14 records are close to realizating symmetric random walks.

These results do not corroborate a widespread opinion that the behavior of most mountain glaciers illustrates the global greenhouse effect. On the contrary, the volume variations in most glaciers (especially large glaciers) demonstrate random or positive changes. Moreover, from our point of view, the rest of the glaciers - with the observed negative trends of glacier volume - are evidence of random regional temperature changes rather than of monotonous, man-made global heating.

4 Stochastic Models for Glacial Cycles

Despite recent achievements in understanding past climatic changes (Adem 1991b; Boulton et al. 1995; Huybrechts and Tsiobell 1995; Marsiat 1995; Berger and Loutre 1996, 1997; Muller and MacDonald 1997, etc.), there are still many difficulties in studying these extremely important natural processes. The causes of this situation are as follows:

1. Data on climatic changes in the past are imperfect and contradictory. We know that even data on climatic changes during the past 10-20 years cannot answer the crucial question of contemporary climatology: is the climate system warming or not? It is clear that paleoreconstructed data are much less accurate and give place to too many uncertainties.
2. Generally speaking, there is no established, unanimously adopted theoretical approach to the problem of paleoclimatic changes. Seemingly, there are hundreds of hypotheses trying to explain glacial cycles (for instance, see Verbitsky 1986). These hypotheses include different mechanisms of feedbacks within and/or between the components of the climate system, also external processes: astronomical, geophysical, processes in the mantle and the core of the earth, etc. As in the case of observational data, existing theories and proposed mechanisms do not cover (convincingly explain and attribute) even recent climatic changes.
3. Finally, many problems arise with respect to choice, identification, and assessment of parameters of mathematical models used for the description of glaciation-deglaciation processes. As in studies of contemporary climate, general circulation models of the atmosphere linked to the models of glaciers and oceans are the most popular in paleoclimatic studies. However, their shortcomings become serious obstacles for the studies of climatic variability at paleo-time scales. Artificial synchronization of oceanic and atmospheric model time and the problem of model drift make the direct using of GCMs in paleostudies questionable. Yet, GCMs cannot be integrated for hundreds of ka in hundreds of variants (usually needed for experiments with a model). Perhaps the most difficult problem for GCMs and detailed models of ice sheets and oceans (connected to these problems) is the difficulty in estimating the numerous parameters needed for these models. Accumulation of errors because of inadequate estimates of parameters can easily compromise large models and give results worse

than those obtained on simpler but better-structured models. Convincing criticism of GCMs with respect to paleoclimate simulations was made recently in many papers: "GCMs seem to be unable to yield a reliable quantitative computation of the net snow accumulation" (Oglesby 1990); "coupling with oceanic GCMs and other complex climatic sub-system models (sea-ice, biosphere) is unrealizable for long simulations" (Marsiat 1995), etc.

The discussion on the existing data, models, and approaches will be continued in the next sections. As a short resume of the above brief overview, in this chapter our goal is to summarize present knowledge in the research area of the last glacial cycles, to try to construct on this basis a simple model which takes into consideration as few mechanisms as it can, and to demonstrate that even a very simple structure like this can, nevertheless, describe basic features of the behavior of the climate system during the past hundreds of ka.

4.1 Stochastic Analysis of Reconstructed Data on Glacial Cycles

4.1.1 Existing Paleoreconstructed Time Series

In our work we use quantified paleoinformation in the form of time series. All series can be divided into four groups:

1. Relatively short, up to 18 ka, series : Northern Hemisphere air temperature (Borzenkova 1992); glacial flow into ocean (Fairbanks 1989); $\delta^{18}O$ global sea level estimations (Bard et al. 1991), and other data of similar type. The typical discreteness interval of these series is several hundred years. The errors within the data are obviously large and unknown (as in almost all other available series used in this work). Moreover, these series usually contain only a small number of points (30 or less), and therefore their statistical and stochastic analysis is difficult. Nevertheless, as will be shown below, they can give important information about the last deglaciation period.
2. Time series with a length of up to 200-300 ka. The typical discreteness interval in these cases is usually several thousand years. Records of this type are relatively numerous and listed in Table 4.1. Because of the large number of points, these series allow more or less accurate

Table 4.1. Paleoclimatic time series and their stochastic models

No.	Series title and author	Length[a]	Model[b]	Coeff.[c]
1	Glacial margin, FESU (Zarrina et al. 1980)	60/30	M	0.3; - 0.4
2	Byrd $\delta^{18}O$ composition (Oeschger et al. 1983)	96/49	W	0.9; +0.0
3	Temperature,Bielorussia (Nazarov et al. 1991)	100/21	M	0.4; - 0.5
4	Precipitation, Bielorussia (Nazarov et al. 1991)	100/21	M	0.4; - 0.6
5	Deep sea core $\delta^{18}O$ (Dansgaard and Duplessy 1981)	128/33	W	0.9; - 0.2
6	Pollen temperatures, California (Adam and West 1983)	128/65	W	0.9; - 0.2
7	North Atlantic sea temperatures (Bradley 1985)	134/68	W	0.8; - 0.2
8	Ocean isotope composition (Shackleton et al. 1983)	134/68	W	0.7; +0.2
9	Pollen precipitation, France (Pons et al. 1991)	140/141	W	0.8; - 0.1
10	Pollen temperatures, France (Pons et al. 1991)	140/141	W	0.9; - 0.1
11	Vostok CO_2 (Barnola et al. 1994)	150/60	W	0.8; - 0.1
12	Vostok Be precipitation (Kotlyakov and Lorius 1992)	155/63	W	0.7; - 0.3
13	Vostok Ca (Legrand and Delmas 1988)	156/79	W	0.8; - 0.2
14	Vostoc methane (Chappellaz et al. 1994)	156/78	W	0.8; 0.0
15	Vostok CO_2 (Kotlyakov and Lorius 1992)	160/41	W	0.8; - 0.1
16	Vostok deuterium (Boutron 1988)	164/83	W	0.9; +0.3
17	Vostok temperatures (Jouzel et al. 1994)	165/67	W	0.9; +0.2
18	Vostok temperatures (Genthon et al. 1987)	165/67	W	0.9; +0.1
19	Vostok dust temperatures (Petit et al. 1980)	180/37	W	0.7; +0.1
20	Siberia temperatures (Zubakov 1986)	200/21	M	0.2; +0.2
21	Deep-sea $\delta^{18}O$ (Martinson et al. 1987)	290/117	W	0.9; +0.2
22	Global mean sea level (Selivanov 1996)	300/31	M	0.3; - 0.2
23	Caspian Sea level (Kliege and Selivanov 1995)	300/61	M	0.5; - 0.4
24	Pollen %, Colombia (Hooghiemstra et al. 1993)	300/29	M	0.3; - 0.2
25	Diatoms in California (Bradbury, 1995)	300/31	M	0.5; - 0.2
26	Mediterranean $\delta^{18}O$ (Parisi et al. 1982)	300/61	M	0.5; - 0.2

[a] Series length in ka/number of points. [b] M - Markov first-order process, W - discrete Wiener process. [c]First trial coefficient of autoregression, first trial autoregression coefficient of series of temporal increments.

statistical (stochastic) analysis of their structure. The length of these registrations is enough for the analysis of features of glacial changes with characteristic periods about 5-50 ka, but is not sufficient for the analysis of the main 100-ka period of glaciation-deglaciation events.

3. A few famous time series with a length of 400-600 ka: deep sea isotopic data (SPECMAP, Ruddiman et al. 1989, etc.), global ice volume (Hays et al. 1976) allow analysis of 100-ka periodicity. The difficulty in this case is that the length of the series is still small in comparison with the period under investigation. Therefore, the significance of results related to the 100-ka period is not very high.
4. Relatively new, high resolution data (Gassman 1994; Koc and Jansen 1994; Dowdeswell and White 1995; Rotcliffe 1995; Ditlevsen et al. 1996; Pavlich 1996; Hughes 1996; Stager and Mayewski 1997, etc.). These authors sometimes claim to provide 1 year resolution for 10-20

ka B.P. and 10-year resolution for hundreds of ka B.P. (Dowdeswell and White 1995). Although the problem of possible errors within these series is far from being solved, such registrations seem to demonstrate important features of paleoclimates: (a) the abrupt, discrete character of climatic changes in the past; (b) the possibility of extremely large increments of climatic parameters during short periods of time, comparable with the range of glaciation cycles themselves; (c) the range of variability of climatic parameters, from maximum to minimum of glaciation, is enormous and can reach dozens of times (Ditlevsen et al. 1996); (d) even more impressively than previous types of paleodata, high resolution records corroborate the random (stochastic) nature of variability: the irregular character of pulsations, their low predictability represent, in general, difficulties in describing the observed variability using deterministic models.

4.1.2 Results of Stochastic Analysis of the Last Deglaciation Period, 0-18 ka B.P.

The most characteristic feature of these registrations is contrasting climate dynamics at the beginning and the end of a deglaciation period. In Figs. 6.1and 6.2 changes in global mean sea level are presented as an example. It was shown in Dobrovolski (1992) that recent instrumentally derived series of annual global mean sea levels (as well as series of instrumentally measured global air temperatures) are satisfactorily described by the discrete Wiener process model [Eq. (2.54)]. The role of the Wiener process model in the study of glaciation cycles will be discussed in more details below. Here, it is important to recall that we can use this model for making a diagnosis of mean deviations (standards) of the global mean sea level in the past moment (t) from its present value (t_o):

$$\sigma[\xi(t) - \xi(t_0)] = \sigma_a \sqrt{t_0 - t}\,, \tag{4.1}$$

where σ_a is the standard of white noise $a(t)$; for instrumentally observed global mean sea level variations σ_a is approximately equal to 1 cm/ya. It is evident from Figs. 4.1 and 4.2 that for the period of time from the present back to approximately 6 ka B.P., Eq. (4.1) gives changes of the same order as reconstructed in Bard et al. (1991), i.e., ± 1 m. These results are also corroborated by global mean sea level reconstructions in Newman et al. (1980).

However, variations of mean sea level earlier than 6 ka B.P. largely (up to 100 times at maximum of glaciation) exceeded those predicted by the extrapolation of recent sea levels using Eq.(4.1). There might be two causes for this: (a) the presence of a strong deterministic trend related to

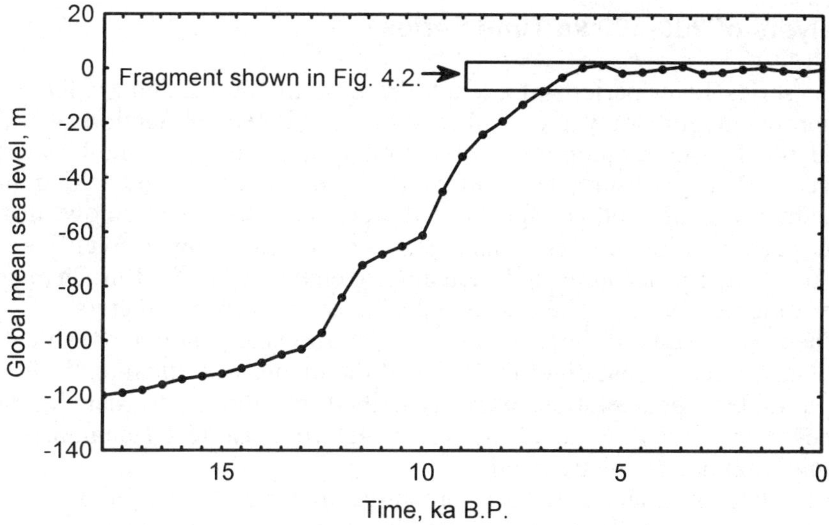

Fig. 4.1. Global mean sea level. (Data from Bard et al. 1991)

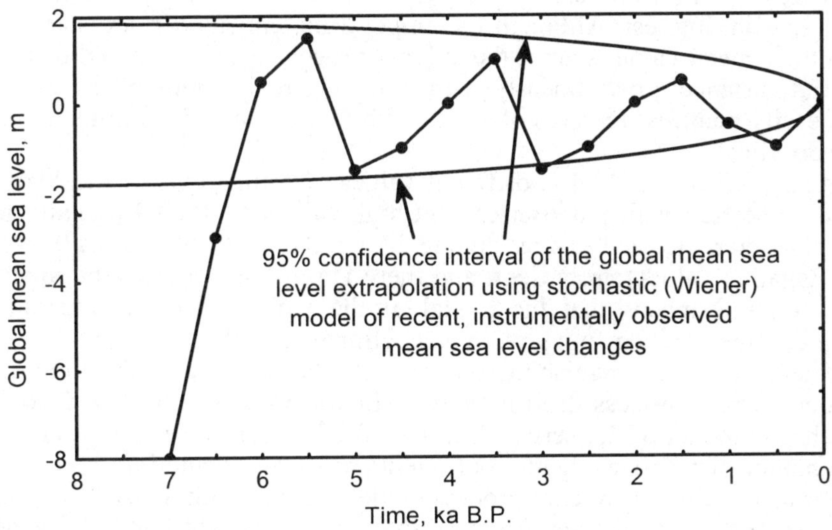

Fig. 4.2. Fragment of Fig. 4.1

the process of glacier melting; (b) changes, in time, of standards of stochastic, white-noise component of processes.

4.1.3 Analysis of 200-300 ka Time Series

These series cover periods of a length from 60 to 300 ka, and are the most numerous in our study (see Table 4.1). As shown in Section 4.3, the standard of random component $a(t)$ is likely to change in time. However, because of large discreteness intervals in registrations, and taking into consideration some other aspects of the problem discussed below, in the first approximation we can consider σ_a as a constant for a given record, and use MEM procedures and their improvements described in Chapter 1 for the identification of the model and estimation of its parameters.

The main results of the analysis of registrations are shown in Table 4.1. In all cases the estimation of the order of the model was equal to 1, i.e., all series under consideration were described by the first-order Markov process [Eq. (2.50)]. None of the series 3-26 from Table 4.1 demonstrated a significant deterministic trend.

As stated in Chapter 1, there are three important types of processes, which can be formally described by Eq. (2.50): the white noise ($k = 0$), the first-order Markov process itself ($0 < k < 1$), and the discrete Wiener process ($k \to 1$). The discrete Wiener process (random walk), which is also close to the Bernoulli symmetric random walk with small grid steps, is of special interest. Although its temporal increments $a(t)$ are stationary and their mean value is zero, the Wiener process itself is not stationary at all: its variance grows linearly with time, and realizations of the process can easily manifest features of series with monotonous deterministic trend or periodical.

As seen from Table 4.1, different values of estimations of the Markov model coefficient fit paleoseries. Yet high values of 0.7-0.9 prevail: they are characteristic for 17 series among 26. On the contrary, values 0.0 - 0.1 (or negative values) are absent, and there is only one coefficient equal to 0.2. All this means that the model of the white noise (widespread in nature, however) does not govern paleoclimatic series.

The task of distinguishing the discrete Wiener model from the first-order Markov process itself is more difficult. As seen from Fig. 2.26, the stochastic model of the series with $c =$ 0.77 (methane record) is close to the model of a process with coefficient $c = 0.91$ (deuterium record). In accordance with the Wiener process model, in bilogarithmic scales spectra of both series are satisfactorily described by the straight line with tangent - 2 (Fig. 4.3) and thus fit the minus two degree law of climatic variability without feedbacks (Hasselmann 1976).

Moreover, even models with lesser coefficients from Table 4.1 fit the discrete Wiener process model if we take into consideration inevitable errors in the data. Consider the discrete Wiener process with additional white noise:

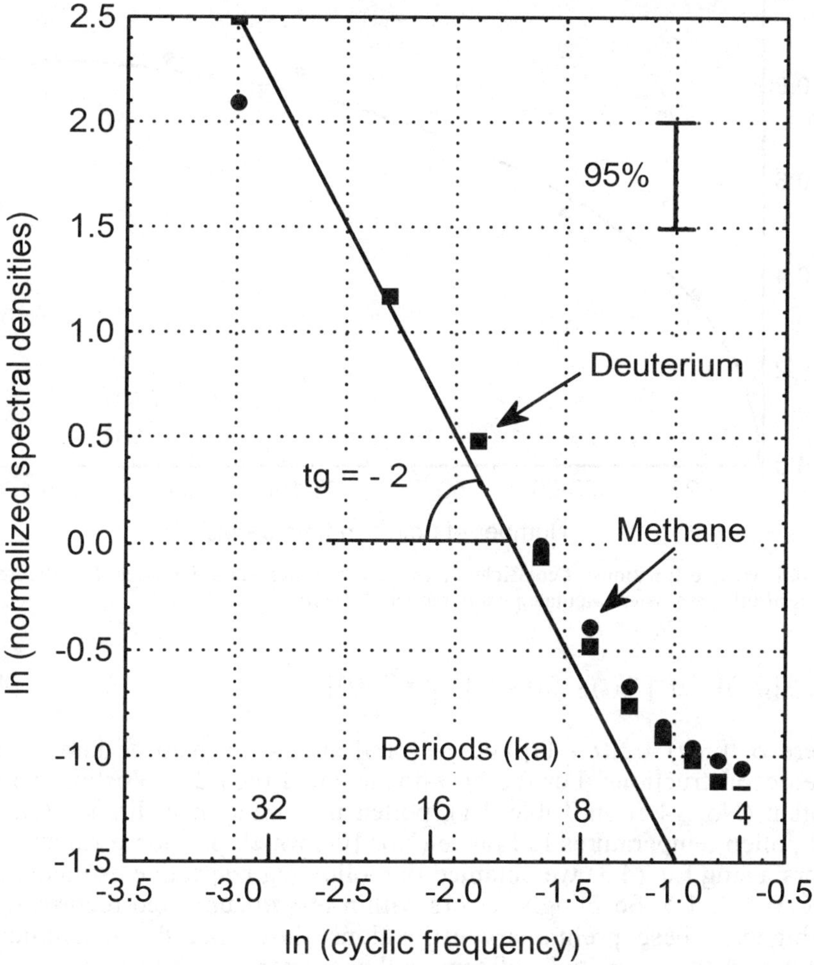

Fig. 4.3. Spectral densities of Vostok methane and deuterium records (as Fig. 2.26 but in bilogarithmic scales)

$$\xi_1(t) = \xi(t) + b(t)\,, \tag{4.2}$$

where $\xi(t)$ is described by Eq. (2.54), and $b(t)$ is a second white noise, which is not correlated with the first white noise $a(t)$. We describe random errors within the data by the component $b(t)$ and, consequently, the whole process by Eq. (4.2).

We can try to estimate the standard of the second white noise σ_b and the ratio $e = \sigma_b / \sigma_a$ using the best-documented series. Indeed, we can write the following equation:

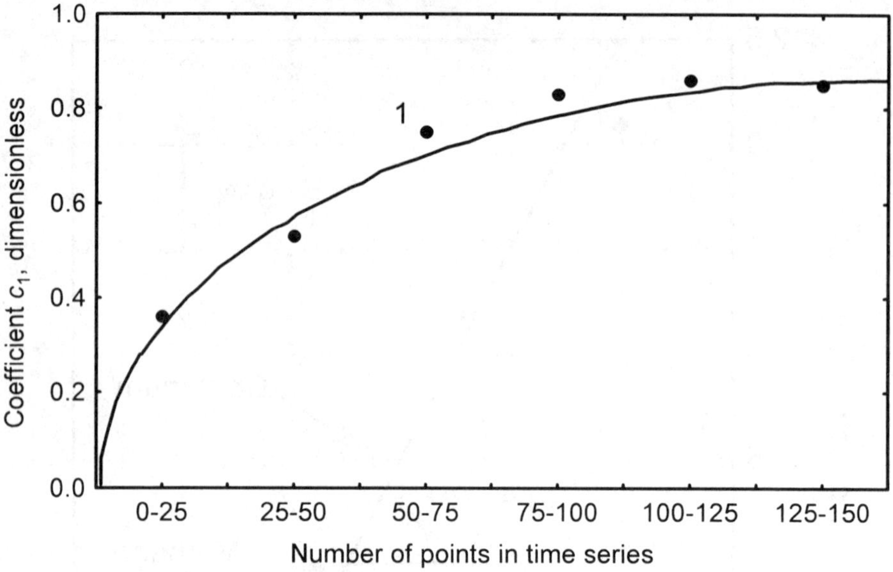

Fig. 4.4. Mean estimations of coefficient c of the first-order Markov process (*1*); theoretical values of this parameter calculated using model (4.2) (*solid line*)

$$\sigma^2 [a(t)] = \sigma^2[\xi_1(t) - \xi_1(t-1)] - 2\sigma^2[b(t)], \tag{4.3}$$

where $\sigma^2[\xi_1(t) - \xi_1(t-1)]$ and $\sigma^2[b(t)]$ are assumed to be known from paleoreconstructions. For the best-documented records - Vostok methane content (No. 14 from Table 4.1), pollen precipitation in France (No. 9), and pollen temperatures in France (No. 10), which provide estimations of errors, using Eq. (4.3) we obtained the following respective estimations of e : 0.3; 1.3; 1.5. Seemingly, errors within less documented records might be higher. These preliminary estimations show that the magnitude of errors and the magnitude of temporal increments in paleoseries can be comparable, and the reasonable estimation of parameter e is around 1. Indeed, Fig. 4.4 demonstrates almost perfect fit, with the use of the Wiener process plus error approach (e = 1.3), of theoretical values of c to the values from Table 4.1 estimated using paleorecords.

Finally, it can be shown analytically that model (4.2) gives negative values of correlation coefficients between neighboring values of the series temporal increments, $\overline{R}_1[\xi_1'(t)]$ where $\xi_1'(t) = \xi_1(t) - \xi_1(t-1)$, using estimated observational data:

$$\overline{R}_1[\xi_1'(t)] = -\frac{e^2}{1+2e^2} \,. \tag{4.4}$$

Table 4.2. Stochastic parameters of the last glaciation trend, from 115 ka B.P. to the last glaciation maximum

Series	Maximum glaciation in ka B.P.	Parameter of linear trend (*LT*)	Error in *LT* (*E*)
Deep sea $\delta^{18}O$ (Martinson et al. 1987)	17	-0.028	0.017
North Atlantic sea temperatures Bradley 1985)	18	-0.143	0.273
Vostok CO_2 (Kotlyakov and Lorius 1992)	16	-1.900	2.140
Deep sea core $\delta^{18}O$ (Dansgaard and Duplessy 1981)	16	-0.008	0.037
Pollen air temperatures (Adam and West 1983)	26	-0.017	0.321
Vostok temperatures (Jouzel et al. 1994)	20	-0.131	0.192
Global ice volume (Hays et al. 1976)	21	-0.041	0.025
Vostok temperatures (Genthon et al. 1987)	20	-0.115	0.186
Pollen temperatures, France (Pons et al. 1991)	15	-0.103	0.154
Pollen precipitation (Pons et al. 1991)	17	-5.610	9.160
Vostok deuterium (Boutron 1988)	24	-0.500	0.810
Vostok Ca (Legrand and Delmas 1988)	14	-0.069	0.124
Vostok methane (Chapellaz et al. 1994)	15	-2.000	7.770
Vostok CO_2 (Barnola et al. 1994)	15	-2.040	1.910

These theoretical considerations are corroborated by the calculations of parameter $\overline{R}_1[\xi_1'(t)]$ from Table 4.1: most of the estimations of this parameter are negative.

The most spectacular feature of the last two glaciation periods is a long visible general decrease in temperature and related parameters (and corresponding increase in parameters which are negatively correlated with temperature) since the beginning of the glaciation period till the maximum of glaciation, i.e., from approximately 115-130 ka B.P. to 14-26 ka B.P. (the two estimations are shown together in Fig. 4.5). In order to investigate the nature of this trend and its parameters, we took appropriate segments of 14 series with the most detailed description of this period (the largest possible number of points within this period) (Table 4.2).

Generally speaking, the best model for the description of these 14 fragments of time series, as well as for the description of most of the complete series in Table 4.1, was the discrete Wiener process. Yet the question remains: do these fragments contain a statistically significant deterministic (linear) trend? At a glance, there is no doubt about the answer to this question because all 14 fragments clearly show general visible long-term, almost linear change of the same sign (if we take graphics for parameters negatively correlated with temperature, with inverse sign).

However, detailed calculations do not corroborate this hypothesis. The natural way to check this assumption is to estimate mean values of

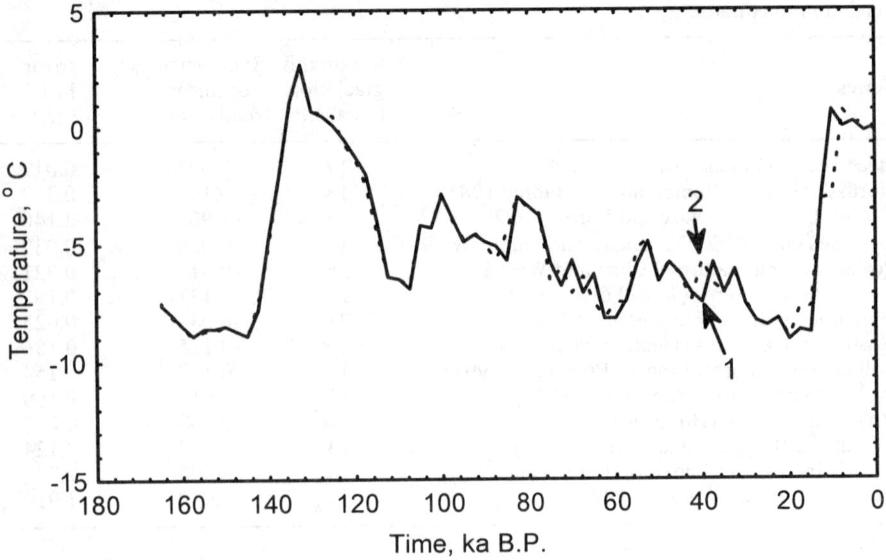

Fig. 4.5. Vostok temperature records. *Solid line 1* data from Jouzel et al. (1994), *dashed line 2* data from Genthon et al. (1987)

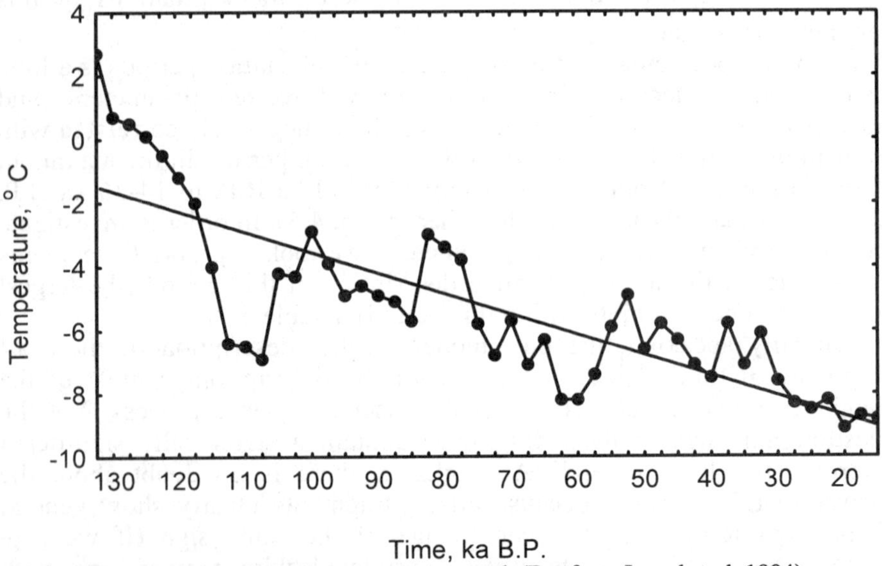

Fig. 4.6. Fragment of the Vostok temperature record. (Dta from Jouzel et al. 1994)

temporal increments of time series and compare these values with errors in this parameter. As an example, a segment of Vostok temperatures series is shown in Fig. 4.6. A linear trend, denoted by a solid line, seemingly exists

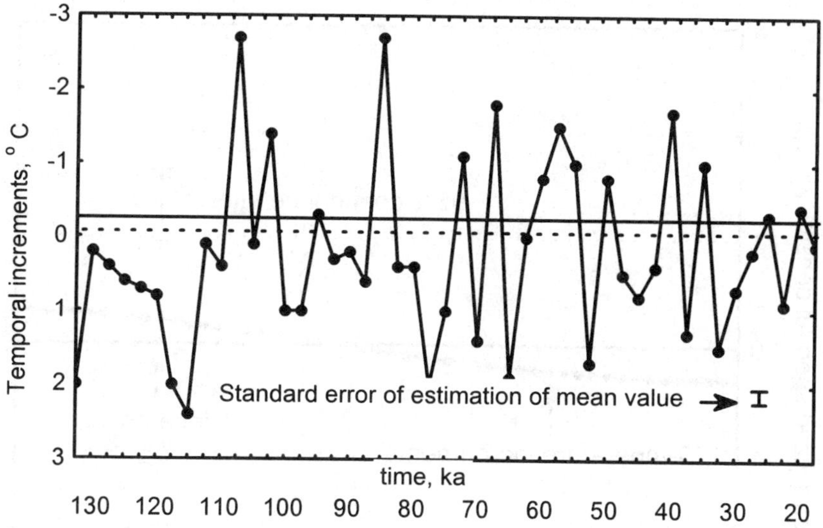

Fig. 4.7. A Fig. 4.6 but for temporal increments

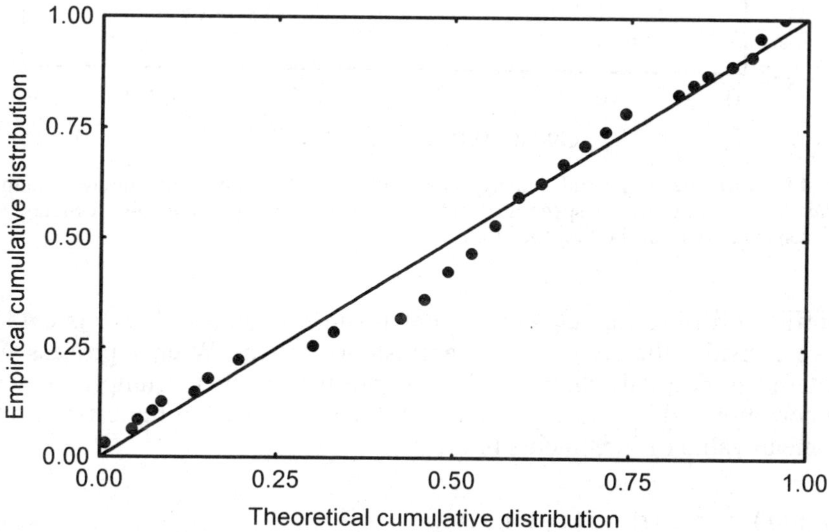

Fig. 4.8. Probability-probability normal distribution plot for Vostok temperature increments (see Fig. 4.7)

within this segment of the series. But if a linear trend really exists, the mean value of the first temporal increments of the series must differ significantly from zero. In Fig. 4.7 first temporal increments for the segment together with their mean value are shown. In this figure no

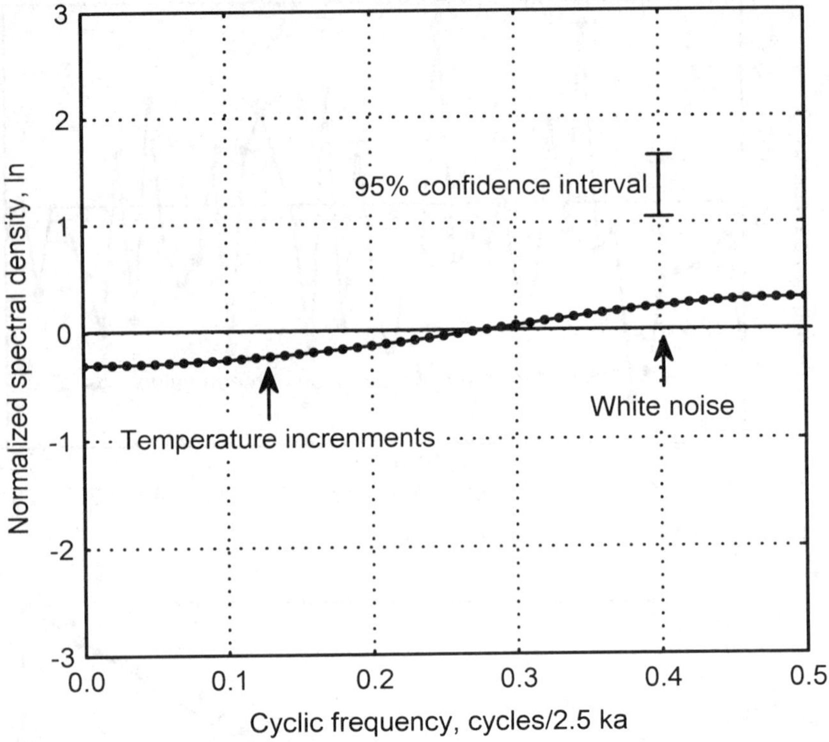

Fig. 4.9. Normalized spectral density (in logarithms) of Vostok temperature increments (*solid line*); white noise spectral density (*dashed line*). *Vertical line* denotes 95% confidence interval for the first spectrum

significant difference between increments mean value and zero is evident. If we consider the series as a manifestation of the Wiener process with white noise normally distributed temporal increments (assumptions which are corroborated by Figs. 4.8 and 4.9), the mean error in the estimation of the mean value of increments is:

$$E(\hat{m}) = \sigma[\xi(t) - \xi(t-1)]\sqrt{N}\,, \tag{4.5}$$

where N is the length of the series (segment) in numbers of points. For the segment of series in Fig. 4.7 the mean value of increments, LT, is equal to - 0.22 °C/2.5 ka. Theoretical estimation of the error of this estimation E is equal to 0.17 °C/2.5 ka. Thus, there is no statistically significant linear trend within the segment of time series investigated. Realizations of the discrete Wiener process with the same number of points (which contain no deterministic trend), on average, give a similar value for the parameter of linear trend, *LT*. We can add that if temporal increments of the random walk are described not by white noise but by another type of stationary process, the method of Yaglom (Yaglom 1987) can be used to assess the error in the mean increment value. Indeed, estimation of the model coefficient of increments is never exactly equal to zero; our estimations using the Yaglom method demonstrated that in this case the significance of linear trend in the investigated series is even less (because of the increase in the integral correlation scale of increments and the less efficient length of the series).

The same calculations have been made for 13 other segments from Table 4.1. (To compare the results, the beginning of the period of glaciation for all registrations was fixed at 115 ka B.P.). In 11 cases of 14, the parameter of linear trend, *LT*, was less than the mean square error in the estimation of this parameter. In the other 3 cases, the order of these values was the same although the absolute value of *LT* was slightly larger than the theoretical estimation of its error. Recall that for the discrete Wiener process with no linear trend, in 33% of cases (i.e., in 4-5 cases out of 14) absolute values of *LT* estimations, on average, must be higher than the mean square root error of their estimation.

We can also add that those of the initial series in Table 4.1 which were less detailed demonstrated an even less statistically significant deterministic trend during the last glaciation. The same calculations have been made for the longest time series that will be discussed in the next section. It was shown that a deterministic monotonous (linear) trend was also not significant for the next to last glaciation.

4.1.4
Longer Time Series. Features of Cyclicity

Famous time series (Hays et al. 1976; Berger et al. 1984; Ruddiman et al. 1989), covering a period of 400 - 600 ka, were analyzed many times. One of them [the SPECMAP series (Berger et al. 1984)] is shown in Fig. 4.10. We would like to draw attention to some features of these records:

1. Their length is still short in comparison with the period of assumed periodical component, 100 ka. That is why, although the spectral peak at 100 ka (Fig. 4.11) is rather sharp, it is comparable with the width of confidence interval of estimations. So all conclusions about 100-ka periodicity must be very cautious. For instance, for the very presence of 100-ka cyclicity during the first 100-130 ka of the registration there is no evidence at all.
2. Peaks at 41 and 23 ka are not statistically significant. The difference between these peaks and spectrum of the first-order Markov model of the same series is less than the confidence interval of reasonable level of probability. Yet, because of the shortness of realization, it is impossible

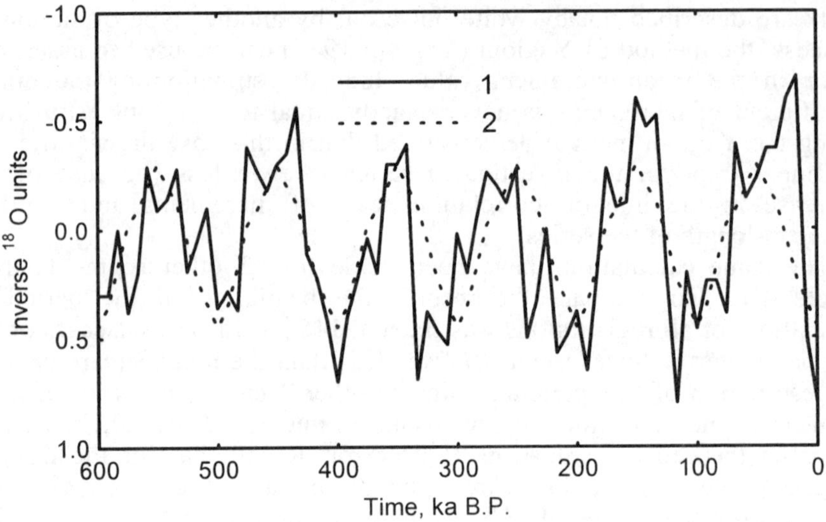

Fig. 4. 10. SPECMAP data (*1*) (from Berger et al. 1984), average 100-ka period dynamics (*2*)

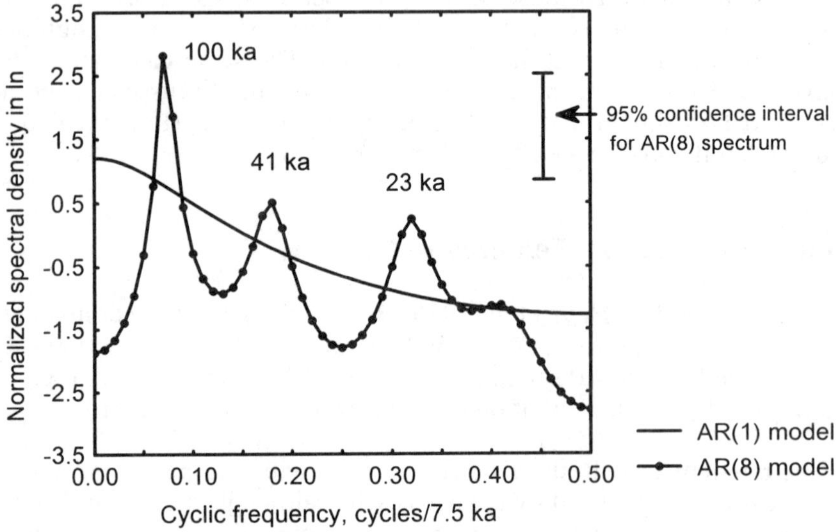

Fig. 4.11. Spectral density estimations. SPECMAP series (see Fig. 2.10)

to accurately determine frequency (period) of these peaks, and to distinguish them from the harmonics of the 100-ka period, 50 and 25 ka, which are close to 41 and 23 ka.

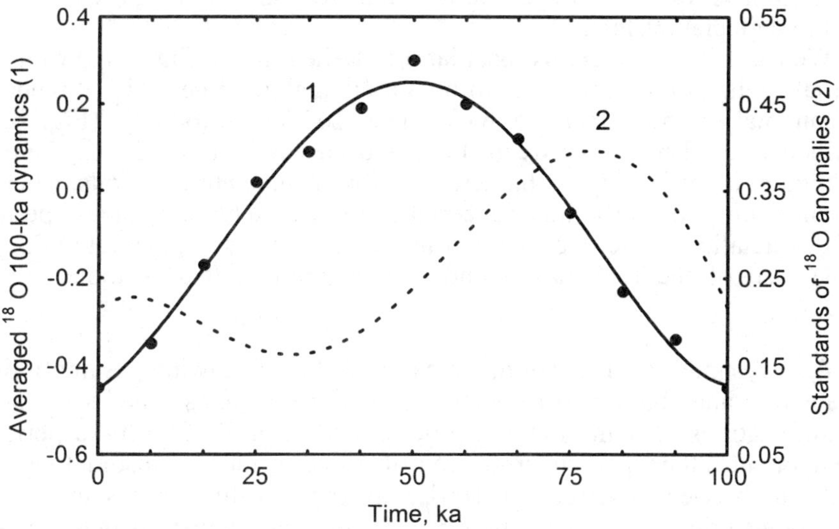

Fig. 4.12. Averaged SPECMAP 100-ka dynamics (*1*), averaged 100-ka standard deviations (*2*)

3. SPECMAP and other registrations demonstrate the essential difference between individual glaciation cycles with respect to their amplitudes. The range of the last two glaciations-deglaciations (200-100 ka B.P. and 100-0 ka B.P.) is about two times larger than the range of the first (600-500 ka B.P.) and fourth (300-200 ka B.P.) glaciation-deglaciation.
4. The durations of cycles are also different. For instance, the length of the fourth cycle (about 150 ka) exceeds by two times or so the length of the third cycle (about 70 ka).
5. The difference in the symmetry-asymmetry properties of individual glacial cycles is startling. The frequently cited asymmetry of the last two cycles (long glaciation, rapid deglaciation) is not evident at all for the previous cycles. Averaged over all six cycles, dynamics (dashed line in Fig. 4.10 and solid line in Fig. 4.12) demonstrates the absence of any asymmetry. Thus it is not at all clear that the last two cycles are typical. Like other cycles, they are rather random manifestations of the general glaciation-deglaciation pattern.
6. Averaged curve (dashed line in Fig. 4.10 and solid line in Fig. 4.12) has a range which is less than that of practically each of the six cycles. This is the result of seemingly random phases of cycles, and the presence of random short-scale anomalies in them. If we use temperature interpretations of the SPECMAP curve, we obtain the average range of cycles 4-4.5 °C against 9-10 °C of the estimated range of the last cycle. Consequently, the average temperature minimum (dashed line) in the cycle is higher than temperature minima in specific cycles, and the average

maximum is lower than specific maxima (by 2-2.5 °C in the case of the last two glacial cycles).
7. We can subtract averaged oscillation (dashed line in Fig. 4.10) from the initial registration (solid line) to obtain anomalous time series. Standards of anomalies thus obtained have clear 100-ka periodicity (Fig. 4.12, dashed line). The lag between these two curves is about 30 ka, i.e., the minimum of variability is characteristic for the glaciation periods, whereas the maximum of variability (uncertainty) is characteristic for the periods of deglaciation. Note that this is true for time scales comparable with the time step of the registration under consideration, that is thousands of years.

As a resume of this section, we can make the following assumption on the nature and character of the last six glaciation cycles. They seem to be manifestations of some random process with a probable 100-ka periodic component rather than superposition of three or more sinusoidal signals. The lag between changes of 100-ka averaged values and standards of anomalies might indicate the presence of some mechanism which provides relative stability during the period of glaciation, and, on the contrary, destabilizes glaciers during the deglaciation periods.

4.1.5 High Resolution Paleorecords

Recent studies on high-frequency variations in the past allow two important conclusions about glaciation-deglaciation processes. First, these registrations manifest abrupt changes in the mean values of climatic parameters. Hundreds and thousands of years can pass without major changes: the past 11,000 were "calm" accordingly to Dowdeswell et al. (1995). However, in a few dozens of years or in a few years (that is, practically in a moment, from the point of view of paleoscales) air temperature can change by several degrees, and precipitation by dozens of percent.

As an example, the record of annual ice accumulation rates in Greenland is presented in Fig. 4.13 (redrawn using data from Dowdeswell et al. 1995). Each of the two halves of this series, using the techniques described in Chapter 2, is satisfactorily modeled by the stationary process of white noise (the first coefficient of autoregression is 0.24 for the first half of the registration, and 0.09 for the second half, which is not far from the mean error of estimation of this coefficient, 0.14). But about 11,640 B.P. the climatic regime suddenly changes, and this shift cannot be explained in terms of stationary stochastic processes. The difference between the mean values of the first and second 50 years of the record, denoted by solid lines in Fig. 4.13 (70 mm/ya), exceeds by more than

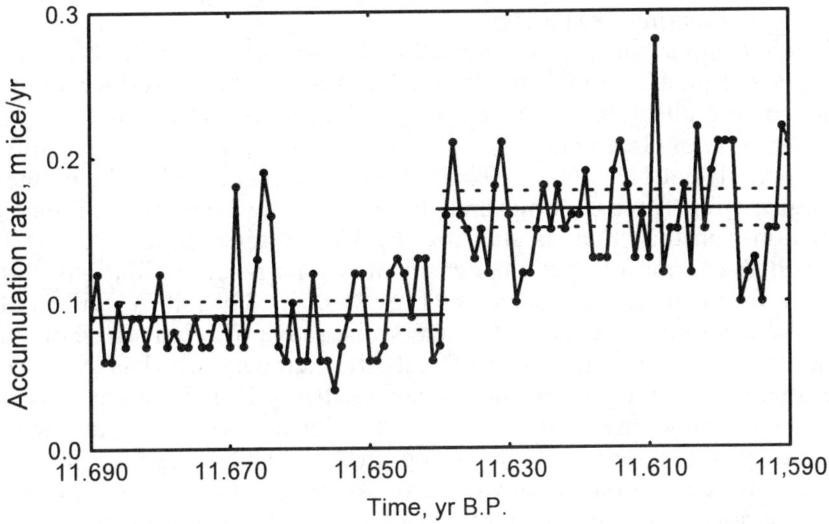

Fig. 4.13. Ice accumulation in Greenland. (Data from Dowdeswell et al. 1995)

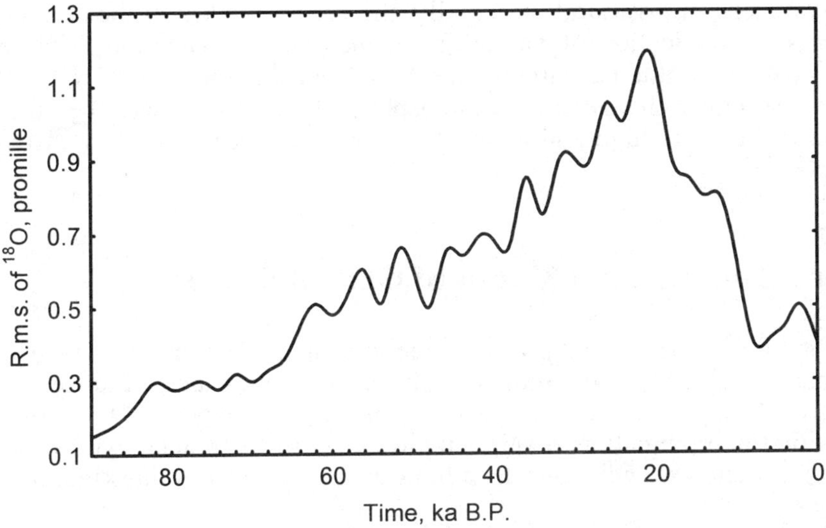

Fig. 4.14. Rms of $\delta^{18}O$ high-frequency variability in Greenland. (Data from Ditlevsen et al. 1996)

three times the 95% confidence interval for the estimations of mean values (denoted by dashed lines, about 20 mm/ya). These data also illustrate the process of transformation of stationary local climatic changes at short time (year-to-year) scales into nonstationary, random walk-like climatic behavior at longtime (ka) scales.

Another important feature revealed by high-resolution data is long-term changes in climatic variability. Figure 4.14 represents the 10-ka root mean square of the absolute value of the high-frequency (150-year high pass) signal in $\delta^{18}O$ in Greenland. The difference between climatic variability at maximum glaciation (around 20 ka B.P.) on one hand, and at the moments of deglaciation (present time and, especially, the beginning of the last glaciation), on the other, is startling: by 10 and more times, i.e., by 100 times in the terms of spectral density and energy of oscillations. From another figure in the cited paper (Fig. 1a-c, p. 810), it is evident that during the period of developed glaciers, the range of relatively short-time variations (secular and ka-to-ka) of the temperature has the same magnitude as the long temperature changes themselves. Physical causes of this phenomenon are evident. At first, global cooling occurs mostly because of the temperature lowering in high and temperate latitudes, whereas tropical latitudes remain relatively warm, so latitudinal gradients at the subpolar atmospheric and hydrological fronts increase, and synoptic activity intensifies. Also, ice sheets are much less stable objects than the ocean, the land surface, and the atmosphere, and their presence must activate temporal variability. Finally, the very geometry of the continents causes intensification of variability at maxima of glaciation. The solid land part of latitudinal circles increases from 82.5°N to 62.5°N by 16.5 times, so equal changes in the atmospheric temperature near the edge of glacier sheets, at high and temperate latitudes, cause absolutely different responses.

4.2 Zero-Dimensional Model of Glacial Cycles

It was demonstrated in previous sections and chapters that facts and mechanisms related to climatic changes at time scales of dozens and thousands of ka are poorly known. This is why it is expedient to construct, as a first approximation, a very simple model of the glacial cycles. In the present section we will consider a 0D stochastic model of this kind.

4.2.1 Hypotheses, Assumptions, and Equations

1. We suppose that changes in the glacier volumes and mean sea level heights can be described by a model with two thresholds (screens) - high threshold, TH_h , and low threshold, TH_l. There are seemingly two limits of glaciation, both related to fundamental planetary-scale limitations and

Table 4.3. Estimates of the maximum volume of continental glaciation in the Late Pleistocene (17-20 ka B.P.) obtained by various kinds of data. (After Dobrovolski and Selivanov 1998)

Mean of estimate	Depression of global mean sea level, m	Global volume of continental glaciation, $10^6 km^3$	Source
Paleoglaciological data	85	58-60	CLIMAP (1976)
	115	67.5	Suetova (1982)
	145	80	Fairbridge (1980)
	160-170	85-87	Denthon and Hughes (1981)
Geological and geomorphological data on ancient sea coasts	120	70	Curray (1964)
	130	74	Bloom (1971)
	100	64	Flint (1971)
	90	61	Kaplin (1973)
	120	70	Shepard (1971)
	114-133	67-73	Carter et al. (1986)
Oxygen isotope data	52-76	50-55	Shackleton and Opdyke (1973)
	131	75	Schoell and Risch (1976)
	105	65	Cronin (1983)
	117	70	Nikolaev et al. (1989)
	123	72	Chappell and Shackleton (1986)
Intensity of gravitational anomalies	117	70	Tarakanov et al. (1992)
	150	81	Lambeck (1990)
Intensity of isostatic rebound	167	87	Morner (1981)
Mean	120	71	

parameters. Indeed, the size of the earth, the distribution of the oceans and land mass, parameters of insolation, and parameters of the earth rotation around its axis and around the sun do not change considerably on the time scale of glacial cycles and thus limit the ice volume. It is useful to take the highest (reconstructed) value of ice volume ever observed in the past as an estimation of TH_h, and the volume of the Antarctic ice sheet (which did not melt considerably during the past 600 ka) as an estimation of TH_l.

As seen from Table 4.3, the problem of estimating TH_h is not easy. The range of estimates of the maximum ice volume is 35 million km^3 (i.e., equal to the contemporary ice sheet volume), and the range of estimations of the sea level depression is 103 m (not far from the mean estimation of sea level deviation itself - 120 m). We suppose that TH_h is a theoretical absolute maximum of glacial volume during the past 600 ka, which is why we have taken the maximum absolute value from Table 4.3, approximately 160 m (Denthon and Hughes 1981; Morner 1981) as an estimation of TH_h.

2. We assume that, roughly speaking, the sea level is a linear function of the ice volume (this assumption is not very tough if we take into consideration the range of estimations in Table 4.3):

$$H(t) - H(t_0) = c_v[V(t) - V(t_0)], \tag{4.6}$$

where t is the time, t_0 the present moment, H the global mean sea level, and V the global ice sheet volume. The present volume of the Antarctic ice sheet is supposed to be about 25 x 10^6 km^3 (Suetova 1987), and the volume of the Greenland ice sheet is equal to approximately 3 x 10^6 km^3 (Suetova 1987; Kotlyakov and Lorius 1992). Thus, estimation of $V(t_0)$ in our model is 28 x 10^6 km^3. Using the above estimations as well as mean values from Table 4.3, we take a linear coefficient $c_v = 2.8$.

3. Within the range denoted in (1), changes in the glacial volume and global mean sea level are supposed to be described in the simplest form, by the model of random walk.

4. It is also evident that standard deviations (variances) in the temporal increments of the random walk are not constant and change with the volume of glaciers. This is a very important point. There are several reasons for such an assumption: on the one hand, it is clear that larger ice sheets are, in principle, subjects of larger random influence from the atmosphere and the oceans. On the other hand, during the cold periods large continental ice sheets move far to the south and find themselves within the zone of intensified atmospheric circulation, thus, within the zone of intensified external (random) impulses. Finally, (as was stated by many authors), there is a fundamental difference between contemporary Antarctic and Greenland ice sheets and ancient continental ice sheets: the latter are much less stable because of their southern position and because they are not islanders (solid ice runoff played a much less important role in their balance). As a result, both ice cores in Fig. 4.14 and sea level reconstructions in Figs. 4.1 and 4.2 manifest enormous increase in external ice sheet forcing (by dozens or hundreds of times).

5. However, we would not agree with the judgment (supported by many scholars) about the existence of a stable state of ice sheets, other than polar, in the form of vast ancient ice sheets. On the contrary, from our point of view, there is evidence of extreme instability of ancient ice

covers, and this is corroborated by the above and many other facts. It is natural to adopt a concept of only one zone of stability - polar ice sheets, which did not change considerably during the past 600 ka. It will be shown below that the existence of only one stable state of ice covers is enough to form or amplify an oscillatory signal with characteristic periods of dozens or 100 ka.

Finally, our simplest model for ice volume [measured in units of global sea level changes in accordance with Eq. (4.6)] looks like this:

$$H(t) = H(t-1) + a_H(t), \tag{4.7}$$

where H is the global mean sea level related to the global ice volume as stated in (4.6), t the discrete time, and $a_H(t)$ the sequence of normal uncorrelated random variables with mean value zero. Formulas for the standard deviation of random temporal increments are as follows

$$\sigma(a_H) = \sigma_S, \qquad H_S < H(t) < TH_l \tag{4.8}$$

$$\sigma(a_H) = \sigma_h + c_h\,[H(t) - H_S], \quad TH_h \le H(t) < H_S, \tag{4.9}$$

where TH_l is a low threshold of glaciation that approximately corresponds to the present volume of the Antarctic ice sheet, diapason of stability (TH_l , H_S) is related to the range of small changes in the stable state of glaciation and corresponds to the volume of Greenland ice sheet plus several meters of uncertainty (we suppose that $|TH_l - H_S| \approx 10$ m), TH_h is a high threshold of glaciation that equals a 160-m deviation of the global sea level, and σ_S, σ_h, c_h are constant values. In accordance with the models of recent changes in global mean sea level (Chap. 3) and paleoclimatic considerations (Sect. 4.1) we have taken σ_S equal to 1 m/10 ka. In accordance with the concept of two states of glaciation, the value of σ_h in our model is much higher and equals approximately half the total diapason of glacial cycles, 60 m/10 ka. In turn, the value of coefficient c_h is the same as the value of σ_S , 1 m/10 ka.

The choice of the model time step Δt is very important. Criteria for estimating Δt were the following. (a) It had to be large enough to avoid the necessity of detailed description of behavior of specific ice sheets and specific mechanisms in the ocean-atmosphere-land system (unknown, however). (b) So the time step has to exceed the characteristic time of ice sheets inertia and melting. Huybrechts and Tsiobell (1995), for instance, suggest that "it may take as little as 5000 - 6000 years to entirely remove the Laurentian and Eurasian ice sheets" (p. 115). (c) At the same time, Δt has to be less than 100, 40, and 20 ka - assumed periods of glacial

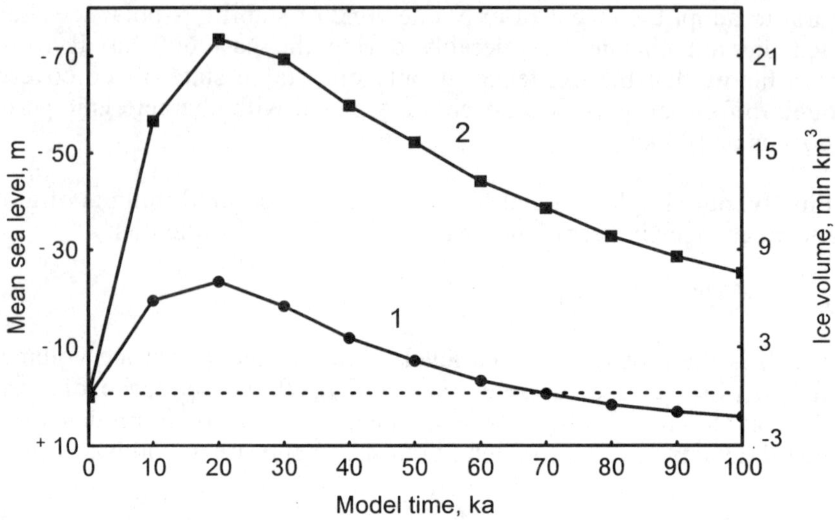

Fig. 4.15. Changes in model mean sea level and mean ice volume (*1*); changes in standard deviations (*2*). Main variant of experiment

glacial oscillation. Thus, it was expedient to choose the order of Δt approximately equal to 10 ka.

Realizations of $H(t)$ in Eq. (4.7) were generated using the Monte-Carlo method. Usually, 10,000 realizations were obtained within each experiment in order to assess statistical parameters of the process. The initial value, $H(t_0)$, in the main variant of experiments was equal to the present global mean sea level, H_S, which, in turn, was supposed to be zero.

4.2.2 Results of Numerical Experiments

A starting result of the main variant of experiment is demonstrated by Fig. 4.15. Here, changes in mean global ice volumes and sea level as well as their standard deviations are presented.

A surprising feature of the curves in Fig. 4.15 is the presence of oscillation: there is a sharp maximum of ice volume (and minimum of sea level) after several dozen ka of model integration; later model glaciers are stabilized. Perhaps changes in Fig. 4.15 are not very realistic (real changes are poorly known, however): they do not reproduce repetitive cycles, perhaps manifest unusual temporal asymmetry. However, from our point of view, patterns in Fig. 4.15 represent an important model prototype of real glacial cycles. They can explain a simple universal mechanism of the emergence of a glacial cycle, without any external periodical forcing or sophisticated nonlinear structures. This experiment can provide answers to the following important and intriguing questions. Why do glacial cycles start? Why, after reaching its maximum, does glaciation enter the stage of degradation? Why, for instance, is the behavior of glaciers after 10 ka of model time (increasing) and after 30 ka (degradation) so different? Recall that the glacier volume at 10 and 30 ka is approximately the same, and all parameters of the model and conditions at the thresholds are the same.

The answers to these questions are related to the very random character of the processes under consideration, and to the presence of different realizations of ice volume variability. Translated into the language of ice sheets, a probabilistic process like this can be explained as follows. Say, at the beginning of integration ($t = 0$) we consider four areas of potential glaciation (Greenland not included): the Laurentide, Cordillerian, Fennoscandian, and Kamchatka areas. Suppose that during first 10,000 years two of them (Nos. 1 and 2) are "lucky": because of random variations of air temperature or other parameters they are covered by growing ice sheets. Two other areas (Nos. 3 and 4) are not lucky and stay without ice; but an average Northern Hemisphere ice sheet demonstrates an overall growth of its total volume. In other words, the mathematical expectation of the Northern Hemisphere ice volume has an inevitable positive trend near the lower threshold of glaciation.

After 10 ka, on average, one of the previously lucky glaciers (say, No. 2) enters the stage of degradation, and No. 1 continues to grow and to move to the south. Why does the mean value of the total NH ice volume continue to increase after 10 ka, taking into consideration the equal probability of positive and negative chances of glaciers No. 1 and No. 2? The answer is: glacier No. 1 has more space and potential for growth to the south than glacier No. 2 for degradation (to the north).

At 20 ka glacier No. 1 is already near the highest limit of its expansion, climatologically and astronomically determined, and cannot provide a large input into the total hemispheric ice volume growth, whereas glaciers Nos. 2-4 continue to be "absorbed" by the northern zone with small temporal increments of ice volume. That is why the behavior of glaciation after 10 and 30 ka of integration is so different: at 10 ka there are more "active" (at least one) glaciers than at 30 ka, although the formal mean ice volume in these two cases can be equal.

The main factor of forming the maximum of glaciation in Fig. 4.15 is the varying standard of temporal increments. In Fig. 4.16, in addition to the basic variant of the model glacier behavior, a pattern corresponding to constant $\sigma(a_H)$ (always equal to 60 m/10 ka) is shown. It is clear that constant $\sigma(a_H)$ gives a curve approaching some constant level that depends on the value of this parameter. In turn, the exact value of the diapason of stability $|TH_l - H_S|$ does not affect considerably the behavior of the

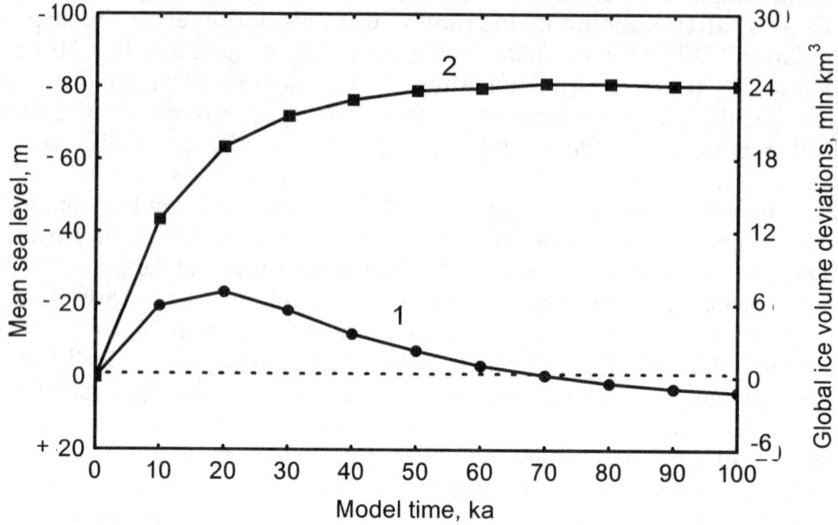

Fig. 4.16. *1* A Fig. 4.15.1; *2* as Fig. 4.15.1 but with constant standard deviations of temporal increments (60 m/10 ka)

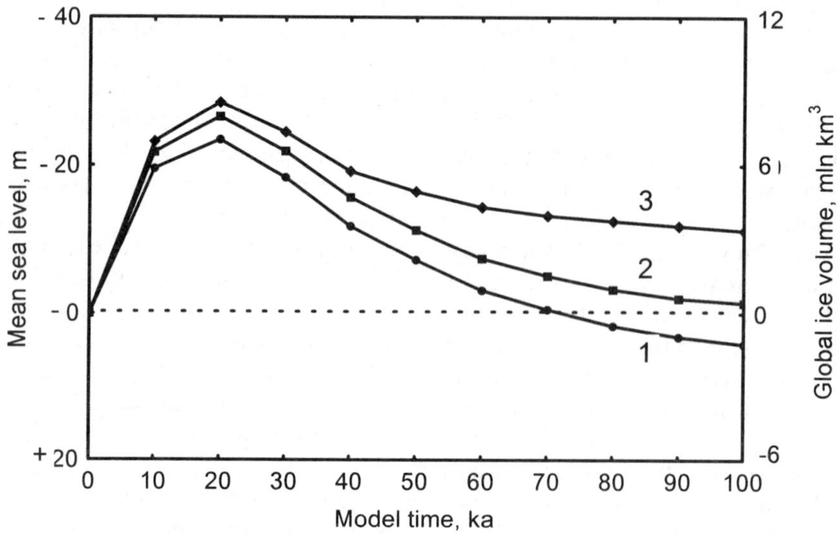

Fig. 4.17. *1* A Fig. 4.15.1; *2* as Fig. 4.15.1 but for $\left|TH_l - H_S\right| = 5$ m; *3* as Fig. 4.15.1 but for $\left|TH_l - H_S\right| = 2$ m

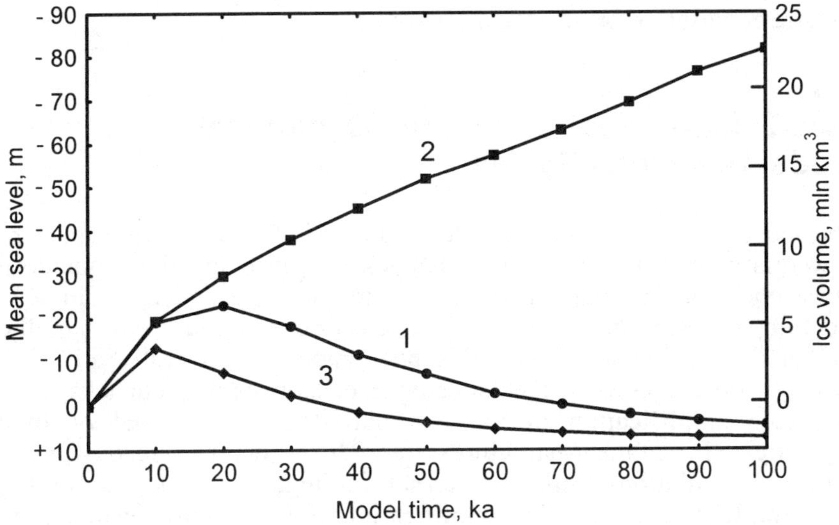

Fig. 4.18. *1* A Fig. 4.15.1; *2* as Fig. 4.15.1 but for $TH_h = \infty$; *3* as Fig. 4.15.1 but for $TH_h =$ 52 m

model glacier cover. Thus, decreasing $|TH_l - H_S|$ by five times (cf. Figs. 4.17.1 and 4.17.2) does not dramatically change the form of the curve.

Finally, Fig. 4.18 demonstrates the sensitivity of the model to the value of the high threshold TH_h. Three cases are presented in this figure: when TH_h = 120 m (basic experiment), TH_h = 52 m (the lowest estimation from Table 4.3), and $TH_h = \infty$. It is interesting that in the latter case the mean value of glaciers does not grow as quickly as could be expected. This behavior of model glaciers is caused by the zone of stability which attracts a large part of model realizations.

We can add to the above short description of a very important phenomenon (of emerging oscillation within the threshold random walk with a varying standard of increments) that similar models and patterns seemingly govern the behavior of other structures on the surface of the earth: ecosystems, specific species, and even ethnic groups and civilizations (Dobrovolski 1998). Three important features unify all the above structures: (1) the presence of natural limits to their expansion or degradation (in living and social systems the lower threshold equals zero, i.e. signifies the disappearance of all representatives of the system), (2) the important role of random impulses affecting the system, (3) the sharp increase in the effect of these impulses corresponding to the growth of the system. For example, it is shown in Dobrovolski (1998) that the average behavior of 40 ethnic groups, characteristic species, and ecosystems is satisfactorily described by equations similar to (4.7) - (4.9). Furthermore, these equations can generate behavior close to that often described by the logistic equation or by strange attractors.

4.3 Two-Dimensional Stochastic Dynamical Model of Glacial Cycles

The most difficult problem in the studies of glacial cycles is to explain how relatively small periodical changes in astronomical parameters can cause enormous changes in the global temperature and ice volumes. As mentioned in Section 4.1, there are dozens or even hundreds of hypotheses and models trying to explain this phenomenon. Practically all of them include, sometimes very sophisticated, mechanisms of feedbacks in order to create amplification of the external astronomic signal or include sophisticated nonlinear structures (strange attractors, etc.), which can generate oscillations without external forcing. Our task was quite the opposite: to demonstrate that amplification of an external signal can occur without any sophisticated deterministic feedbacks or deterministic nonlinear attractors. From our point of view, the most simple and natural assumptions suffice to explain glacial cycles. Below, we will try to demonstrate this, using a more realistic model than the zero-dimensional one.

4.3.1 Mathematical Model, Parameters, and Experiments

4.3.1.1 Computational Area

In Fig. 4.19 model land and oceans are shown. We do not describe internal ice sheet processes in the model, so that (also because of large time steps - 10 ka, see Sect. 4.2), spatial resolution can not be very detailed: 5° x 10°. Cylindrical rectangular cartographic projection was adopted; its distortions, evidently, were taken into account. Ice sheets were allowed to occupy not only the land but also parts of the oceanic shelf with depths less than 200 m (denoted by dashed lines in Fig. 4.19).

Also, in Fig. 4.19 regions where calculations were made for specific ice sheets are shown: Taimyr (1), Cordillerian (2), Laurentide (3), Fennoscandian (4), Kamchatka (5), and Tibet (6).

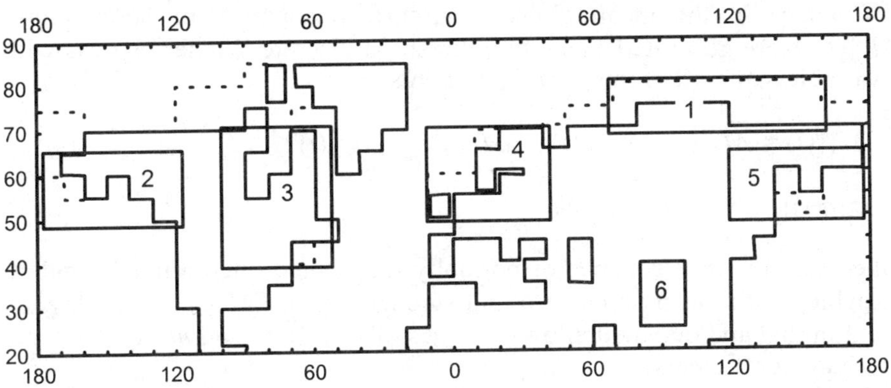

Fig. 4.19. Grid of the two-dimensional stochastic dynamical model. Computational areas of specific ice sheets: *1* Taimyr; *2* Cordillerian; *3* Laurentide; *4* Fennoscandian; *5* Kamchatka; *6* Tibet

4.3.1.2
Equations and Parameters of the Model

Processes of ice sheet growth and melting are related mostly to the summer air temperatures. Therefore, the main independent variable in our simple 2D model was near-surface summer air temperature $T_s(x,y,t)$. It was calculated in the following way:

$$T_s(x,y,t) = T_{s,o}(x,y) + \Delta T_{s,g}(t) . \qquad (4.10)$$

Here, x, y are plane coordinates; t is time; $T_{s,o}(x,y)$ the field of present-day summer temperatures; $\Delta T_{s,g}(t)$ changes in global summer temperatures. Data on $T_{s,o}(x,y)$ - mean July temperatures - were taken from Bol'shoi Sovetskii atlas mira (1964-).

Evidently, Eq. (4.10) is a very rough approximation: it does not take into account internal dynamics and feedbacks of the ocean-atmosphere-ice sheet system. However, from our point of view, it is better than attempts to use large GCMs connected with oceanic and ice sheet models because details of many processes and detailed parameters of the climatic system in the past are not known (see discussion in Sect. 4.1). Seemingly, using GCMs we cannot estimate realistic 2D spatial and temporal distributions of climatic parameters of dozens and hundreds of ka ago. Again, we stress that we are trying to construct a most simple structure that could reproduce realistic glacial cycles with a minimum of a priori assumptions and prescribed parameters. For this purpose, also taking into consideration a very large time step in the model, an equation like (4.10) seems appropriate, as a first approximation.

Changes in the mean global near-surface summer air temperature, $\Delta T_{s,g}(t)$ were generated as the two-threshold discrete Wiener process with a variable standard of temporal increments:

$$\Delta T_{s,g}(t) = \Delta T_{s,g}(t-1) + a(t), \quad \Delta T_{s,g,min} \leq \Delta T_{s,g}(t) \leq \Delta T_{s,g,max} \tag{4.11}$$

$$\sigma[a(t)] = m_1 * \Delta T_{s,g}(t) + m_2, \tag{4.12}$$

where $a(t)$ is the sequence of normally distributed uncorrelated, random variables with mean value zero and standard $\sigma(a)$; $\Delta T_{s,g,min}$ and $\Delta T_{s,g,max}$ are minimal and maximal allowed values of $\Delta T_{s,g}(t)$; and m_1 and m_2 are constant coefficients.

Different values of parameters $\Delta T_{s,g,min}$, $\Delta T_{s,g,max}$, m_1 , m_2 were tested in the experiments. Basic adopted values were correspondingly: -20 °C, +1 - +2°C, -4.9, 10.2 °C . Estimation for $\Delta T_{s,g,min}$ was taken from Salamatin et al. (1997); this is a compromise between much larger values (for instance, Stepanov 1996), on the one hand, and modest estimations (Boulton et al. 1995; Fabre et al. 1995; Rhode et al. 1995, etc.), on the other. Estimation of $\Delta T_{s,g,max}$ was taken from Velichko (1982); similar assessments are given by Boulton et al. (1995), Fabre et al. (1995) (averaged estimation), and others. Values of m_1 and m_2 were calculated using cited results by Ditlevsen et al. (1996) on changes in climatic variability, combined with the above estimations of temperature maximum and minimum, and estimations by Dobrovolski (1992) of the present-day global temperature variability, recalculated for time scales of the order of 10 ka (time step of the model).

A very simple equation for the precipitation rate $P(x,y,t)$ was:

$$P(x,y,t) = P_o(x,y) * [1 + k_p \Delta T_{s,g}(t)], \tag{4.13}$$

where $P(x,y,t)$ is the present-day annual precipitation taken from Jaeger (1976). The value of coefficient k_p was taken from Huybrechts and Tsiobell (1995) and represented 0.04 $(°C)^{-1}$; in accordance with the cited work, Eq. (4.13) does not take into account the elevation of ice sheets.

Changes in ice sheet thickness, $H(x,y,t)$, were calculated using formulas:

$$H(x,y,t) = H(x,y,t-1) + \Delta t * P(x,y,t-1) / 0.91, \quad T_s(x,y,t-1) < 0 \tag{4.14}$$

$$H(x,y,t) = 0, \quad T_s(x,y,t-1) \geq 0, \tag{4.15}$$

where Δt is the time step; and 0.91 the density of ice (see Fabre et al. 1995, for instance).

Finally, external forcing of the model was created by adding to the values of $\Delta T_{s,g}(t)$, $\Delta T_{s,g,min}$, $\Delta T_{s,g,max}$ in Eq.(4.11) the periodical term $EX(t)$:

$$EX(t) = A \sin(\omega t + \varphi) . \tag{4.16}$$

Different values of the amplitude, phases, and frequencies in Eq. (4.16) were tested. Evidently, frequencies corresponding to periods around 100, 40, and 20 ka represented the primary interest.

4.3.1.3 Numerical Experiments

Equation (4.11) was integrated using the Monte Carlo method. The time step was always equal to 10 ka (see Sect. 4.2). Results of experiments were usually averaged over a large number of realizations - 1,000, and sometimes 3,000 and more, although experiments with only one or a few realizations were also carried out. Integrations started from the present conditions, i.e., from the value $\Delta T_{s,g}(t) = 0$, and from present Greenland ice thickness (taken using data from Fabre et al. 1995). The length of each realization was 600 ka in order to generate six glacial cycles. In special cases, for the purposes of exact estimation of spectra, longer realizations were obtained.

The problem of the number of realizations in the numerical experiment is important. Averaging over a large number of realizations gives an ideal theoretical behavior of the mean values of climatic and glaciological parameters - air temperature, precipitation, and ice thickness, and enables us to obtain three-dimensional statistics and cross-correlations of the above parameters. This is an important advantage of a simple model such as the present one, in comparison with large ocean-atmosphere-glaciers-land surface models, which cannot give the necessary number of realizations (thousands) with the length of hundreds of ka each. However, experiments with a fewer number of realizations are also interesting. Although spatial correlations were not introduced explicitly into the model by combining the number of realizations, we can implicitly reproduce situations with various levels of correlation between different glacial areas. Averaging over the large number of realizations corresponds to the situation when different glaciers are poorly correlated with each other. On the contrary, one realization reflects the situation with a high spatial correlation of glaciation.

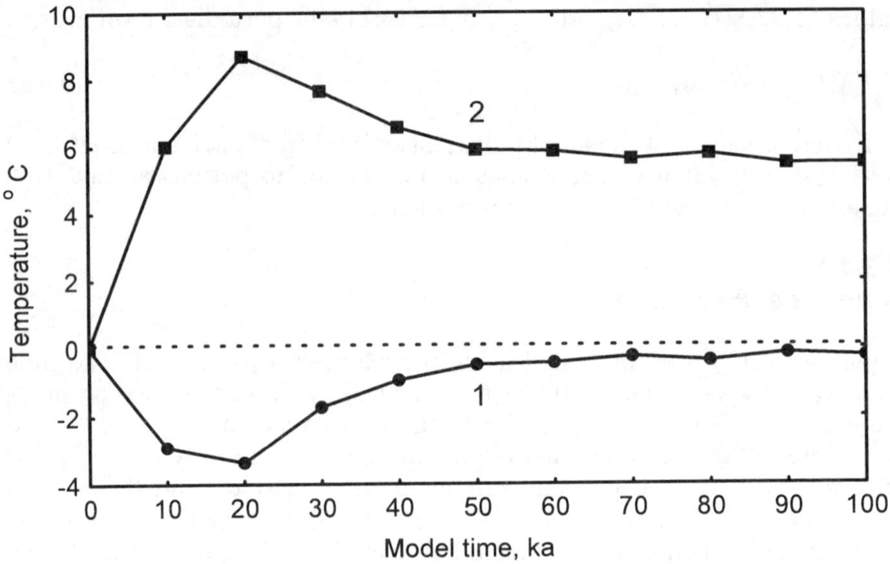

Fig. 4.20. Mean temperatures (*1*), standard deviations of mean temperatures (*2*). Experiment without external forcing

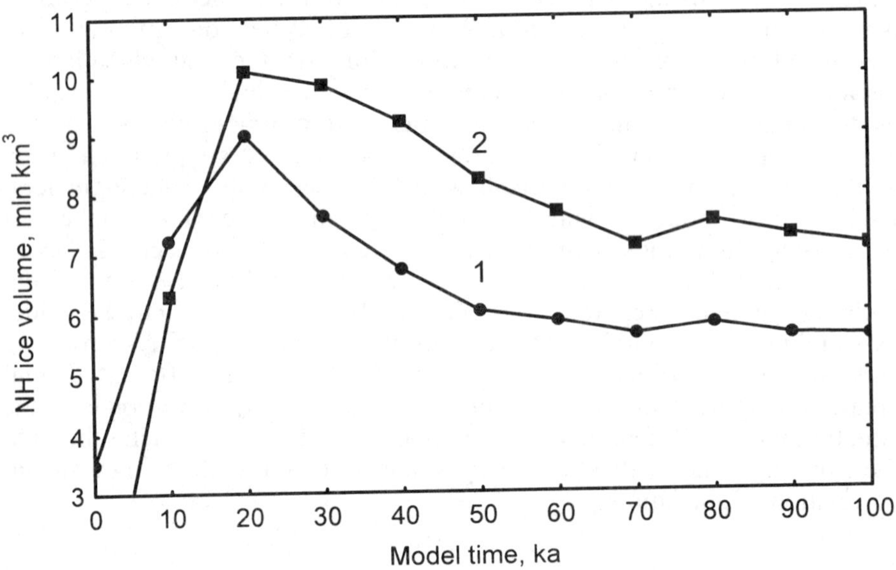

Fig. 4.21. Mean model global ice volumes (*1*), standard deviations of ice volumes (*2*). Experiment without external forcing

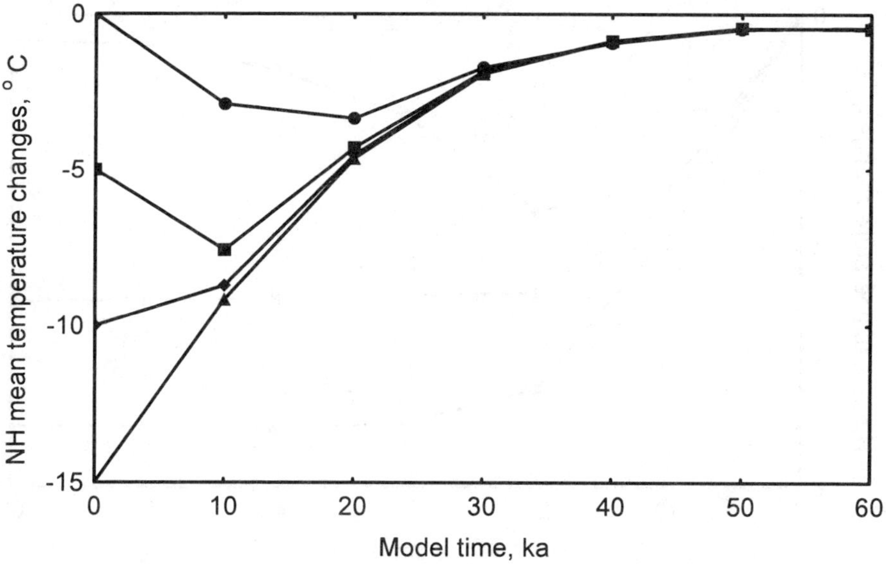

Fig. 4.22. Sensitivity of model Northern Hemisphere temperatures to initial conditions

4.3.2 Results

4.3.2.1 Experiments Without External Forcing

In Figs. 4.20 and 4.21 changes in mean Northern Hemisphere air temperatures and ice volumes as well as their standard deviations are presented. Parameters of the model correspond to the above estimated values. Parameter $\Delta T_{s,g,max}$ is equal to + 2 °C in this experiment, parameter A from Eq. (4.16) was zero. Temperature and ice volume were averaged over 1,000 realizations.

Results shown in Figs. 4.20 and 4.21 are analogs of results of the 0D model (Sect. 4.2). The very existence of the oscillation in the experiment without external forcing depends on three parameters of the model: initial conditions $T_{s,o}$, maximal global temperature $\Delta T_{s,g,max}$, and the parameter describing relations between global temperature and temperature random increments m_1. In Fig. 4.22 the sensitivity of the model results to various initial conditions are presented. Extremely low initial global temperature (-10 to -15 °C) is not transformed into oscillation, whereas present and intermediate (+5 °C) temperature conditions result in temperature minimum at 10 or 20 ka. This result is corroborated by observational data: transformation of extremely cold conditions into the warm climate is

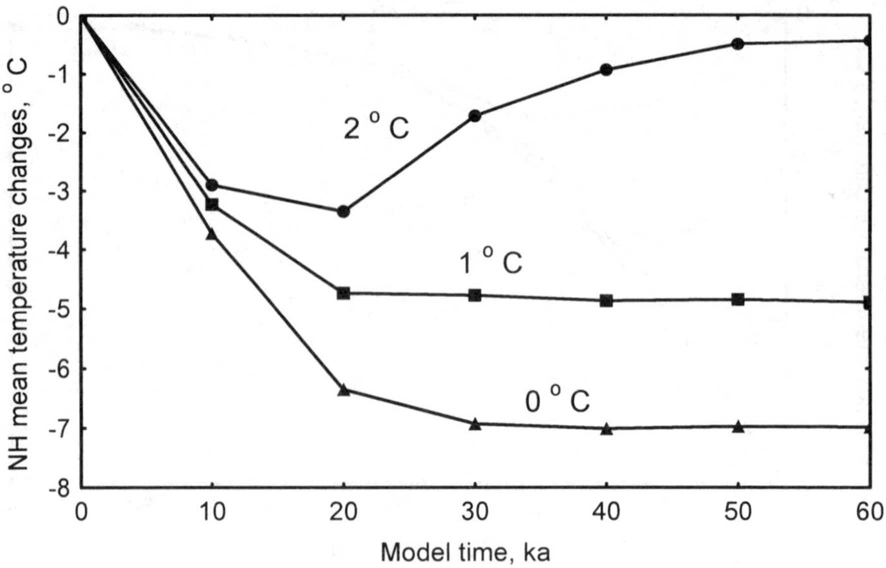

Fig. 4.23. Northern Hemisphere model temperature sensitivity to the high temperature limit: 0, 1, 2 °C

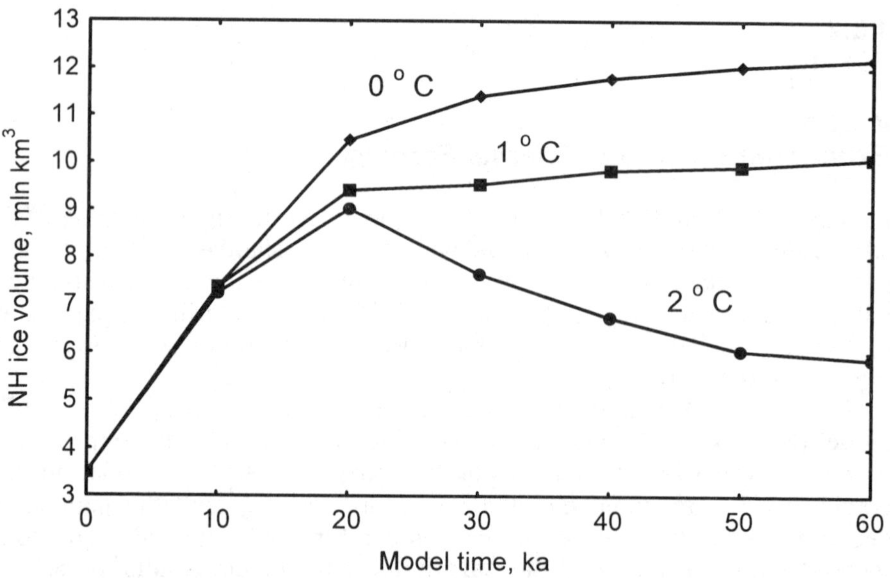

Fig. 4.24. Model ice volume sensitivity to initial conditions

accompanied by much faster destruction of glaciers than transfer from intermediate glacial conditions to warmer conditions.

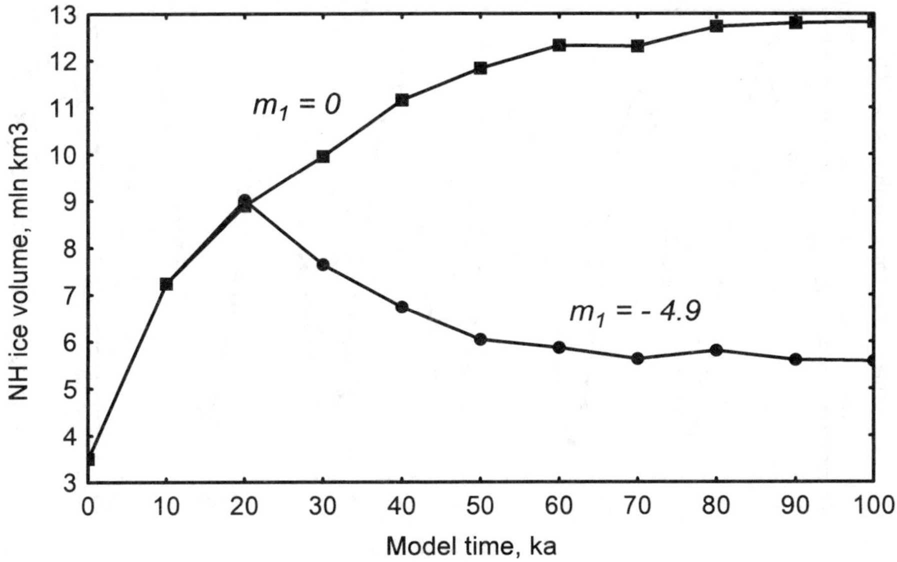

Fig. 4.25. Northern Hemisphere model ice volume sensitivity to parameter m_1

Another parameter which governs the entering into oscillation, is the maximum global temperature (Figs. 4.23 and 4.24). Although 1 and 0 °C do not seem very realistic initial conditions, these figures demonstrate that small changes in the maximum global temperatures can play an important role.

Evidently, the regime of oscillation is controlled by the parameter m_l . Two variants of glacier behavior are presented in Fig. 4.25: with constant (equal to the present one) standard deviations of global temperature increments ($m_l = 0$), and with standard deviations linearly changing with global temperature (m_l = - 4.9). Random temperature increments with constant standards do not create glaciation maximum at 20 ka. It is interesting that Fig. 4.25 gives an almost perfect pattern of bifurcation with the point of bifurcation at 20 ka and almost perfect symmetry of curves after this point with a horizontal line at 9 x 10^6 km^3 as the axis of symmetry. In this way, Fig. 4.25 resembles the famous patterns of Feigenbaum mapping. Perhaps it would be useful in the future to investigate in more detail the behavior of the model in the vicinities of the point of bifurcation.

Another interesting result appearing in Fig. 4.25 is that larger, on average, temperature increments (when m_l = - 4.9) eventually lead to a much lower steady level of glaciation than lesser increments. The cause of this paradoxical feature is the much quicker transfer of the glacier regime to the stable zone in the first case.

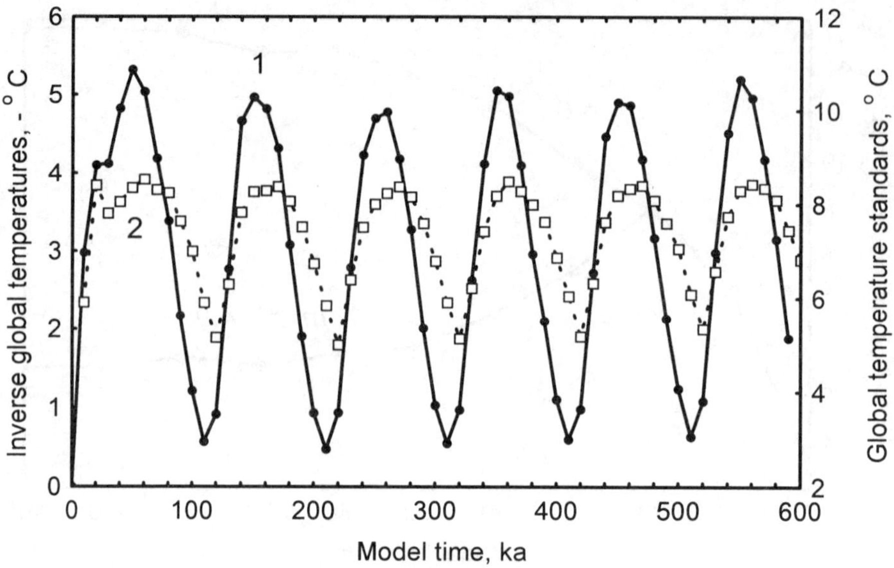

Fig. 4.26. Northern Hemispheric model temperatures (*1*), standard deviations (*2*). Experiment with external 100-ka periodical forcing

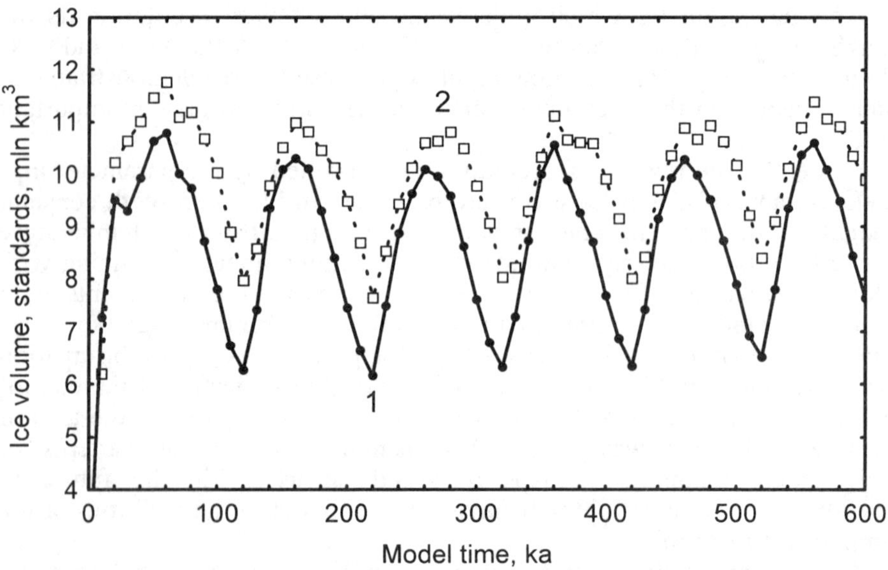

Fig. 4.27. Northern Hemispheric model ice volume (*1*), standard deviations (*2*). Experiment with external 100-ka periodical forcing

4.3.2.2 Experiments with External Forcing. Globally Averaged Results

In Figures 4.26 and 4.27 results of experiments with external periodical (with period 100 ka) forcing are presented. The amplitude of additional temperature harmonic A was taken as 0.5 °C in these calculations; φ was equal to $\pi/2$ [see Eq. 4.16]. Curves in Figs. 4.26 and 4.27 are averaged over 3,000 realizations each, i.e., totally equations of the model were integrated for 1,800,000,000 years for each experiment. Other parameters of the model were as indicated in previous sections.

These results show that external periodic forcing with minor amplitude (0.5 °C) causes oscillations of considerable range: about 4.4 °C for the NH temperature; 3.9×10^6 km^3 for the NH ice volume. Standard deviations of NH temperature and ice volume manifest oscillations as well.

The above results can be compared with existing paleoreconstructed time series (Figs. 4.10, 4.12). The comparison between modeled and observed glacial cycles demonstrates the following features of similarity:

1. Averaged modeled and observed cycles, the latter being recalculated from δ^{18}O into temperature units, have a similar range: about -4 or -5 °C.
2. Unlike the two last glacial cycles, averaged (both observed and modeled) cycles have almost symmetric form.
3. The range of averaged 100-ka oscillations is less than the range of specific actual cycles (the range between maximum and minimum within the specific cycle).
4. The minimum of the averaged 100-ka cycle is higher than the observed minimal values, and the maximum of the averaged cycle is lower than the observed maximal values.
5. There is a shift (time lag) between the averaged temperature curve and averaged deviation standards: the changes in mean temperature precede changes in standards. Thus periods of ice sheet growth are more stable than periods of ice sheet degradation in both observation and model results.
6. The maximum of external forcing precedes the minimum of modeled averaged glaciation cycles by 20 ka. Recall that these results are similar to those obtained by Muller and MacDonald (1997) from both paleoreconstructions and investigations of orbital inclination forcing.

We would like to make some comments concerning results (5) and (6). It is important to stress that in our model we did not take into account internal dynamics of glaciers, their horizontal movements, inertia, etc.: glaciers were allowed to completely disappear (and in a moment - in terms of the model time step) in the case of positive summer air

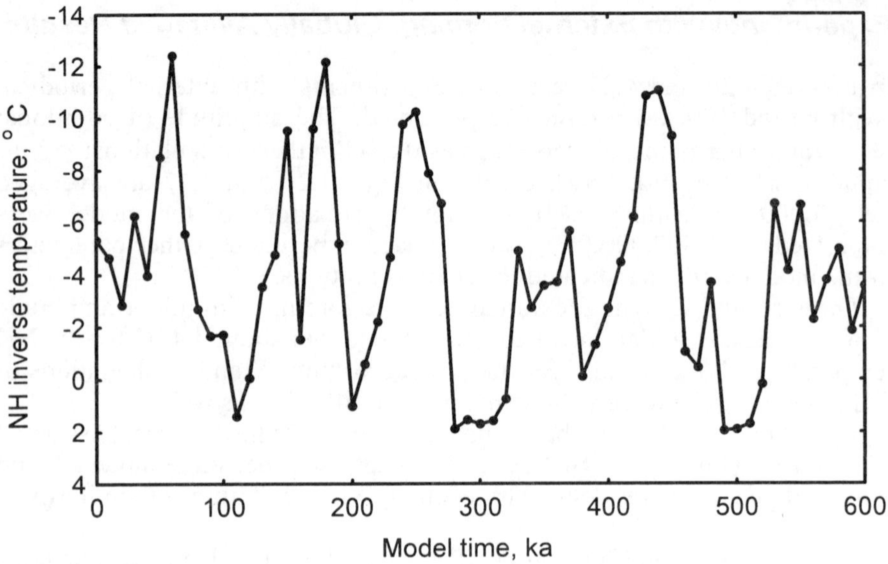

Fig. 4.28. Northern Hemispheric six-realization-average temperatures

temperatures. The cause of the above time lags was more general and more profound - long periods of time (lags) were needed in order to accumulate positive or negative anomalies of air temperature.

Evidently, real glacial cycles differ by their form from the model cycles averaged over a large number of realizations. To compare model and observed time series, it is expedient to obtain model registrations averaged over a fewer number of realizations. In Figs. 4.28 and 4.29 model glacier dynamics, averaged over six realizations, are presented. These results can be treated as follows: they can be considered as an average dynamics of six regions of glaciation, which are only partly correlated with each other. Such an assumption is rather realistic, at least more realistic than the assumption of absolutely synchronous changes in all glacier regions. Recall that correlation between different glaciers is created in this case in the model by the synchronous effect of external periodic forcing, whereas asynchronous features are caused by non-correlated random temperature increments.

To be sure, the glacial dynamics in Figs. 4.28 and 4.29 is more realistic in comparison with the observed one (Fig. 4.10). Glacial cycles are not all of the same length: the fastest cycle (the 3rd one, 200-280 ka of model time) is about 1.5 times shorter than the 5th model cycle). Also glacial cycles are often interrupted by the periods of deglaciation. As well as in reconstructed records, some of the model cycles are sharp (real last two glacial cycles in Fig. 4.10, and model cycles No 3, 5 in Fig. 4.28), and

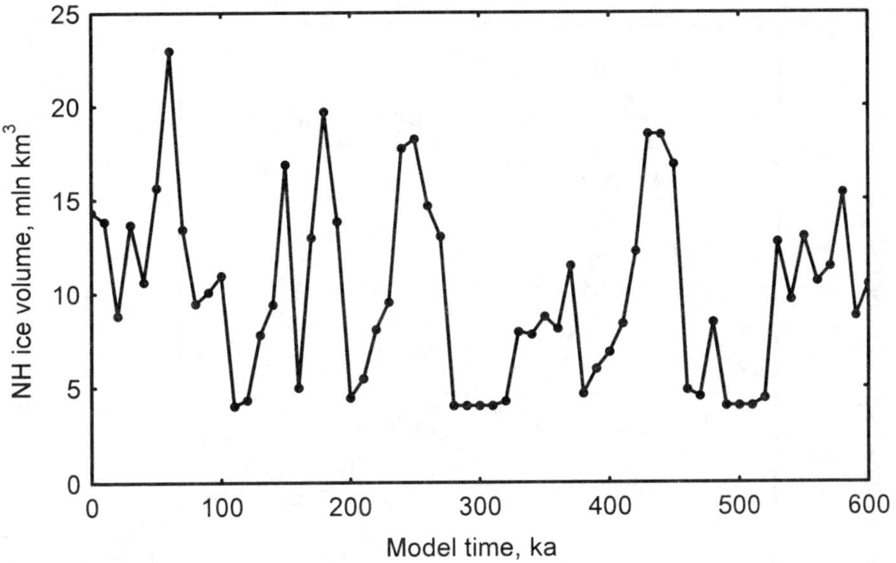

Fig. 4.29. Northern Hemispheric six-realization-average ice volume

some are not so evident (600-500 ka B.P. in Fig. 4.10, cycle No. 6 in Fig. 4.28).

It is important that, unlike patterns averaged over a large number of realizations, six-realizations-averaged curves might not be symmetric. For the sharpest model glacial cycles (Nos. 3 and 5 in Fig. 4.28), as well as for the last two real glacial cycles, the stage of increasing glaciation is much longer than the stage of deglaciation.

In general, model changes in glacier volume (Fig. 4.29) are well correlated with changes in the air temperature (Fig. 4.28). Some differences are remarkable, however, especially when the glacial cycles are not sharp. For instance, in the 6th model cycle, temperature minimum (Fig. 4.28) occurs 50 ka earlier than the glacial maximum.

As a result of the above considerations we can propose the following hypothesis. Glacial cycles, both observed and modeled, can be divided into two groups: sharp glacial cycles (the last two real cycles, and model cycles No. 3, 5) on one hand, and less evident cycles, on the other. A more or less steady and long increase in glacial volume as well as quick melting is characteristic for the first group of cycles, whereas enormous interruptions or smoothed changes and the absence of visible asymmetry are characteristic for the second group. The cause of the difference, from our point of view, can be the different correlation between glacier regions: in the first case glaciation seems to occur synchronously because of the coincidence of random increments in the temperatures.

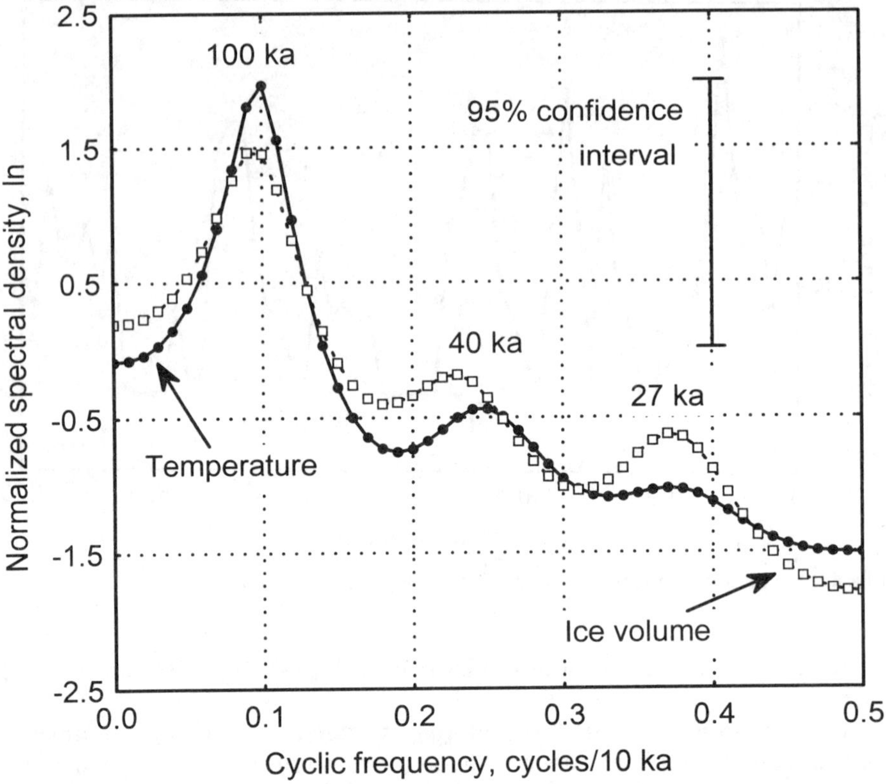

Fig. 4.30. Spectral densities of model temperatures and ice volumes estimated using realizations from Figs. 4.28 and 4.29

It is interesting to compare spectra of modeled and reconstructed time series. In Fig. 4.30 spectra of model temperature and glacier changes are presented. As in the case of real glacial cycles (Fig. 4.11), the peak at 100 ka is evident and rather sharp. Peaks at periods of 40 and 27 ka are also here, although they are smaller than peaks at 41 and 23 ka in the reconstructed data. Because of the proximity of these two small peaks to 50 and 25 ka respectively, their existence might be related to harmonics of the 100-ka main period. In turn, 50- and 25-ka harmonics reflect, correspondingly, the asymmetry of the cycles and the relatively flat form of their maxima and minima.

Paradoxical is the pattern of probability distribution of glacier volumes in Fig. 4.31. Here, category 0-5 mln km^3 is not presented because it is related mostly to the Greenland ice sheet, which is always present in experiments and reconstructions, and does not say much about features of the NH glacier changes; yet it is difficult to compare probabilities near

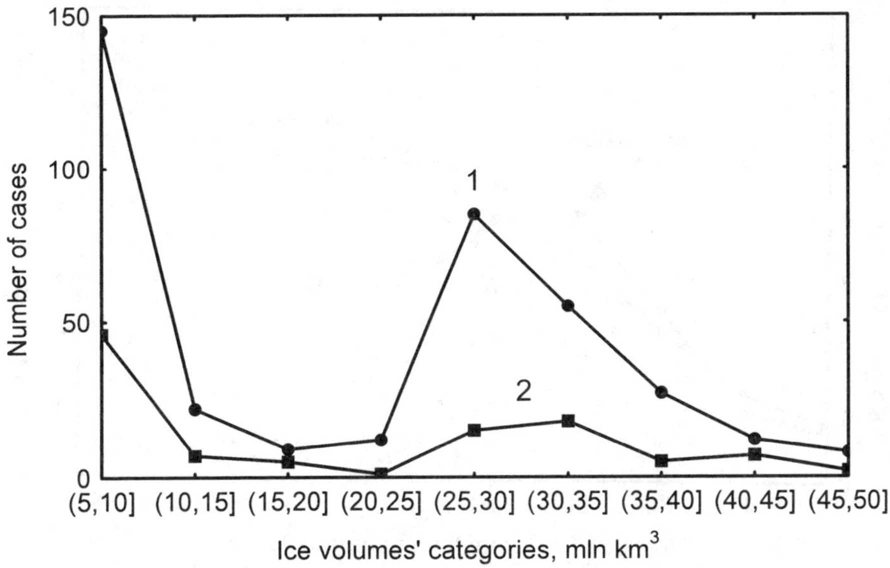

Fig. 4.31. Ice volume probability distribution densities at 60 ka (*1*) and 120 ka (*2*) of model time

zero and probabilities within categories of 20-35 x 10^6 km^3 because the latter correspond to enormous volumes of glaciers. Nevertheless, it is seen from this figure that the largest difference in probabilities during deglaciation on the one hand and glaciation, on the other, corresponds to two zones of attraction: near zero ice volumes, and between 25-30 x 10^6 km^3. Thus, the most surprising is that the external forcing of only 0.5 °C in amplitude, resulting in the average range of glacier volume changes of 4.5 x 10^6 km^3, gives the range of zones of glacier volume attraction of 20-25 x 10^6 km^3. Perhaps this explains the nature of the enormous amplification of the external astronomic signal. Also these results demonstrate that the numerous attempts to deduce the very large range of deterministically described glaciation cycles from small changes in astronomical parameters are not necessary. Very small changes in external forcing and relatively small changes in averaged glacial volumes born very large changes in the probabilistic structure of glaciation cycles. In other words, the very probabilistic (stochastic) approach to the problem with a minimum of natural a priori assumptions can easily replace numerous sophisticated mechanisms and processes (feedbacks, attractors, GCMs) being usually fitted in order to reproduce amplification of external forcing.

Experiments were conducted on the model sensitivity to the key parameters of external periodical forcing: amplitude, period, and phase. It is evident from Fig. 4.32 (the amplitude of the forcing is always 0.5 °C

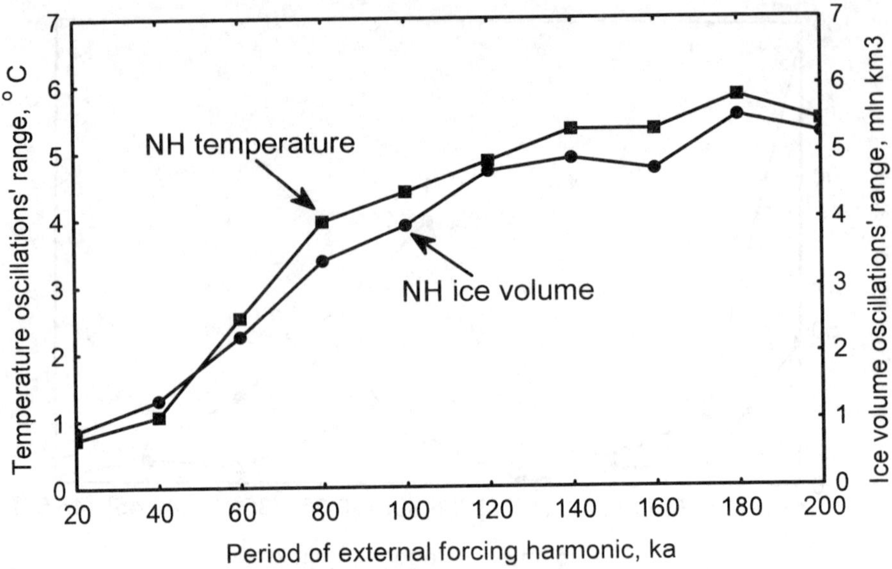

Fig. 4.32. Model sensitivity to the period of external oscillatory forcing

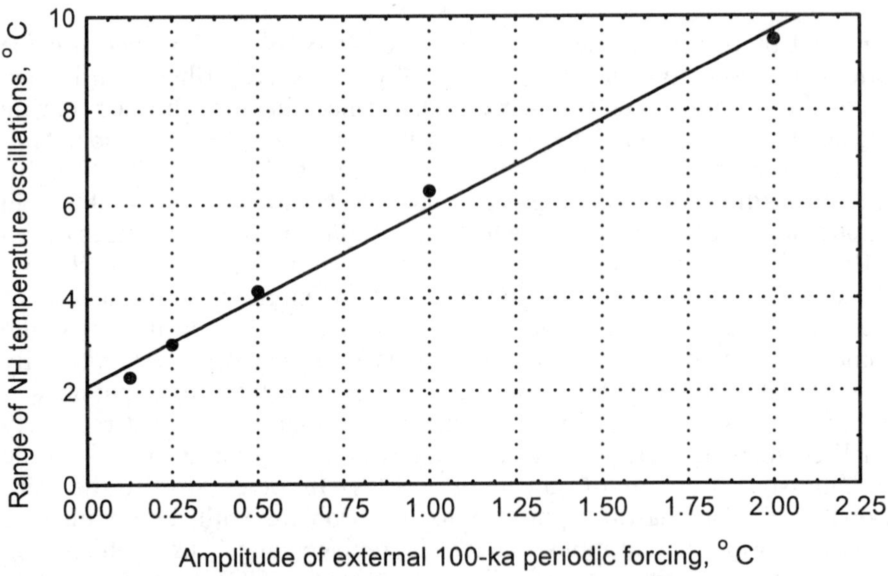

Fig. 4.33. Model sensitivity to the amplitude of external oscillatory forcing

here) that the model amplification of external signal is practically absent at the period of 20 ka, and constantly grows with the period of forcing,

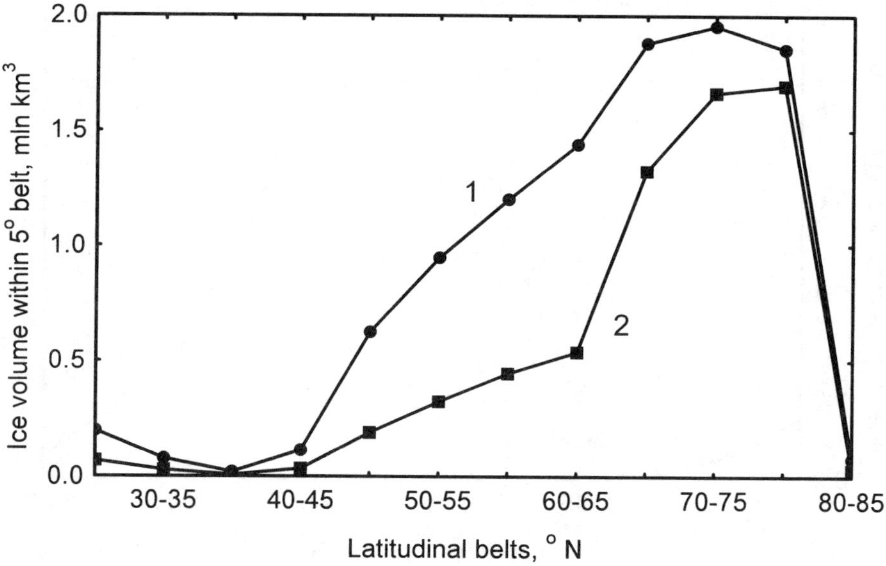

Fig. 4.34. Latitudinal distributions of the model ice volumes at maximum (*1*) and minimum (*2*) of the 100-ka glacial cycle

both for temperature and glacier volume. Perhaps this is the explanation of the outstanding role of 100-ka peak in the spectra: even if the amplitudes of forcing at the periods around 20 and 40 ka were comparable with that of 100 kaa, their effect would be four to five times less in amplitude, and 16-25 times less in terms of spectral density.

Finally, sensitivity to the amplitude of external periodic forcing is practically linear (Fig. 4.33). The phase of external forcing did not influence the results save for a short period of time just after the start of integrations.

4.3.2.3 Zonally Averaged Results

In Fig. 4.34 changes in ice volume (for both maximum and minimum of the model 100-ka glacial cycle) versus latitude are presented. The difference between curve 1 and curve 2 represents the distribution of the range of 100-ka oscillations over latitudes and is shown in Fig. 4.35. The latitude of the maximum in the range of ice volume oscillations (60-65°N) is far to the south of the latitude of the maximum of glacial volumes (70-75°N), which, in turn, is caused by the constant presence of the Greenland ice sheet. As seen from Fig. 4.19, the maximum of the range of glacial cycles coincides with the zone of glaciers of moderate latitudes.

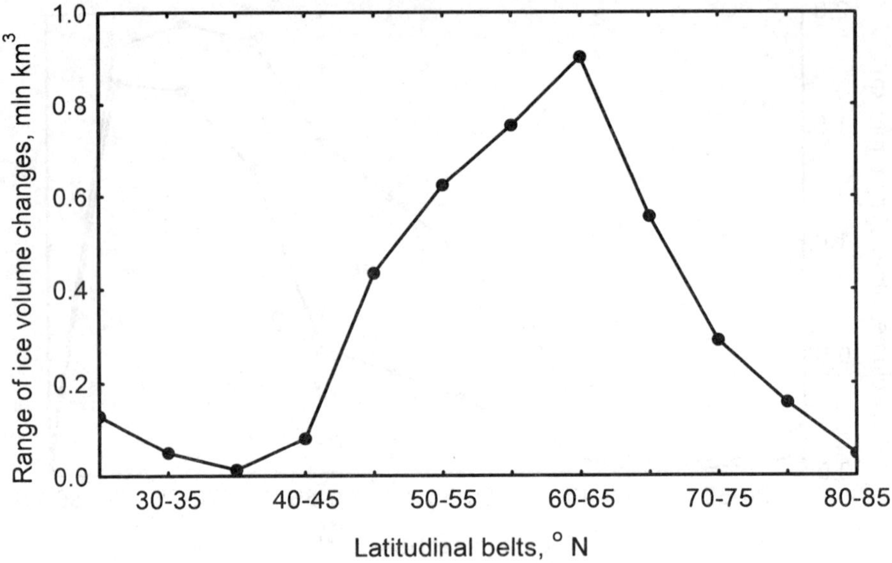

Fig. 4.35. Latitudinal distribution of the range of 100-ka ice volume cycles

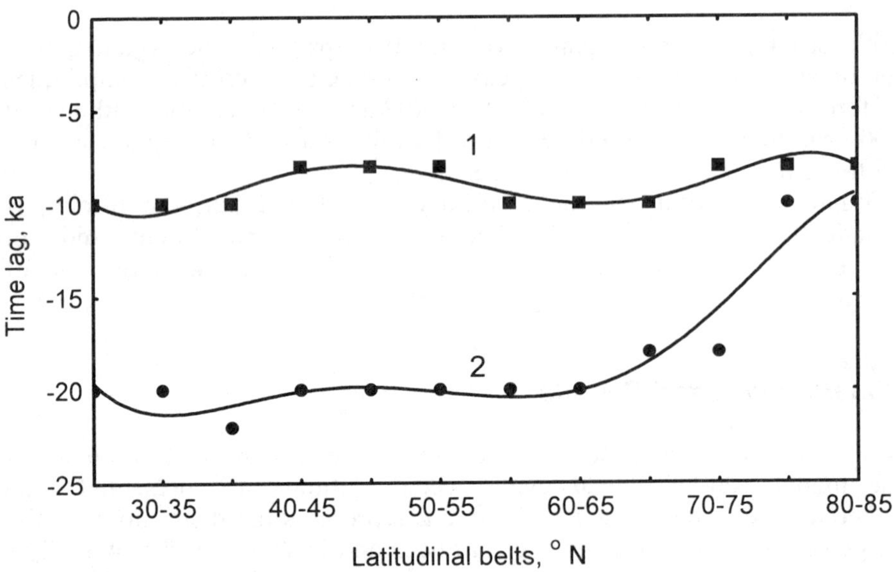

Fig. 4.36. Latitudinal distribution of the phases of glaciation maxima (*1*) and minima (*2*)

The zonal distribution of phases of glaciation is shown in Fig. 4.36. In general (as was mentioned in previous sections), there is a time lag

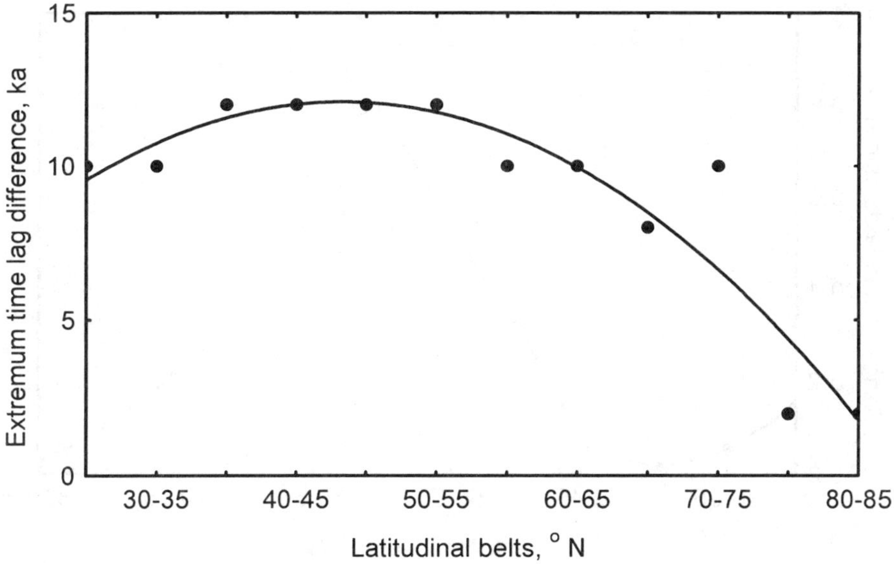

Fig. 4.37. Changes in the asymmetry of 100-ka glacial cycles

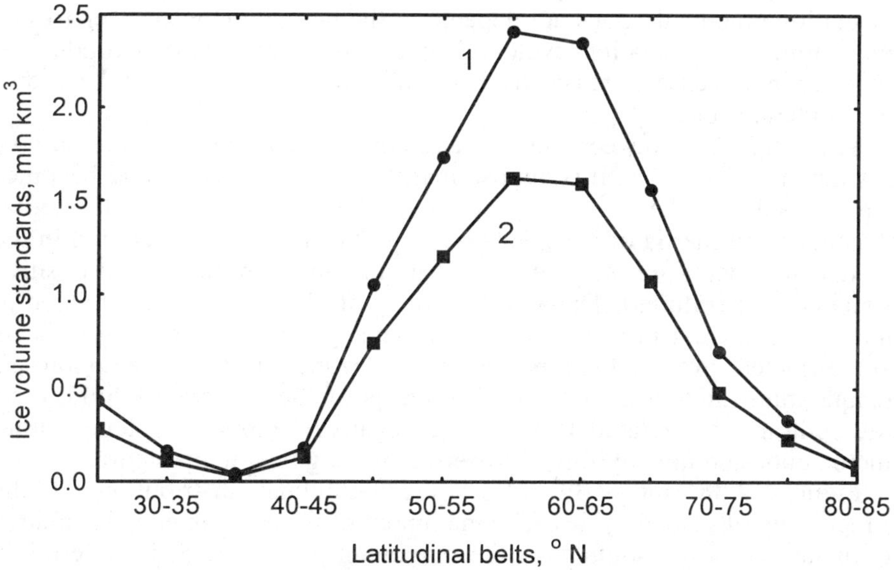

Fig. 4.38. A Fig. 4.34 but for standard deviations

between the external periodical forcing and changes in glacial volumes. It is evident from this figure that the phase of maxima of zonally averaged

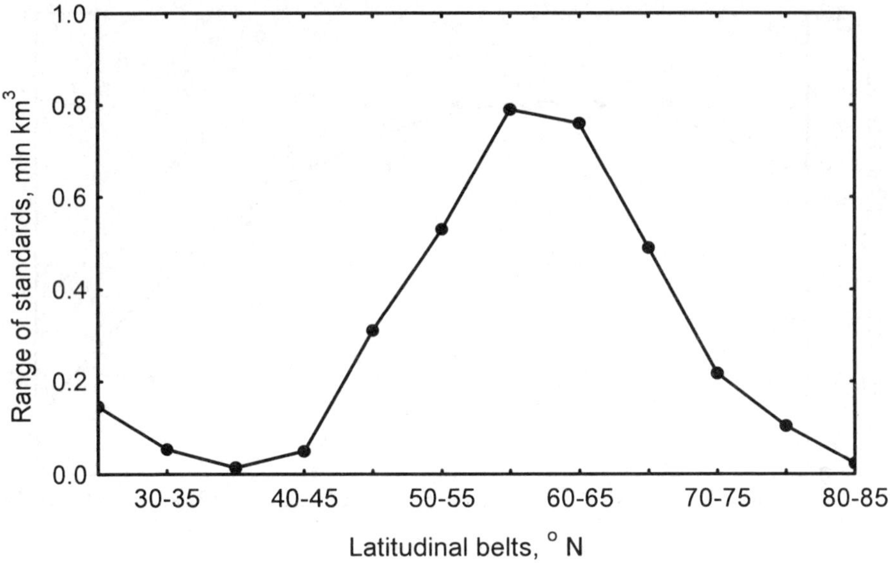

Fig. 4.39. A Fig. 4.35 but for standard deviations

glaciation practically does not depend on the latitude. However, phases of the minima of glaciation cycles change considerably with latitude: the time lag in this case increases from 10 ka in the arctic zone to 20 ka within the moderate zone.

The difference between curve 1 and curve 2 represents the measure of asymmetry (Fig. 4.37). It is almost absent in the arctic zone, and becomes considerable to the south. It should be noted that details of internal dynamics and inertia of ice sheets were not taken into consideration in the model (ice sheets are allowed to melt in a moment in the case of positive summer temperatures). The above features of glacial cycles are related to more fundamental processes in the climate system: "unlucky" ice sheets are attracted over a long period by the zone of minor temporal air temperature random increments. Ice sheets of the moderate latitudes, on the contrary, are related to the large negative temporal air temperature increments, and quickly follow lowering of the global temperature.

Figures 4.38 and 4.39 reflect the latitudinal distribution of the irregularity of glacial cycles. The maximum of these parameters is situated even farther to the south than the maximum in Fig. 4.35; it is certainly related to the center of the latitudes of «moderate zone» ice sheets. For instance, this maximum coincides with the latitude of the Valdai glaciation, the most famous within the territory of the former USSR,.

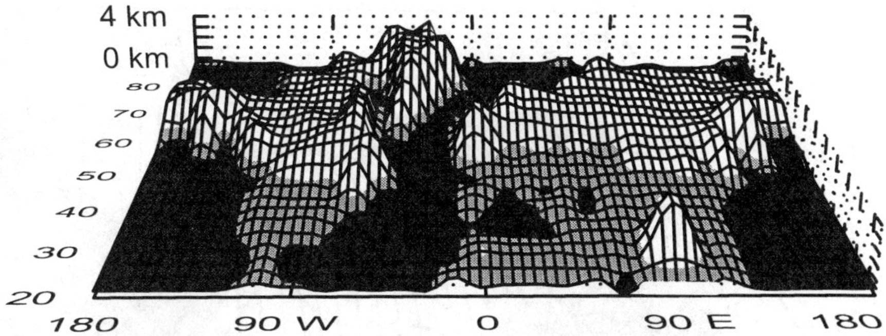

Fig. 4.40. Maximum model ice thickness, 10,000 realizations. Areas with ice thicker than 200 m are shown *white*

4.3.2.4 Regional Results

As mentioned above, the present model provides four-dimensional statistics of ice sheets (three spatial dimensions plus time) as a result of generating a large number of model realizations. In the series of Figs. 4.40-4.45 three-dimensional patterns of ice sheets for different probabilities and time moments are shown. In Figs. 4.40-4.42 maximum ice thickness at each gridpoint of the model, i.e., thickness with zero probability, is shown. Corresponding NH total ice volume was equal in this case to about 50 x 10^6 km^3. A model time of 60 ka (model maximum of glaciation) was chosen for these calculations, and 10,000 realizations were generated for this purpose. The resulting field of maximum ice thickness was not very sensitive to the moment of time or number of realizations. Maximum values in Figs. 4.40-4.42 are characteristic for the Laurentide and Kamchatka ice sheets (about 3 km), also for the Fennoscandian and Cordillerian sheets (2.5-2.7 km), for the Tibet region (about 2 km); thick ice evidently covers Greenland (more than 3 km). It is interesting that there was no artificial limitation of the ice thickness in the model (as well as there was no horizontal movements of ice or formation of icebergs). Nevertheless, experiments with even a low number of realizations (several hundreds) quickly reached a maximum ice thickness. The fundamental cause of the ice thickness limitation in the model was the statistical structure of random air temperature pulsations together with the limitation of low air temperatures. A cross-section through North America and Eurasia along 57.5°N, corresponding to Figs. 4.40-4.42, is shown in Fig. 4.46. This pattern seems natural: even if assumptions about ancient conditions in the atmosphere are uncertain, the concentration of ice in midlatitudes near sea shores looks reassuring.

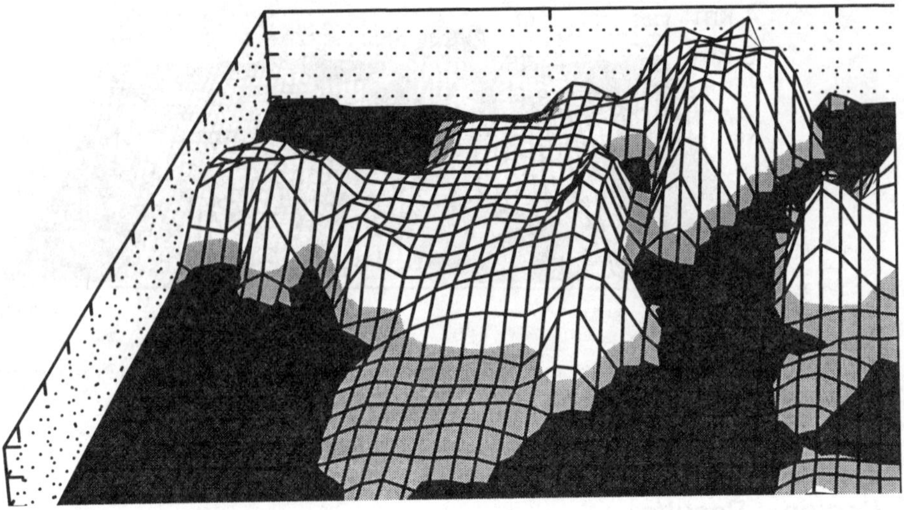

Fig. 4.41. Fragment of Fig. 4.40. North American ice sheet

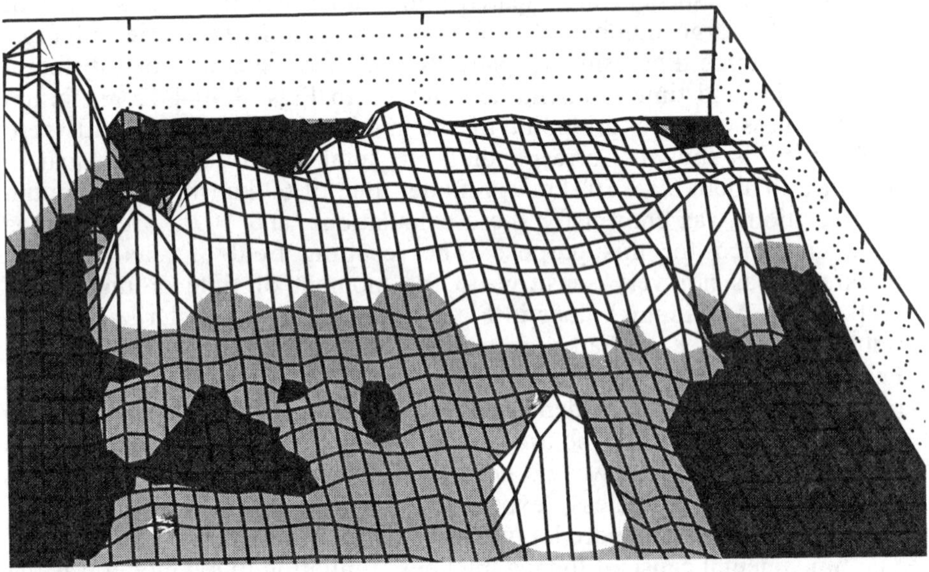

Fig. 4.42. Fragment of Fig. 4.40. Eurasian ice sheet

In Fig. 4.43 an ice thickness at 60 ka (maximum glaciation in the model) with 33% probability is shown. Here, we considered areas with ice thickness of more than 200m as relatively stable. The area covered by ice in Fig. 4.43 is practically the same as in Figs. 4.40-4.42 although ice

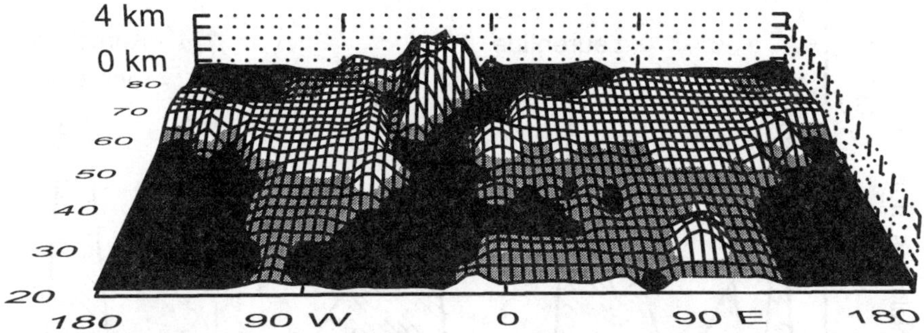

Fig. 4.43. Ice thickness at 60 ka model time. 33% probability. Areas with ice thickness more than 200 m are *white*

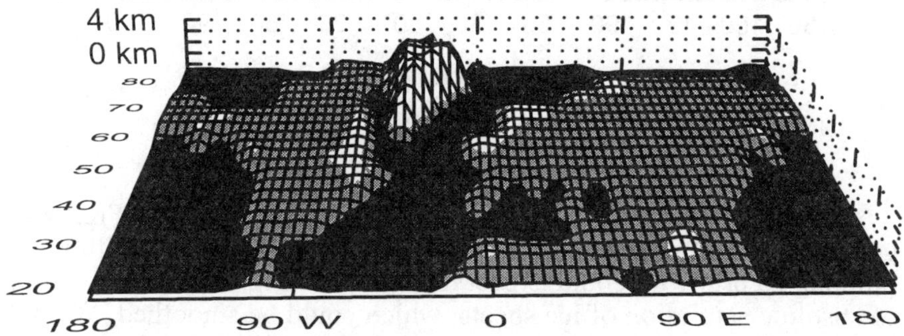

Fig. 4.44. Mean ice thickness at 60 ka model time

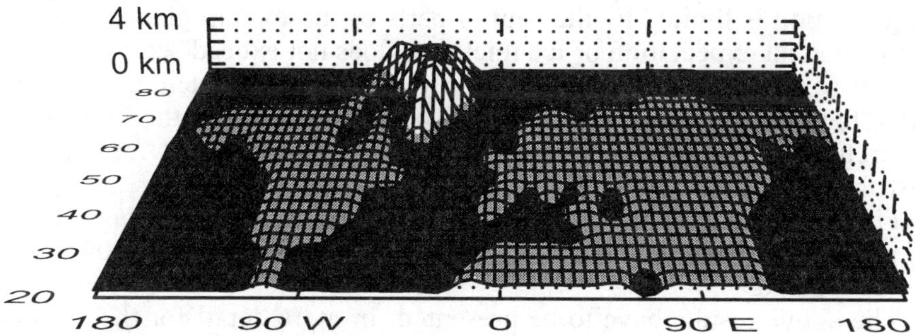

Fig. 4.45. Mean ice thickness at 120 ka model time

thickness in the centers of glaciation was less than the maximum thickness: correspondingly 1.2 - 1.3, 1.1, 1 km for the above regions.

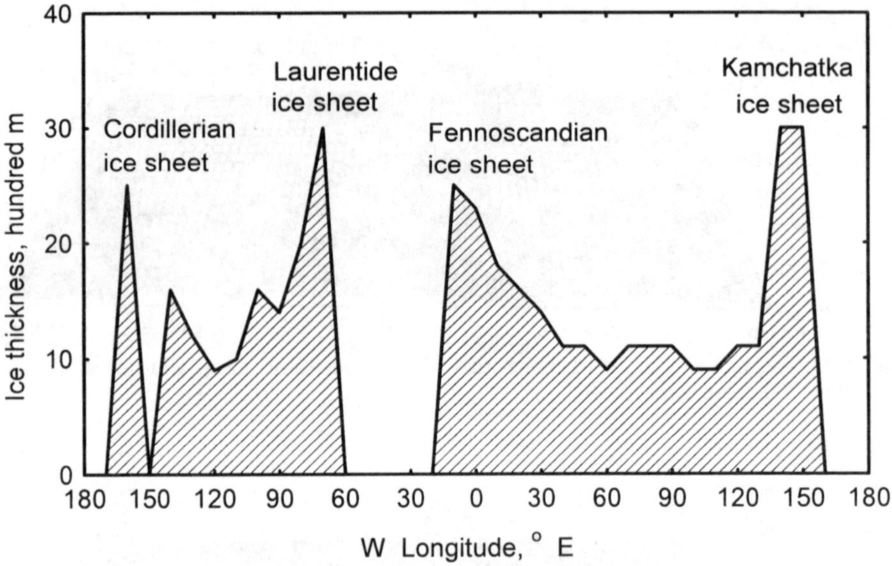

Fig. 4.46. Maximum model ice thickness. 57.5°N cross-section

An appropriate west-east cross-section is presented in Fig. 4.47. In general, ice peaks are less sharp in this case, in comparison with Fig. 4.46 (although a priori we could expect the opposite: Fig. 4.46 reflects the stage of absolute saturation of ice sheets, which could be smoothed. We did not include effects of horizontal movements of ice, however. Thus we speak here rather of potential ice thickness.

However, mean glaciers at the model maximum of glaciation (60 ka, for instance, Fig. 4.44) differ considerably from the above two cases. The area of glaciation is limited by the central parts of these main glaciers, and the mathematical expectation of ice thickness does not exceed several hundred meters (except for Greenland, evidently). These results demonstrate once more the extremely unstable character of glaciers even in the moments of theoretical glaciation maximum, also a large asymmetry of ice thickness distributions.

Finally, the pattern for the minimum of glaciation (in our case, 120 ka of model time, Fig. 4.45) does not reproduce ice thicker than 200, except for Greenland.

The above results have to be presented in more detail for the behavior of specific ice sheets (regions). Evidently, it is difficult to determine the exact areas of these regions; approximate corresponding areas are shown in Fig. 4.19. Thus, it is difficult to compare absolute values of different parameters for these regions. Nevertheless, we can calculate an important relative parameter - the phase of maximum and minimum of each glacier -

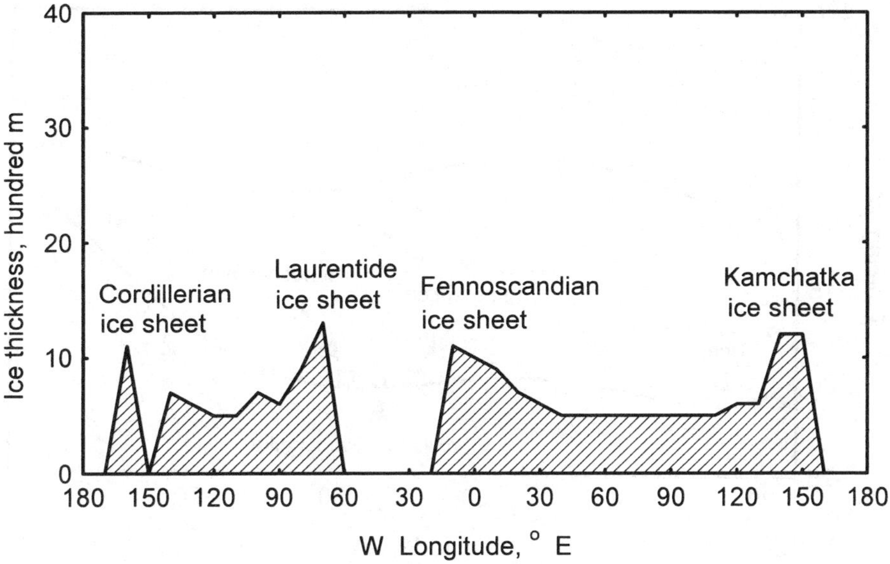

Fig. 4.47. 66% probability ice thickness at 60 ka model time. 57.5°N cross-section

and compare it with the zonally averaged phase (Fig. 4.48). The first interesting result was that, unlike the zonally averaged pattern, for each of six regions the minimum time lag (in comparison with the external forcing maximum) was the same as the maximum time lag. Thus, for these specific glaciers a temporal asymmetry (at least, on average) of oscillation does not exist. The time lag of both maximum and minimum of glaciation was almost the same as the time lag of minimum of glaciation in the zonally averaged pattern. Perhaps the cause of such behavior is the need for much time for the main glaciers to accumulate negative air temperature anomalies for their growth. Taimyr is an exception: this is not a main glacier area in the model; it was taken as an example of an ordinary arctic territory. The absence of asymmetry in the case of Taimyr is related to its extreme northerly situation, where the same feature is characteristic for the entire latitudinal zone.

Finally, the following conclusions with respect to model results can be formulated. A stochastic approach to the studies of climatic variations during the last 600 ka can give a realistic description and explanation of the observed features of glacial cycles. Some of the above model results are directly corroborated by existing paleoreconstructions. Other, more detailed, results, which do not have direct analogs in paleoreconstructions, can be of interest for possible comparison with future reconstructions (some of them are corroborated at expert level, e.g., those related to the phases of glacier growth and melting). We would like to stress that model

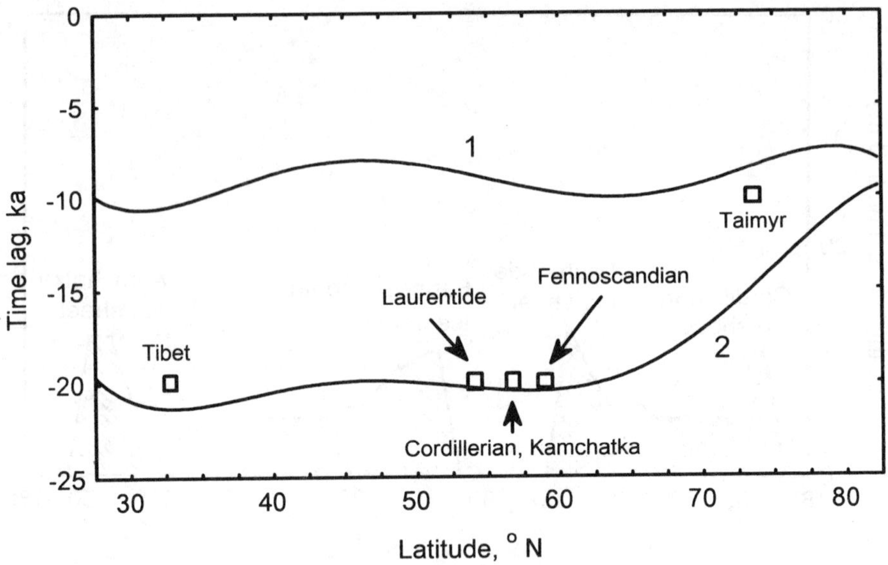

Fig. 4.48. A Fig. 4.36 but with model results for specific glaciers

results were obtained using very simple assumptions and mathematical models. Thus only few but fundamental facts about paleoclimates enable us to clarify many interesting features of glacial cycles.

Taking into consideration these results, some suppositions can be formulated with respect to the behavior of the climate system in the future (here we do not discuss the problem of man-made heating). At the present time we are close to the high-temperature natural limit of global temperature. Therefore, manifestations of a general natural negative trend of global temperature in the next thousands of years are inevitable. It is characteristic for the future entry into glaciation that it can be relatively stable, more stable than the stage of deglaciation. Perhaps the time of creation of temperate zone glaciers near oceanic coasts will exceed the time interval needed for the forming of ice cover in the internal continental regions. Evidently, the above projections have a statistical character, i.e., reflect probable future changes on average.

Conclusion

Results of the present work can be summarized as follows. Stochastic analysis of climate records shows that the theoretical and model approaches which prevail in contemporary climate change science are not consistent with existing observational data:

1. Neither global nor local climatic parameters (thousands of time series of several dozens of types of variables, except for several records of atmospheric gas content) manifest any detectable monotonous deterministic trends.
2. Existing coupled deterministic models of the climate system - GCMs plus oceanic and land surface models - do not reproduce realistic spectra of global climatic variabilities, i.e., these models are not capable of simulating the most important features of climate change. For instance, coupled models underestimate, by an order of magnitude, the long-term natural variability of the climate system. Apparently, the coupled models are unable to describe the basic climate process adequately: accumulation of random portions of water and heat within climatic subsystems.
3. Hypotheses on the presence of low-dimensional nonlinear attractors within the climate system (that are intended to explain ENSO and similar events, and their global climatic effect) are not corroborated by an analysis of regional and global climatic records. The existence of such a mechanism would inevitably cause the dominant frequencies in the spectra, constrained to specific frequency bands. However, this is not true.

Our studies demonstrate that climatic variability has a strongly stochastic nature. Random impulses and processes are not "errors" or unpleasant "noise", but the very essence of climatic variability. Analysis of new types of data, using new methods of time series processing, allows us to check the "classical" global hypotheses of stochastic climate theory and to introduce new hypotheses:

1. In general, our results corroborate the -2 degree law of climatic variability. At the same time, we can now say much more about the applicability of this law: it is true mostly for accumulation (primary, or state), globally averaged parameters of thick climatic subsystems: global air and sea temperatures, and global mean sea level. On the contrary, local accumulation (temperature, water content) and transfer (heat and

mass fluxes) variables of thick subsystems and both local and some global variables of thin climatic subsystems (vegetation spatial density, area of snow and ice cover) demonstrate different stochastic patterns.

2. Thus, two important natural phenomena take place with respect to stochastic properties of recent climate changes: local-global polarization of stochastic models of the thick climatic subsystems, and a local-global convergence of stochastic models of thin subsystems. A stochastic dynamical model of these phenomena, on the basis of Hasselmann's two-scale climate separation concept, is proposed here.
3. The most fundamental stochastic models are usually good first approximations for the description of recent climatic variabilities: the discrete Wiener process, the first-order Markov process, and the white noise.

Stochastic analysis of paleorecords shows that the following model can play an important role in the description of climatic variability on paleoscales: the two-threshold (two-screen) discrete Wiener process with varying standards of temporal increments. Our study demonstrates that such a stochastic mechanism can largely amplify external 100-ka periodical forcing. Thus, we hypothesize that there is no need for sophisticated feedback mechanisms or sophisticated mechanisms within the components of the climate system to explain and describe the most important features of glacial cycles.

References

Adam DP, West GJ (1983) Temperature and precipitation estimates through the last glacial cycle from Clear Lake, California, pollen data. Science 219: 168-170

Adem J (1991a) Review of the development and applications of the Adem thermodynamic climate model. Clim Dyn 5: 145-160

Adem J (1991b) Numerical experiments on the simulation of sea surface temperature for the last 18,000 years. Ann Geophys 9: 42-49

Adem, J, Mendoza VM, Villanueva EE (1994) Numerical prediction of the sea surface temperature in the Pacific and Atlantic oceans. Geofis Int 34: 149-160

AGU (1995) Determining global mean sea level rise. US National Report to IUGG, 1991-1994. Rev Geophys 33 Suppl

Ahmad Z, Sutton J, Mitchell W (1985) Multi-channel sea surface temperature (MCSST) from the TIROS-N advanced very high resolution radiometer (AVHRR) for the FGGE period. Ocean Eng Environ 1: 314-319

Ariel' IZ, Girdyuk GV, Egorov BN, Malevskii-Malevich SP, Murashova AV (1984) Kharakteristiki energoobmena okean-atmosfera. (Characteristics of ocean-atmosphere energy fluxes). Gidrometeoizdat, Leningrad, 80 pp (in Russian)

Arkin PA, Janowiak JE (1993) Tropical and subtropical precipitation. In: Gurney, R.J., Foster, J.L., Parkinson, C.L. (eds). Atlas of satellite observations related to global change. Cambridge University Press, Cambridge, pp 165-181

Astaf'eva NM (1996) Interaction of the atmosphere and the ocean: analysis of multi-scale structure of data of long-term observations on the El Niño-Southern Oscillation. Preprint. Institute of Space Information (IKI) of the Russian Academy of Sciences, No 1945: 1-32 (in Russian)

Babkin VI, Selyakov KI (1995) Planets, solar activity, and Volga runoff. In: Conditions of forming and methods of forecasting of Volga Runoff. State Hydrological Institute (GGI), Sankt Petersburg, pp 21-26 (in Russian)

Baggeroer AB (1976) Confidence intervals for regression (MEM) spectral estimates. IEEE Trans Inf Theory, IT-22: 218-224

Baker T, Vassie I, Woodworth P (1989) Taking the measure of the World's sea level. Earth Sci Progr Newsl 82: 8-10

Balling RC Jr (1996) Geographic analysis of differences in trends between near-surface and satellite-based temperature measurements. Geophys Res Lett 23: 2939-2941

Bard E, Arnold M, Duplessy J-C (1991) Reconciling the sea level record of the last deglaciation with the $\delta^{18}O$ spectra from deep-sea cores. Quat Proc 1: 113-133

Barnett TP (1984) Long-term trends in surface temperature over the oceans. Mon Weather Rev 112: 303-312

Barnett TP (1988) Global sea level change. In: Climate variations over the past century and the greenhouse effect. A report based on the First Climate Trend Workshop, 7-9 Sept 1988, Washington, DC, National Climate Program Office, Maryland, 57 pp

Barnola JM, Raynaud D, Lorius C, Korotkevich YS (1994) Historical CO_2 record from the Vostok ice core. In: Trends'93: a compendium of data on global change. ORNL/CDIAC-65. Carbon dioxide information Analysis Center, Oak Ridge National Laboratory, Oak Ridge, Tenn, USA, pp 7-10

Barnsley MF, Devaney RL, Mandelbrot BB, Peitgen HO, Saupe D, Voss RF (1988) The science of fractal images. Springer, Berlin, Heidelberg, New York

Berger A (ed) (1984) Milankovitch and climate: understanding response to astronomical forcing. Reidel, Dordrecht

Berger A, Loutre MF (1996) Modeling the climate response to astronomical and CO_2-forcing. CR Acad Sci Ser 2, Fasc. A 323: 1-16

Berger A, Loutre MF (1997) Paleoclimate sensitivity to CO_2 and insolation. Ambio 26: 32-37

Birman BA, Balashova EV (1986) Structure and variability of contemporary climate of the Northern Hemisphere. Atlantic and Pacific Ocean. Sea surface temperature, 1957-1983. Vsesoyuznyi Nauchno-Issledovatel'skii Institut Gidrometeorologicheskoi Informatsii - Mirovoi Tsentr Dannykh, Obninsk, 195 pp (in Russian)

Blanke B, Neelin J, Gutzler D (1997) Estimating the effect of stochastic wind stress forcing on ENSO irregularity. Int Sci Conf Trop Oc Glob Atm (TOGA) Programme: Melbourne, 2-7 Apr 1995, WCRP/WMO, No 717, pp 633-637

Bloom AL (1971) Glacial-eustatic and isostatic controls of sea level since the last glaciation. In: Turekian, K.K., (ed) The Late Cenozoic glacial ages. Yale Univ Press, New Haven, pp 355-380

Boer GJ (1985) Modelling the atmospheric response to the 1982/1983 El Niño. In: Coupled ocean-atmosphere models. Elsevier, Amsterdam, pp 7-17

Bol'shoi Sovetskii atlas mira (1964-) (Big Soviet atlas of the World) Nauchno-Redaktsionnaya Katrosostavitel'skaya Chast', Gosudarxtvennoe Upravlenie Geodezii i Kartografii pri Sovete Ministrov SSSR, Moscow

Borzenkova II (1992) Climate change during the Cenozoic. Gidrometeoizdat, Sankt-Petersburg, 248 pp (in Russian)

Bottomley MC, Folland CK, Hsiung J, Newell RE, Parker DE (1990) Global ocean surface temperature atlas. The Meteorological Office, Bracknell, UK

Boulton GS, Hulton N, Vautravers M (1995) Ice-sheet models as tools for paleoclimatic analysis: the example of the European ice sheet through the last glacial cycle. Ann Glaciol 21: 103-110

Boutron C (1988) Plomb dans l'atmosphère. La Recherche 198; 19: 446-455

Box GEP, Jenkins GM (1970) Time series analysis: forecasting and control, vol 1. Holden Day, San Francisco, 408 pp

Box GEP, Jenkins GM, Reinsel GC (1994) Time series analysis: forecasting and control. Prentice Hall, Englewood Cliffs

Bradbury JP (1995) A diatom-based paleohydrologic record of climate change for the past 800 kyr from Owens Lake, California. Geological Society of America. Special Paper. Past Global Changes (PAGES) Publ Ser 1995, No 95-1, 92 pp

Bradley RS (1985) Quaternary palaeoclimatology. Wiley, Boston, 234 pp

Bradley RS (1988) The explosive volcanic eruption signal in Northern Hemisphere continental temperature records. Clim Change 12: 221-243

Brayan K, Griffies S (1996) Climatic variability of the Northern Atlantic on the decadal time-scale: is it predictable? Phys Atm Oc 32: 591-599

Bunimovich VI, Leontovich MA (1946) On the distribution of the number of large deviations in electrical fluctuation. Dokl Acad Nauk SSSR 53: 21

Burg JP (1967) Maximum entropy spectral analysis. Paper presented at the 37th Ann Int Meet Soc of Explor Geophys, Oklahoma City, Oklahoma, Oct 31, 1967

Burg JP (1968) A new analysis technique for time series data. Paper presented at Advanced Study Institute on Signal Processing, NATO, Enschede, Netherlands

Butler C, Johnson D (1996) A provisional long mean air temperature series for Armagh observatory. J Atmos Terr Phys 58: 1657-1672

Bye JAT (1996) Coupling ocean-atmosphere models. Earth-Sci Rev 40: 149-162

Byshev VI, Ivanov YA (1969) The time spectra of some characteristics of the atmosphere above the ocean. Izv Acad Nauk SSSR, Fizika Atmosfery i Okeana 5: 17-28

Cane MA, Clement AC, Kaplan A, Kushnir I, Pozdnyakov D, Seager R, Zebiak SE, Murtugudde R (1997) Twentieth-century sea surface temperature trends. Science 275: 957-960

Carl P, Worbs KD, Tschentscher I (1995) On a dynamic systems approach to atmospheric model intercomparison. WCRP/WMO, 732: 445-450
Carsey FD, Pihos G (1989) Beaufort-Chukchi seas summer and fall ice margin data from Seasat: conditions with similarities to the Labrador Sea. IEEE Trans GRS-27: 541-551
Carter RM, Johnson PP, Weersma J (1986) Sea-level controls on the postglacial development of the Great Barrier Reef, Queensland. Mar. Geol. 71: 137-164
Carton J A (1997) See-saw sea. Nature 385: 487-88
Cavalieri DJ, Gloersen P, Campbell WJ (1984) Determination of sea ice parameters with the Nimbus-7 SMMR. J Geophys Res 89: 5355-5369
Cavalieri DJ, Crawford JP, Drinkwater MR, Eppler DT, Farmer LD, Jentz RR, Wackerman CC (1991) Aircraft active and passive microwave validation of sea ice concentration from the Defence Meteorological Satellite Program Special Sensor Microwave Imager. J Geophys Res 96: 21,989-22,008
Chang ATC, Foster JL, Hall DK (1990) Satellite estimates of Northern Hemisphere snow volume. Remote Sens Lett 11: 167-172
Chaos Data Analyzer. The professional version (1995) pack of computer programs. American Institute of Physics, New York
Chappell J, Shackleton NJ (1986) Oxygen isotopes and sea level. Nature 324: 137-140
Chappellaz, J., Barnola, J.M., Reynaud, D., Korotkevich, Y.S., Lorius, C. (1994): Historical CH_4 record from the Vostok ice core. In: Trends'93: a compendium of data on global change. ORNL/CDIAC-65. Carbon dioxide information Analysis Center, Oak Ridge National Laboratory, Oak Ridge, Tenn, USA, pp 229-232
Chen RS (1985a) Surface air temperatures over land and ocean, 1949-1972. In: 3rd Conf Clim Var Symp Contemp Clim, 1850-2100, Los Angeles, Calif, Jan 8-11. Boston, Mass, pp 39-40
Chen TC (1985b) Global water vapour flux and maintenance during PGGE. Mon Weather Rev 113: 1801-1819
Chistyakov VF (1996) Global variations of insolation and oscillations in the earth's climate. Vestnik Dal'nevostochnogo Otdeleniya Rossiiskoi Akademii Nauk, No 2: 75-85 (in Russian)
Choudhury BJ (1989) Monitoring global land surface using Nimbus-7 37 GHz data. Theory and examples. Int J Remote Sens 10: 1579-1605
Choudhury BJ (1993) Desertification. In: Gurney, R.J., Foster, J.L., Parkinson, C.L. (eds) Atlas of satellite observations related to global change. Cambridge University Press, Cambridge, pp 313-325
Choudhury BJ, DiGirolamo NE (1994) Relations between SMMR 37 GHz polarization difference and rainfall, and atmospheric carbon dioxide concentration. Int J Remote Sens 15: 3643-3661
Chui CK (1992) An introduction to wavelet analysis. Academic Press, Boston
CLIMAP Project Members (1976) The surface of the ice age EARTH. Science 191: 1131-1137
Connolley W, King J (1996) A modeling and observational study of East Antarctic surface mass balance. J Geophys Res D. 101: 1335-1343
Cramer H (1946) Mathematical methods of statistics. Princeton Univ Press, Princeton, NJ
Cronin TM (1983) Rapid sea level and climatic change: evidence from continental and island margins. Quat Sci Rev 3: 177-214
Curray JR (1964) Transgressions and regressions. In: Papers in marine geology. Miller, R.L.(ed) Macmillan, NewYork, pp 173-203
Currie R (1996) M_n and S_c signals in North Atlantic tropical cyclon occurrence. Int J Climatol 16: 427-439
Currie R, Vines R (1996) Evidence for luni-solar M_n and solar cycle S_c signals in Australian rainfall data. Int J Climatol 16: 1243-1265
Dansgaard W, Duplessy J-C (1981) The Eemian interglacial and its termination. Boreas 10(2): 219-228
Denthon GH, Hughes TJ (eds) (1981) The last great ice sheets. Wiley, New York., 477 pp

Ditlevsen PD, Svensmark H, Johnsen S (1996) Contrasting atmospheric and climate dynamics of the last-glacial and Holocene periods. Nature 379: 810-812

Dobrovolski SG (1982) Simple dynamic-stochastic model of large sea surface temperature anomalies. Meteorol Gidrol, 6: 68-75 (in Russian)

Dobrovolski SG (1991) Anomalies of global heat and water exchange. Stochastic models. Soviet Geophysical Committee, Moscow, 128 pp (in Russian)

Dobrovolski SG (1992) Global climatic changes in water and heat transfer-accumulation processes. Elsevier, Amsterdam, 280 pp

Dobrovolski SG (1994) On the applications of thermodynamic climate models of J.Adem in Russia. Atmosfera 7: 139-142

Dobrovolski SG (1998) Life as a random walk. 42 pp (unpublished)

Dobrovolski SG, Rybak OO (1992) Incorporation of hydrological cycle elements into the dynamic-stochastic climate model. GeoJournal 27.3: 247-254

Dobrovolsi SG, Rybak OO, Selivanov AO, Yushkov VP (1995) Report on the RFFI Project Study of the phenomenon of local-global polarization of water redistribution between the components of the earth hydrosphere. Moscow, 30 pp

Dobrovolski SG., Choudhurt BJ (1996) Temporal climatic variabilities of global atmospheric, oceanic, and land surface parameters. Atmosfera 9, 4: 273-289

Dobrovolski SG, Yarosh ES (1980) Synoptic and interannual variability in sea surface temperature in the North Atlantic Ocean. Vestnik MGU, Geogr Ser, Dep VINITI, No 4071-80, 14 pp (in Russian)

Dovgalyuk V, Klimenko V (1996) On long-term variations in the intensity of El Niño occurrences. Geophys Res Lett 23: 3771-3774

Dowdeswell JA, White JWC (1995) Greenland ice core records and rapid climate change. Philos Trans R Soc Lond A 352: 350-371

Dugam SS (1995) ENSO, equatorial Pacific SST, NAO and sea ice extent in the Northern Hemisphere. WCRP/WMO, No 717, pp 825-829

Dymnikov VP, Gritsun AS (1996) Barotropic instability and structure of low frequency changes in circulation simulated by two-layer baroclinic atmospheric model. Phys Atmos Ocean 32: 724-736 (in Russian)

Dzhaparidze KO, Yaglom AM (1983) Spectrum parameter estimation in time series analysis. Developments in statistics. Academic Press, New York 4: 1-96

Edward JA, Fitelson MM (1973) Notes on maximum entropy processing. IEEE Trans Inform Theory IT-19: 232

Efimov VV, Prusov AV, Shokurov MV (1995) Classification of interannual surface temperature anomalies of the world ocean. Oceanology 35: 505-513 (in Russian)

Elsen M, Rotmans J (1993) Modelling climate-related feedback processes. J Environ Sci Health. A. 28: 2095-2151.

EPA (1984) Potential climatic impacts of increasing atmospheric CO_2 with emphasis on water availability and hydrology in the United States. United States Environment Protection Agency, Washington, DC, 96 pp

Esbensen SK, Reynolds RW (1981) Estimating averaged air-sea transfer of heat and momentum using the bulk aerodynamic methods. J Phys Oceanogr 11: 457-465

Fabre A, Letreguilly A, Ritz C, Mangeney A (1995) Greenland under changing climates: sensitivity experiments with a new three-dimensional ice-sheet model. Ann Glaciol 21: 1-7

Fairbanks RG (1989) A 17,000-year glacio-eustatic sea level record: unfluence of glacial melting rates on the Younger Dryas event and deep-ocean circulation. Nature 342: 637-642

Fairbridge RW (1980) Holocene sea-level oscillations. Striae 14, 1-4: 23-60

Fairbridge RW, Jelgersma S (1990) Sea level. In: R.Paepe (ed) Greenhouse effect, sea level, and drought. Kluwer, Dordrecht, pp 117-141

Fairbridge RW, Krebs OA Jr (1962) Sea level and the Southern Oscillation. Geophys. J. 6: 532-545

Farmer JD, Sidorowich J (1988) Evolution, learning and cognition. World Scientific, Singapore

Feder J (1988) Fractals. Plenum, New York

Feigenbaum M (1984) In: Chaos. World Scientific, Singapore

Feng G (1996) Fokker-Plank equation for long-term climatic fluctuations and its solutions. Acta Meteorol Sin 54: 753-758

Fennessy MJ, Shukla J (1988) Numerical simulation of the atmospheric response to the time-varying El-Niño SST anomalies during May 1982 through October 1983. J Clim: 1, 195-211

Fennessy MJ, Marx L, Shukla J (1985) General circulation model sensitivity to 1982-1983 equatorial Pacific sea surface temperature anomalies. Mon Weather Rev 113: 858-864

Fifth annual climate assessment (1994) United States Department of Commerce, Washington, DC, 111 pp

Flint RF (1971) Glacial and Quaternary geology. Wiley, New York, 892 pp

Folland CK, Parker DE, Kates FE (1984) Worldwide marine temperature fluctuations, 1856-1981. Nature 310: 670-673

Foster JL, Chang ATC (1993) Snow cover. In: Gurney, R.J., Foster, J.L., Parkinson, C.L. (eds) Atlas of satellite observations related to global change. Cambridge University Press, Cambridge, pp 361-370

Frankignoul C (1985) Sea surface temperature anomalies, planetary waves, and air-sea feedback in the middle latitudes. Rev Geophys 23: 357-390

Frankignoul K (1995) Climate spectra and stochastic climate models. In: von Storch H, Navarra A (eds) Analysis of climate variability: applications of statistical techniques. Proc Autumn School Comm Europ Comm, Elba, Oct 30 - Nov 6, 1983, Springer, Berlin, pp 30-40

Frankignoul C, Hasselmann K (1977) Stochastic climate models. Part II. Application to sea-surface temperature anomalies and thermocline variability. Tellus 29: 359-370

Frost MR (1975) Stress, evaporative heat flux and sensible heat flusx distributions of the North Atlantic (mid-latitude) and their contribution to the production of large scale sea surface temperature anomalies. Master Thesis. Southampton University, Southampton

Garduño R, Adem J (1995) Adem model's results on global warming by CO_2 with emphasis in Central Mexico. World Res Rev 7: 246-253

Garduño R, Adem J, Mendoza B (1997) Contributions of preindustrial CO_2 and climate feedbacks on the Maunder minimum cooling. World Res Rev 9: 240-247

Gassmann F (1994) Abrupt change scenarios: global change scenarios revisited with respect to emerging scientific insights. PSI Ber. 10: 3-9

Gates WL, Cubasch U, Meehl GA, Mitchell JFB, Stouffer RJ (1993) An intercomparison of selected features of the control climates simulated by coupled ocean-atmosphere general circulation models. World Climate Research Programme WCRP-82, WMO/TD No 574, World Meteorological Organization, Geneva, 46 pp

Genthon C, Barnola JM, Raynaud D (1987) Vostok ice core: climatic response to CO_2 and orbital forcing changes over the last climatic cycle. Nature 329: 414-418

Giddings L, Choudhury BJ (1989) Observation of hydrological features with Nimbus-7 37 GHz data, applied to South America. Int. J. Remote Sens 10; 1673-1686

GISS (1998) Global temperature anomalies. http://www.giss.nasa.gov/data/gistemp/GLB.Ts.txt

Glacier mass balance bulletin (1993) Haeberli W, Herren E, Hoelzle M (eds) IAHS/UNEP/UNESCO, World Glacier Monitoring Service, ETH, Zurich

Gleckler PJ, Randall DA, Boer G, Colman R, Dix M, Galin V, Helfand M, Kiehl J, Kitoh A, Lau W, Liang X-Z, Lykossov V, McAvaney B, Miyakoda K, Planton S (1994) Cloud-radiative effects on implied oceanic energy transports as simulated by atmospheric general circulation models. Report No. 15, PCMDI. Lawrence Livermore National Laboratory, Livermore, CA, 13 pp

Global Climate Observing System Newsletter (1996) No 5, p 4

Gloersen P, Barath FT (1977) A scanning multichannel microwave radiometer for Nimbus-G and Sea Sat-A. IEEE J. OE-2: 172-178

Gloersen P, Campbell DJ, Cavalieri DJ, Comiso JC, Parkinson CL, Zwally HJ (1992) Arctic and Antarctic sea ice, 1978-1987: satellite passive-microwave observation and analysis. NASA, Washington, DC, 290 pp

Gnedenko BV (1962) The theory of probability. Chelsea Publ, New York

Gornitz V, Lebedeff S (1987) Global sea level changes during the past century. In: Nummdal D. (ed) Sea level fluctuations and coastal evolution. SEPM Spec Publ No 41

Gorodetsky AK (1989) Estimating sea surface temperature using remote sense methods on the basis of spectra and angular measurements of infra-red waves. In: Oceanographic investigations using remote-sensing methods. Morskoi Gidrofizicheskii Institut, Sevastoppol', Dep VINITI 07.06.89, No 3765-B89

Gotz G (1995) Predictability of nonlinear dynamical systems. Idojaras 99: 1-32

Grassberger P, Procaccia I (1983) Strange attractors. Phys Rev Lett 50:346

Gu D, Philander S (1997) Interdecadal climate fluctuations that depend on exchanges between the tropics and extratropics. Science 275:805-807

Gualdi S, Navarra A, von Storch H (1997) Tropical interseasonal oscillation in operational analysis and in a family of general circulation models. J. Atmos Sci 54: 1185-1202

Haeberli W, Hoelzle M (1995) Application of inventory data for estimating characteristics of and regional climate change effects on mountain glaciers - a pilot study with the European Alps. Ann Glaciol 21: 206-212

Haykin S (1979) Non-linear methods of spectral analysis. Top Appl Phys 34: 247

Hasselmann K (1966) Feinman diagrams and interaction rules of wave-wave scattering processes. Rev Geophys 4: 1-32

Hasselmann K (1967) Non-linear interactions treated by the methods of theoretical physics (with application to the generation of waves by wind). Proc R Soc A299: 77-100

Hasselmann K (1976) Stochastic climate models. Part 1. Theory. Tellus 28: 473-485

Hasselmann K (1981) Construction and verification of stochastic climate models. In: Berger A (ed) Climatic variations and variability: facts and theories. D Reidel, Dordrecht pp 481-497

Hasselmann K (1982) An ocean model for climate variability studies. Progr Oceanogr 11: 69-92

Hays JD, Imbrie J, Shackleton NJ (1976) Variations in the Earth orbit: pacemaker of the ice ages. Science 194: 1121-1132

Hecht Z (1996) Cold water pored on climate models. New Sci 150: 17

Hendricks JR, Leben RR, Born GH, Koblinsky CJ (1996) Empirical orthogonal function analysis of global TOPEX/POSEIDON altimeter data and implications for detection of global sea level rise. J Geophys Res 101: 14,131-14,145

Hoelzle M, Trindler M (1998) Data management and application. Stud Rep Hydrol 56: 53-72

Hooghiemstra H, Melice JL, Berger A, Shackleton NJ (1993) Frequency spectra and paleoclimatic variability of the high-resolution 30-1450 ka Funza 1 pollen record (Eastern Cordillera, Columbia). Quat Scie Rev 12: 141-156

Houghton JT, Meira Filho LG, Callande BA, Harris N, Kattenberg A, Maskell K (eds) (1996) Climate Change 1995. The science of climate change. Cambridge University Press, Cambridge

Hughes T (1996) Can ice sheets trigger abrupt climatic change? Arct Alp Res 28: 448-465

Hurrell J (1996) Influence of variations in extratropical wintertime teleconnections on Northern Hemisphere temperature. Geophys Res Lett 23: 665-668

Huybrechts P, T'siobell S (1995) Thermomechanical modelling of Northern Hemisphere ice sheets with a two-level mass-balance parametrization. Ann Glaciol 21: 111-116

IAHS (ICSI)/UNEP/UNESCO (1988) Fluctuations of glaciers 1980-1985 (Haeberli W, Müller P, eds) UNESCO, Paris

IAHS (ICSI)/UNEP/UNESCO (1991) Glacier mass balance bulletin, No 1. World Glacier Monitoring Service, ETH, Zurich

IAHS (ICSI)/UNEP/UNESCO (1993a) Fluctuations of glaciers 1985-1990 (Haeberli W, Hoelzle M, eds) UNESCO, Paris

IAHS (ICSI)/UNEP/UNESCO (1993b) Glacier mass balance bulletin, No 2. (Haeberli W, Herren E, Hoelzle M, eds) World Glacier Monitoring Service, ETH, Zurich

IAHS (ICSI)/UNEP/UNESCO (1994) Glacier mass balance bulletin, No 3. (Haeberli W, Hoelzle M, Bösch H, eds) World Glacier Monitoring Service, ETH, Zurich

IAHS (ICSI)/UNESCO (1967) Fluctuations of glaciers 1959-1965 (Kasser P, ed) UNESCO, Paris

IAHS (ICSI)/UNESCO (1973) Fluctuations of glaciers 1965-1970 (Kasser P, ed) UNESSCO, Paris

IAHS (ICSI)/UNESCO (1977) Fluctuations of glaciers 1970-1975 (Müller F, ed) UNESCO, Paris

IAHS (ICSI)/UNESCO (1985) Fluctuations of glaciers 1975-1980 (Haeberli W, ed) UNESCO, Paris

IOC (Intergovernmental Oceanographic Commission) (1990) Global Sea Level Observing System (GLOSS) implementation plan. Intergovernmental Oceanographic Commission, Technical Series, No. 35 90 pp

IPCC (1993) Climate change 1992 - the supplementary report to the IPCC Scientific Assessment. WMO and UNEP, Cambridge University Press, Cambridge, 218 pp

Jaeger L (1976) Monatskarten des Niederschlags fur die ganze Erde. Ber Dtsch Wetterdienst 139: 38 p.

Janssen J (1997) Does wind stress depend on sea-state or not? A statistical error analysis of HEXMAX data. Boundary-Layer Meteorol 83 C: 479-503

Jaynes ET (1968) Prior probabilities. IEEE Trans. Syst Sci Cybern SEC-4: 227-241

Johnson J, Hanson C, Hardegree S (1996) Stochastic weather simulation: overview and analysis of two commonly used models. J Appl Meteorol 35: 1878-1896

Jones PD (1994) Recent warming in global temperature series. Geophys Res Lett 21: 1149-1152

Jouzel J, Lorius C, Petit JR, Barkov NI, Kotlyakov VM (1994) Vostok isotopic temperature record. In: Trends'93: a compendium of data on global change. ORNL/CDIAC-65. Carbon dioxide information Analysis Center, Oak Ridge National Laboratory, Oak Ridge, Tenn, USA, pp 90-106

Jovanovic B (1993) Has the sun activity any influence to terrestrial phenomena? Bull Appl Math 66, No 870: 1-12

Kachiola K, Reck R (1997) Comparison of global climate change simulations for 2xCO_2 induced warming. Clim Change 35: 53-69

Kalinin GP (1968) Problems in global hydrology (Problemy global'noi gidrologii). Gidrometeoizdat, Leningrad, 377 pp (in Russian)

Kane R (1989) Relationship between the southern oscillation/El Niño and rainfall in some tropical and midlatitude regions. In: Proc. Ind. Acad. Sci. Earth Planet Sci 89: 223-235

Kane R (1997) Prediction of droughts in north-east Brazil: role of ENSO and use of periodicities. Int J Climatol 17, 6: 655-665.

Kane R, De Paula E (1996) Atmospheric CO_2 changes at Mauna Loa, Hawaii. J Atmos Terr Phys 58:673-1681

Kaplin PA (1973) Recent history of the world ocean coasts. Nauka, Moscow, 265 pp (in Russian)

Karoly DJ, Kestin TS (1998) Time-frequency variability of ENSO and stochastic simulations. http://www.ldeo.columbia.edu/cullen/po_seminar/karoly_abstract.html

Kerenyi J (1993) Surface temperature derived from METEOSAT infrared data using atmospheric correction. Idojaras 97: 251-257

Kerr R (1994) Climate modeling's fudge factor under fire. Science 265: 1528

Kerr R (1997) Greenhouse forecasting still cloudy. Science 276: 1040-1042

Khinchin AY (1934) Korrelationstheorie der stationaren stochastischen Prozesse. Math. Ann 109: 604

Kishtawal C, Bsu C, Pandey P (1996) An algorithm for retrieving vertical wind profiles from satellite-observed winds over the Indian ocean using EOF analysis. J Appl Meteorol 35: 532-540

Kliege RK Selivanov AO (1995): Budget of sedimentary material in the Caspian Sea and its possible role in water-level changes. Water Resour (Moscow) 22 (3): 330-335 (in Russian)

Kliege RK, Danilov ID, Konishchev VI (1997) Istoriya gidrosfery (The history of the hydrosphere). Nauchnyi Mir, Moscow, 368 pp (in Russian)

Koc N, Jansen E (1994) Response of the high-latitude Northern Hemisphere to orbital climate forcing: Evidence from Nordic seas. Geology 22: 523-526

Kolmogorov AN (1956) Foundations of the theory of probability, Chelsea Publishing, New York

Konda M, Imasato N, Shibata A (1996) A new method to determine near-sea surface air temperature by using satellite data. J Geophys Res C 101: 14349-14360

Kotlyakov VM, Lorius C (1992) Global changes during the last glacial-interglacial cycle. Izv Acad Sci USSR Geogr ser 1: 5-22

Kritskii SN, Menkel' MF (1950) Gidrologicheskie osnovy rechnoi gidrotekhniki (Hydrological basis for river hydrotechnics). Izdatel'stvo Akad Nauk SSSR, Moscow, Leningrad, 377 pp (in Russian)

Krupchatnikov V (1995) Global attractors for the Lorenz model on the sphere. Bull Novosibirsk Comput Cent Ser Numer Model Atmos, Ocean Environ Stud 2: 31-40

Kruze HA, Hasselmann K (1986) Investigation of processes governing the large-scale variability of the atmosphere using low-order barotropic models as a statistical tool. Tellus A38: 12-24

Kurgansky MV, Dethloff K, Pisnichenko, IA, Gernandt H, Chmielewski F-M, Jansen W (1996) J. Geophys Res D 101: 4299-4314

Ladoy P, Lovejoy S, Schertzer D (1984) Une étude de bifurcation locale/globale pour les températures météo/climatologiques. 25e Congr Int De Géographie, Paris, 1984. Resumés de communications 1: p 3.17

Lambeck K (1990) Late Pleistocene, Holocene and present sea levels: constraints on future change. Palaeogeogr Palaeoclimatol Paaleoecol 9: 205-217

Larin DA (1982) On the accuracy of mean monthly estimations of meteorological parametres using ship data. Meteorol Gidrol 8: 109-115 (in Russian)

Larin DA (1986a) On the sensitivity of the error of averaging meteorological fields to the changes in ship data probability distributions. In: Trudy Vsesoyuznogo Nauchno-Issledovatel'skogo Instituta Gidrometeorologicheskoi Informatsii, Moscow, Gidrometeoizdat, Vyp 122, pp 84-90 (in Russian)

Larin DA (1986b) On the real accuracy of ship meteorological observations. In: Trudy Vsesoyuznogo Nauchno-Issledovatel'skogo Instituta Gidrometeorologicheskoi Informatsii, Moscow, Gidrometeoizdat, Vyp 122, pp 90-105 (in Russian)

Legrand MR, Delmas RJ (1988) Continental Ca from the Vostok kern. Nature 334: 418-420

Lemke P (1977) Stochastic climate models. Part III. Application to zonally averaged energy models. Tellus 29: 385-392

Loève M (1960) Probability theory. Van Nostrand, Princeton, NJ

Lorenz EN (1984) Irregularity: a fundamental property of the atmosphere. Tellus A36: 98-110

Lorenz EN (1990) Can chaos and intransitivity lead to interannual variability? Tellus A42: 378-389

Lvovich MI (1986) Voda i zhizn' (Water and life). Mysl', Moscow, 256 pp (in Russian)

Lyapunov AM (1954) Sobranie Sochinenii (Collected works), vol 1. Nauchno-Tekhnicheskoe Izdatel'stvo, Moscow, 26-33, 33-44, 45-100, 101-122 (in Russian)

Malkevich MS, Gorodetsky AK (1988) Determination of the ocean surface temperature taking account of atmospheric effects by measurements of the angular IR-radiation distribution of the ocean-atmosphere system made from the satellite Cosmos-1151. Remote-Sens Rev 3: 137-161

Manabe S, Stouffer R (1988) Two stable equilibria of a coupled ocean-atmosphere model. J Climate 1: 841-866

Mandelbrot BB (1982) The fractal geometry of Nature. Freeman, San Francisco, 402 pp

Mann ME, Park J, Bradley RS (1995) Global interdecadal and century-scale climate oscillations during the past five centuries. Nature 378: 266-270

Markov AA (1948) Collected works (Izbrannye trudy). Nauka, Moscow

Marple SL, Jr (1987) Digital spectral analysis with applications. Prentice Hall, Englewood Cliffs, NY, 584 pp

Marsiat I (1995) The waxing and waning of the Northern Hemisphere ice sheets. Ann Glaciol 21: 96-102

Martinson DG, Pisias NG, Hays JD (1987) Age dating and orbital theory of ice ages: development of a high-resolution 0 to 300,000 year chronostratigraphy. Quat Res 27: 1-29

Materialy Nablyudenii na Vodnobalansovykh Stantsiyakh (Materials of observations at water-balance stations) (1965-1990) Gidrometeoizdat, Leningrad (in Russian)

Matson M, Ropelewski CF, Varnadore MS (1986) An atlas of satellite-derived Northern Hemisphere snow cover frequency. Natl Weather Serv, Washington, DC

McCravy A (1997) Coupled model intercomparison project (CMIP).CMIP Power spectra of global mean temperatures. http://www-pcmdi.llnl.gov/cpvey/cmip/Fig1.htm

McPhaden MJ (1993) TOGA-TAO and the 1991-93 El Niño-Southern Oscillation event. Oceanography 6: 36-44

Measuring sea surface temperature from space (1987) Meerestechn. 18, 3: 92-93

Middleton D (1960) An introduction to statistical communication theory. McGraw-Hill, New York

Milman AS, Wilheit TT (1985) Sea surface temperatures from the scanning multichannel microwave radiometer on Nimbus 7. J. Geophys. Res. C90: 11631-11641

Mirovoi Vodnyi Balans i Vodnye Resursy Zemli (1974) (World water balance and water resources of the earth). Gidrometeoizdat, Leningrad, 638 pp (in Russian)

Mitchell JM (1966) Stochastic models of air-sea interaction and climate fluctuations. Symp Arctic Heat Budget Atm Circ, Lake Arrowhead, California, Santa Monica, 1966. Rand Corp., pp 40-56

Monin AS (1969) Prognoz pogody kak zadacha fiziki. (Weather forecasting as a task of physics). Nauka, Moscow, 184 pp (in Russian). English translation: Weather forecasting as a problem in physics. MIT Press, Cambridge, Mass, 1972, 199 pp

Morner NA (1981) Eustasy, paleoglaciation and paleoclimatology. Geol. Rundschau 70: 691-702

Muller RA, MacDonald CJ (1997) Glacial cycles and astronomical forcing. Science 277: 215-218

Mutlow CT, Zavody AM, Barton IJ, Llewellyn-Jones DT (1994) Sea surface temperature measurements by the along-track scanning radiometer on the ERS-1 satellite: early results. J Geophys Res C 99: 22575-22588

Muzylev SV, Privalsky VE, Ratkovich DYa (1982) Stokhasticheskie modeli v inzhenernoi gidrologii (Stochastic models in engineering hydrology). Nauka, Moscow, 184 pp (in Russian)

Namias J (1985) Extra-tropical connections. WCRP Publ. Series, WMO and WCRP, Geneva, No 4, pp 11/13-11/23

Namias J (1987) Where are we going in the study of short-period climate fluctuations? In: Quo Vadimus: Where are we going? A collection of contributions to IUGG Symp. At the XIX Gen Ass Graz, Austria, pp 130-132

NASA (1995) Global data sets for land-atmosphere model. 5 CD-ROMs, NASA-GEWEX, GDAAC, GSFC, Greenbelt, MD

Nazarov VI, Velichkevich FYu (1991) Paleogeography, paleoclimate, paleotemperature. Inst. Geochemistry and Geophysics Acad Sci BSSR, Mensk, 28 pp (in Russian)

Newman WS, Markus LF, Pardi RR (1980) Eustasie and deformation of the geoid. In: Earth reology and its isostasie. Wiley, New York, pp 555-567

Nicholls N, Lavery B, Frederiksen C (1996) Recent apparent changes in relationships between the ENSO and Australian rainfall and temperature. Geophys Res Lett 23: 3357-3360

Nikolaev SD, Blyum NS, Nikolaev VA (1989) Paleogeography of oceans and seas in the Cenozoic (by isotopic and micropaeontological data). VINITI, Moscow, 196 pp

NOAA (1998) Southern Oscillation Index and TAO SST anomaly time series. TAO Project Office, PMEL. Http://www.pmel.noaa.gov/toga-yao/gif/faily/soi_110w

NOAA (1998) Global land+sea surface temperature index. http://www.ncdc.noaa.gov/images/climate/research/1998/anomalies

Oeschger H, Beer J, Siegenthaler V, Stauffer B, Dansgaard W, Langway C (1983) Late-glacial climate history from ice cores. Paleoclimate Research and Models (NATO ASI Ser). Reidel, Dordrecht, pp 95-107

Oglesby RJ (1990) Sensitivity of glaciation to initial snow cover, CO_2, snow albedo, and oceanic roughness in the NCAR GCM. Climate Change 4; 219-235

Oort AH, Maher MAC (1985) Observed long-term variability in the global surface temperature of the atmosphere and oceans. In: Coupled ocean-atmosphere models, Elsevier, Amsterdam, pp 183-198

Ostrovsky AG (1983) Features of variability of sea surface temperature anomalies in the North Atlantic. Doctorate (Candidate) Thesis, Moscow State University, Moscow, 170 pp (in Russian)

Ostrovsky AG, Piterbarg LI (1986) On adaptation of a numerical model of sea surface temperature anomalies in the ocean to observational data. In: Modelirovanie gidrofizicheskikh polei i protsessov v okeane. Nauka, Moscow, pp 133-140 (in Russian)

Owen JA, Palmer TN (1987) The impact of El Niño on an ensemble of extended-range forecasts. Mon Weather Rev 115: 2103

Parisi E, Chita MB (1982) Late Quaternary paleoceanographic changes recorded by deep-sea benthos in the Western Mediterranean Ridge. Geogr Fisique Dynamique Quaternaire 5: 102-114

Parker DE, Jones PD, Folland CK, Bevan A (1994) Interdecadal changes of surface temperature since the late nineteenth century. J Jeophys Res 99: 14373-14399

Parkinson CL, Gloersen P (1993) Global sea ice coverage. In: Gurney RJ, Foster JL, Parkinson CL (eds). Atlas of satellite observations related to global cange. Cambridge Univ Press, Cambridge, pp 371-383

Parkinson CL, Comiso JC, Zwally HJ, Cavalieri DJ, Gloersen P, Campbell WJ (1987) Arctic sea ice, 1973-1976: satellite passive-microwave observations. NASA SP-489, National Aeronautics and Space Administration, Washington, DC, 296 pp

Pate-Cornell E (1996) Uncertainties in global climate change estimates. Clim Change 33: 145-149

Pavlich MJ, (1996) Marine-terrestrial correlation as a means of testing global clmate change. US Geol Surv Circ, No 1119: 61-66

Pearce F (1994) Fire and flood greet El-Niño's third year. New Sci 141: 9

Peltier WR, Tushingham AM (1989) Global sea level rise and the Greenhouse Effect: might they be connected? Science 244: 806-810

Permanent Service for Mean Sea Level (1998) Global, regional and national level data sets. http:// biudc.nbi.ac.uk./psmsl/sea_level.html

Petit JR, Mounier L, Jouzel J, Korotkevich YS, Kotlyakov VM, Lorius C (1980) Paleoclimatological and chronological implications of the Vostok core dust record. Nature 343: 56-58

Piterbarg LI (1985) On the stochastic character of oceanic SSTA generation. Dokl Akad Nauk SSSR 282: 1473-1477 (in Russian)

Piterbarg LI (1989) Dinamika i prognoz krupnomasshtabnykh anomalii temperatury poverkhnosti okeana. Statisticheskii podkhod (Dynamics and forecast of large-scale sea

surface temperature anomalies. Statistical approach. Gidrometeoizdat, Leningrad, 200 pp (in Russian)
Pons A, De Beaulieu J-L, Guiot J (1991) Le pollen remonte le temps Quaternaire et les implications climatique. La Recherche 231: 518-520
Portman DA (1993) Identifying and correcting urban bias in regional time series: surface temperature in China's northern plains. J Climate 6: 2298-2308
Privalsky VE (1985) Climatic variability (Stochastic models, predictability, spectra). (Klimaticheskaya izmenchivost'(Stochasticheskie modeli, predskazuemost', spektry)). Nauka, Moscow, 184 pp (in Russian)
Privalsky VE, Jensen DT (1993) Time series analysis package. Autoregressive time and frequency domain analysis of scalar and multi-variate time series. Utah Climate Center, Utah University, Logan, UT, 61 pp
Privalsky VE, Panchenko VA, Asarina EYu (1992) Time series models (Modeli vremennykh ryadov). Gidrometeoizdat, Sankt-Petersburg, 226 pp (in Russian)
Puchkovsky SV (1996) On the probability of a "greenhouse catastrophe" and the nature of feedbacks. Biofizika 416: 520-522 (in Russian)
Ramanathan V, Vogelmann AM (1997) Greenhouse effect, atmospheric solar absorbtion and the Earth's radiation budget: from the Arrhenius-Langley era to the 1990s. Ambio 26: 38-46
Ratkovich DYa (1976) Mnogoletnie kolebaniya rechnogo stoka (Multi-year variations of river runoff. Gidrometeoizdat, Leningrad, 256 pp (in Russian)
Ratkovich DYa (1993) Gidrologicheskie osnovy vodoobespecheniya (Hydrological basis for water supply). Water Problems Institute, Moscow, 429 pp (in Russian)
Ratkovich DYa, Bolgov MV (1997) Stokhasticheskie modeli kolebanii sostavlyayushchikh vodnogo balansa rechnogo basseina (Stochastic models of variations of components of a river watershed water balance). Water Problems Institute, Russian Foundation for Fundamental Research, Moscow, 263 pp (in Russian)
Reynaud L, Dobrovolski SG (1998) Statistical analysis of glacier mass balance data. Stud Rep Hydrol 56: 73-83
Reynolds RW (1988) A real time global sea surface temperature analysis. J Clim 1: 75-86
Reynolds RW, Folland CK, Parker DE (1989) Biases in satellite-derived sea surface temperature data. Nature 341: 728-731
Rhode H, Charlson R, Crawford E (1995) Svante Arrhenius and the greenhouse effect. Ambio 26, 1: 2-5
Roach D, Mitchum GI, Wyrtki K (1989) Length scales of interannual sea level variations along the pacific margin. J Phys Oceanogr 19: 122-128
Ropelewski CF (1989) Monitoring large-scale cryosphere/atmosphere interactions. Adv Space Res 9: 213-218
Rotcliffe R (1995) Back to basics: is our climate changing? Weather 50, 2: 54-57
Ruddiman W, Raymo M, Martinson D, Clement B, Blackman J (1989) Paleoceanography 4: 353-412
Rybak OO (1992) Dynamic-stochastic model of anomalies of heat and moisture exchange in large regions. Doct (Candidate) Thesis, Moscow State University, Moscow, 180 pp (in Russian)
Sahay A, Sreenivasan KR (1996) The search for a low-dimensional characterization of a local climate system. Philos Trans R Soc, Lond A 354: 1715-1750
Salamatin AN, Lipenkov VYa, Blinov KV (1994) Vostok (Antarctica) climate record time-scale deduced from the analysis of a borehole-temperature profile. Ann Glaciol 20: 207-218
Salas JD, Obeysekera JTB, Smith RA (1981) Identification of streamflow stochastic models. J Hydraul Div Proc Am Soc Civ Eng 107: 853-866
Sausen R, Barthel K, Hasselmann K (1988) Coupled ocean-atmosphere models with flux corrections. Clim Dyn 2: 154-163
Schlesinger ME, Ramankutty N (1994) An oscillation in the global climate system of period 65-70 years. Nature 367: 723-726

Schoell DW, Risch M (1976) Recent submergence of southern Florida: a comparison with adjacent coasts and other eustatic data. Geol. Soc. Am. Bull. 78 (4): 682-691

Self S, Rampino MR (1988) The relationship between volcanic eruptions and climate change. EOS Trans Am Geophys Union 69, 66: 74-75, 85-86

Selivanov AO (1996) Global sea-level changes and coastal evolution during the Pleistocene and Holocene. Water Problems Institute, Schwartz Publ, Moscow, 278 pp (in Russian)

Shackleton NJ, Opdyke HD (1973) Oxygen isotope and paleomagnetic stratigraphy of equatorial Pacific core V28-238: Oxygen isotope temperatures and ice volumes. Quat Res 3: 39-55

Shackleton NJ, Hall MA, Line J, Shaxi C (1983) Carbon isotope data in core V19-30 confirm reduced carbon dioxide concentration in the ice age atmosphere. Nature 306: 319-322

Shannon CE (1948) A mathematical theory of communication. Bell Syst Tech J 27: 379-423

Shepard FP (1971) Sea level rise during the past 20,000 years. Ann Geomorphol 13: 285-291

Shrirer HN, Fosmire SJ, Wells R, Suciu L (1997) Estimating the correlation dimension of atmospheric time series. J Atmos Sci 54: 211-229

Simmonds I, Hope P (1997) Persistence characteristics of Australian rainfall anomalies. Int J Climatol, 17: 597-613

Sirutis J, Reynolds R, Lu Y-H (1986) Monthly forecast experiment: Part II. With sea--surface temperature anomalies forcing. Preliminary report. In: Long-range forecast, Res Rep Ser WMO, Geneva No 6/2, 555 pp

Slutski EE (1938) Sur les fonctions aléatoires presque periodiques et sur la décomposition des fonctions aléatoires stationnaires en composantes. Actualités scientifiques et industrielles, No 738, Hermann, Paris, 33

Smylie DE, Clarke GKC, Ulrych TJ (1973) Analysis of irregularities in the earth's rotation. In: Methods in computational physics 13, Academic Press, New York, pp 391-430

Sonechkin DM, Ivashchenko NN (1996) Contemporary approach to the problem of existence of rithms in the atmosphere dynamics. Trudy Gidrometeorol Nauchno-Issled Tsentr Rossiiskoi Federatsii, 329: 72-79 (in Russian)

Sonechkin DM, Datsenko NM, Ivashchenko NN (1997) Estimation of the global heating trend using the wavelet analysis. FAO (Physics of the atmosphere and the oceans) 97: 184-194 (in Russian)

Song N, Starr D, Wuebbles D (1996) Volkanic aerosols and interannual variation of high clouds. Geophys Res Lett 23: 2657-2660

Spencer NE (1992) Joint IAPSO-IOC workshop on sea level measurements and quality control. Paris, 12-13 October 1992. Intergovernmental Oceanographic Commission Workshop Report No 81, 167 pp

Spencer NE, Vassie JM (1985) Comparison of sea-level measurements obtained from deep pressure sensors. In: Advances in underwater technology and offshore engineering, volume 4. Evaluation, comparison and calibration of oceanographic instruments. Graham and Trotman, for SUT, London, pp 183-207

Spencer NE, Woodworth PL (1993) Data holdings of the permanent service for mean sea level (November 1993). Bidston, Birkenhead: Permanent service for mean sea level, 81 pp

Sprott JC, Rowlands G (1995) Chaos data analyzer. The professional version. User's manual. American Institute of Physics, New York, 70 pp

Stager J, Mayewski Z (1997) Abrupt early to mid-Holocene climatic transition registered at the equator and the poles. Science 276: 1834-1836

Steffen K, Maslannik JA (1988) Comparison of Nimbus-7 scanning multichannel microwave radiometer radiance and derived sea ice concentrations with Landsat imagery for the North Water area of Baffin Bay. J Geophys Res 93: 10,769-10,781

Stepanov VN (1996) Role of the ocean in formation of long-term climate fluctuations. Nature (Moscow) 10: 23-35 (in Russian)

Stone L, Saparin PI, Huppert A, Price C (1998) El Nino chaos: the role of noise and stochastic resonance on the ENSO cycle. Geoph Res Lett 25: 175-178

Suetova IA (1982) Areas and volumes of ancient ice sheets of the Quaternary period. In: Rychagov GI (ed) Geographical investigations of the Quaternary period (Geograficheskie Issledovaniya Chetvertichnogo Perioda). Moscow University Press (Izdatel'stvo Moskovskogo Universiteta), Moscow, pp 22-33 (in Russian)

Suetova IA (1987) Morphometrical characteristics of Antarctida (by the new cartographic materials). In: Kapitsa, AP (ed) Theoretical and methodological problems of palaeogeography. Moscow University Press (Izdatel'stvo Moskovskogo Universiteta), Moscow, pp 203-215 (in Russian)

Susskind J (1993) Water vapor and temperature. In: Gurney RJ, Foster JL, Parkinson CL (eds) Atlas of satellite observations related to global change, Cambridge University Press, Cambridge, pp 89-129

Takahashi M, Zhao N (1997) Equatorial waves in a general circulation model simulating a quasi-biennial-oscillation. J Meteorol Soc Jap 75: 529-540

Takata R, Tokioka T, Noda A (1997) Study on interannual variation of frozen ground simulated by climate models. Res Rep IAHS 3: 13-20

Tarakanov YuA, Kliege RK, Selivanov AO (1992) Marine terraces as indicators of global geoidal and hydrological changes. GeoJournal 27: 493-497

Thomas RH (1993) Ice sheets. In: Gurney RJ, Foster JL, Parkinson CL (eds) Atlas of satellite observations related to global change. Cambridge University Press, Cambridge, pp 385-400

Timofeev YuM, Martynov AA (1996) An assessment of the temperature and brightness of the earth surface from space. Issledovaniya Zemli Iz Kosmosa 4: 12-17 (in Russian)

Todd PM (1996) A COADS-based Southern Oscillation index (SOI). JISAO Data. http://tao.atmos.washington.edu/pacs/additional_analysis/soi2.html

Tokioka T, Kitoh A, Katayama A (1986) Atmospheric response to the SSTA in the mature phase of El Niño. J Meteorol Soc Jpn 64: 347-362

Trenberth K, Hoar T (1996) The 1990-1995 ENSO event: longest on record. Geophys Res Lett 23: 57-60

Trishchenko AP, Khokhlova AV (1996) Use of satellite data of radiometer SCARABE for the assessment of sea surface temperature. Meteorol Gidrol 12: 27-41

Tsonis AA (1992) Chaos: from theory to applications. Plenum Press, New York

Ulrych TJ, Bishop T (1975) Maximum entropy spectral analysis and autoregressive decomposition. Rev Geophys Space Phys 13: 183-200

USDC (1994) Fifth annual climate assessment 1993. United States Department of Commerce, Washington, DC, 111 pp

Van den Bos A (1971) Alternative interpretation of maximum entropy spectral analysis. IEEE Trans Inform Theory IT-17: 493-494

Van der Sluijs JP, De Bruyn GJ, Westbroo P (1996) Biogenic feedbacks in the carbonate-silicate geochemical cycle and the global climate. Am J Sci 296: 932-953

Vasechkina EF, Timchenko IE, Yarin VD (1996) Forecasting of hydrological processes using the methods of self-organizing mathematical models. Morskoi Gidrofizicheskii Zhurnal 3: 48-57 (Russian)

Vecchia AV (1985) Periodic ARMA (PARMA) model with application to water resources. Water Res Bull 21: 721-730

Velichko AA (1982) Paleogeografiya Evropy za poslednie 100 tysyach let: atlas i monografiya. (Paleogeography of Europe during the last 100 thousand years: atlas and monograph). Nauka, Moscow, 156 pp 15 sep Maps (in Russian)

Verbitsky MYa (1986) One important mechanism of glacial cycles' generation. In: Monin AS (ed) Modelirovanie gidrofizicheskikh polei i protsessov v okeane. Nauka, Moscow, pp 48-54 (in Russian)

Walsh J (1986) Long-range forecasting: truth of consequences. Science 231: 1362

Wang B, Fang Z (1996) Chaotic oscillations of tropical climate: a dynamic system theory for ENSO. J Atmos Sci 53: 2786-2802

Warrick RA, Le Provost C, Meier MF, Oerlemans J, Woodworth PL (1996) Changes in sea level. In: Houghton JT, Meira Filho LG, Callander BA, Harris N, Kattenberg A, Maskell K (eds) Climate change 1995. The science of climate change. Cambridge University Press, Cambridge, pp 361-405

Waylen P, Quesada M, Caviedes C (1996) Temporal and spatial variability of annual precipitation in Costa Rica and the Southern Oscillation. Int J Climatol 16: 173-193

Wimbush M (1990) Inferring sea level variation from acoustic travel time and bottom pressure measurements. In: Eden HF (ed) Towards an integrated system for measuring long term changes in global sea level. Report of a workshop held at Woods Hole Oceanographic Institution, May 1990, Joint Oceanographic Institutions (JOI), Washington, DC, pp 147-153

World water balance and water resources of the earth (1974) Gidrometeoizdat, Leningrad, 638 pp (in Russian)

World Weather Disc Climate data for the planet earth (1994). Weather Disc Associates, Seattle, WA

Wright WJ, Smith JN, Allan RJ (1995) The 1991-95 extended warm Pacific event: Southern Hemisphere effects and predictability. Int Sci Conf Trop Oc Glob Atm (TOGA) Programme, Melbourne, 2-7 Apr 1995, WCRP and WMO, No 717, pp 875-879

Woodworth PL (1991) The permanent service for mean sea level observing system. J Coastal Res 7: 699-710

Wunsch C (1972) Bermuda sea level in the relation to tides, weather and baroclinic fluctuations. Rev Geophys Space Phys 10: 1-49

Wyrtki K (1989) Sea level: the rising tide. UNESCO Sources, No 6, p 13

Yaglom AM (1952) An introduction to the theory of stationary random functions. Uspekhi Matematichaskikh Nauk 7, Vypusk 5(51): 3-168 (in Russian)

Yaglom AM (1962) An introduction to the theory of stationary random functions. Prentice-Hall, Englewood Cliffs, NJ, 235 pp

Yaglom AM (1981) Correlation theory of stationary random functions (Korrelyatsionnaya teoriya statsionarnykh sluchainykh funktsii). Gidrometeoizdat, Leningrad, 280 pp (in Russian)

Yaglom AM (1987) Correlation theory of stationary and related random functions, vol I. Basic results. Springer, Berkin, Heidelberg, New York, 526 pp

Yarosh ES (1986) Issledovanie sezonnykh anomalii kharakteristik atmosfernoi vetvi gidrologicheskogo tsikla (Study of seasonal anomalies of the atmospheric water cycle parametres). Doctorate (Candidate) Thesis. Water Problems Institute, Moscow, 137 pp (in Russian)

Yokoyama R, Tanba S (1988) Estimation of sea surface temperature via NOAA/AVHRR sensor: comparison with sea data by fixed buoys. In: IGARSS'83: Int Geosci Remote Sens Symp: Remote Sens, Paris 1988, 1: 275-280

Yuan Y, Yane M, Martinson D (1996) Cycling around the South Pole. Nature 380: 673-674

Zarrina EP, Krasnov II, Spiridonova EA (1980) Late Quaternary climatostratigraphic correlation and chronology of the north-western and central parts of the Russian Plain. In: XXYI Int Geol Congr Rep of Soviet Geologists. Quaternary Geology and Geomorphology. Nauka, Moscow, pp 46-50 (in Russian)

Zebiak SE (1986) Atmospheric convergence feedback in a simple model for El Niño. Mon Weather Rev 114: 1263-1271

Zhurbenko IG, Kozhevnikova IA (1990) Stochastic processing (Stokhasticheskoe modelirovanie protsessov). Moscow University Press (Izdatel'stvo Moskovskogo Universiteta), Moscow, 148 pp (in Russian)

Zubakov VA (1986) Global climatic events during the Pleistocene. (Global'nye klimaticheskie sobytiya na protyazhenii Pleistotsenovogo perioda) Gidrometeoizdat, Leningrad, 288 p (in Russian)

Zwally HJ, Comiso JC, Parkinson CL, Campbell WJ, Carsey FD, Gloersen P (1983) Antarctic sea ice, 1973-1976: satellite passive-microwave observations. NASA SP-459, National Aeronautics and Space Administration, Washington, DC, 206 pp

Index